AF787326

Global Warming - A Concerning Component of Climate Change

Edited by Vinay Kumar

Published in London, United Kingdom

Global Warming - A Concerning Component of Climate Change
http://dx.doi.org/10.5772/intechopen.1000331
Edited by Vinay Kumar

Contributors
Abubakar Danmaigoro, Adina Berbecea, Anthonia Ifeyinwa Achike, Asma Abahussain, Başak Kiliç Taşeli, Bernadette J. Holmes, Camellia Moses Okpodu, Chinasa Onyenekwe, Damtew Degefe Merga, Dennis Mark Onuigbo, Eric Eboh, Isidora Radulov, Jayanti Pal, Joseph Akpan, Julian Schlubach, Kevin J. Duffy, Mahmood Saeed Mustafa Alalawi, Mariam Al-Mannai, Mohammed Gedefaw, Nahed Salman, Nischal Sharma, NwaJesus Anthony Onyekuru, Obiora C. Collins, Oludolapo A. Olanrewaju, Oludolapo Olanrewaju, Oluwafemi E. Ige, Paul A. Iji, Phillip Sagero, Prince Owusu-Wiredu, Pınar Rüya Uludağ, Raju Attada, Razia Rashid, Robert Ross, Rohtash Saini, Royford Bundi Magiri, Sayoni Sarkar, Shivani Singh, Sophia Chirongoma, Stelios Andreadakis, Terence Mupangwa, Vinay Kumar, Walter Okello, Wati Mocevakaca

© The Editor(s) and the Author(s) 2024

The rights of the editor(s) and the author(s) have been asserted in accordance with the Copyright, Designs and Patents Act 1988. All rights to the book as a whole are reserved by INTECHOPEN LIMITED. The book as a whole (compilation) cannot be reproduced, distributed or used for commercial or non-commercial purposes without INTECHOPEN LIMITED's written permission. Enquiries concerning the use of the book should be directed to INTECHOPEN LIMITED rights and permissions department (permissions@intechopen.com).

Violations are liable to prosecution under the governing Copyright Law.

Individual chapters of this publication are distributed under the terms of the Creative Commons Attribution 4.0 License which permits commercial use, distribution and reproduction of the individual chapters, provided the original author(s) and source publication are appropriately acknowledged. If so indicated, certain images may not be included under the Creative Commons license. In such cases users will need to obtain permission from the license holder to reproduce the material. More details and guidelines concerning content reuse and adaptation can be found at http://www.intechopen.com/copyright-policy.html.

Notice

Statements and opinions expressed in the chapters are these of the individual contributors and not necessarily those of the editors or publisher. No responsibility is accepted for the accuracy of information contained in the published chapters. The publisher assumes no responsibility for any damage or injury to persons or property arising out of the use of any materials, instructions, methods or ideas contained in the book.

First published in London, United Kingdom, 2024 by IntechOpen
IntechOpen is the global imprint of INTECHOPEN LIMITED, registered in England and Wales, registration number: 11086078, 167-169 Great Portland Street, London, W1W 5PF, United Kingdom

British Library Cataloguing-in-Publication Data
A catalogue record for this book is available from the British Library

Additional hard and PDF copies can be obtained from orders@intechopen.com

Global Warming - A Concerning Component of Climate Change
Edited by Vinay Kumar
p. cm.
Print ISBN 978-1-83769-859-2
Online ISBN 978-1-83769-858-5
eBook (PDF) ISBN 978-1-83769-860-8

An electronic version of this book is freely available, thanks to the partial funding support of libraries working with Knowledge Unlatched. KU is a collaborative initiative designed to make high quality books Open Access for the public good. More information about the initiative and links to the Open Access version can be found at www.knowledgeunlatched.org

We are IntechOpen, the world's leading publisher of Open Access books

Built by scientists, for scientists

6,900+
Open access books available

186,000+
International authors and editors

200M+
Downloads

156
Countries delivered to

Our authors are among the
Top 1%
most cited scientists

12.2%
Contributors from top 500 universities

WEB OF SCIENCE™

Selection of our books indexed in the Book Citation Index in Web of Science™ Core Collection (BKCI)

Interested in publishing with us?
Contact book.department@intechopen.com

Numbers displayed above are based on latest data collected.
For more information visit www.intechopen.com

For EU product safety concerns:
IN TECH d.o.o., Prolaz Marije Krucifikse Kozulić 3, 51000 Rijeka, Croatia,
info@intechopen.com or visit our website at intechopen.com.

Meet the editor

Vinay Kumar holds a Ph.D. from the Indian Institute of Tropical Meteorology, Pune, India, with a research focus on monsoonal droughts. He is currently an Assistant Professor of Meteorology at the University of the Incarnate Word, Texas, USA. Beyond his teaching responsibilities, he actively engages in research, exploring topics related to natural hazards, risk, uncertainty, and health implications arising from extreme weather and climate conditions. He has authored more than fifty research articles, contributed to several book chapters, and produced numerous research reports.

Contents

Preface

Global Warming – A Concerning Component of Climate Change is a comprehensive exploration of the multifaceted aspects of global warming, a critical element within the broader context of climate change. As the world grapples with the escalating challenges posed by environmental degradation, understanding the intricate dynamics of global warming is paramount for informed decision-making and sustainable action. Our planet's climate is a complex and interconnected system, and even slight variations can have significant consequences.

This book comprises sixteen chapters, each offering unique insights into various dimensions of global warming and its repercussions on our planet. From the intricate water balance components affected by land use to the quantification of carbon footprints in unexpected domains such as high school operations, and the examination of climate variability in diverse ecosystems, this compilation encompasses a wide spectrum of research and analysis.

This book is not simply a collection of scientific observations; it is a multifaceted exploration of climate change's causes, impacts, and potential solutions.

The first section ("Climate Change: Land Use and Human Influence") establishes a firm foundation. Chapters 1 and 2 delve into the technical aspects of climate change, examining how land-use and human activity affect the Earth's water balance and carbon footprint. Chapters 3 and 4 focus on specific regions, analyzing how global warming alters weather patterns in the Indian summer monsoon and the Western Himalayas. Chapter 5 then shifts the focus to the human dimension, exploring community resilience and the stories of climate champions on the frontlines.

The second section ("Climate Change: Soil Health and Anthropogenic Role") examines the intricate relationship between climate change and human actions. Chapter 6 highlights the often-overlooked role of soil health in mitigating climate change. Chapter 7 explores the delicate balance between agricultural practices, ecosystem services, adaptation, and mitigation strategies. Chapter 8 takes a critical look at the fashion industry, unveiling the environmental costs of "fast fashion" and its growing impact on our planet.

The human element remains central in the third section ("Climate Change: Eco-Regions and Other Influences"). Chapter 9 examines young people's knowledge, attitudes, and practices regarding climate change, highlighting the crucial role of education and awareness. Chapter 10 delves into a fascinating aspect: the intersection of religion, climate change, and gender in Africa. Chapter 11 brings us back to the realm of climate variability, analyzing trends in two distinct Ethiopian eco-regions.

The fourth section ("Climate Change: Food Production and Environmental Factors") showcases the wide-ranging effects of climate change on a global scale.

Chapter 12 examines the combined stresses of air pollution and climate change on cocoyam production in Nigeria. Chapter 13 explores the impact of climate change on dairy production in Fiji and the Pacific Islands, highlighting adaptation strategies. Shifting focus to the other side of the globe, Chapter 14 analyzes recent temperature and snow cover changes in the Northern Hemisphere, with a particular focus on Alaska.

The final section ("Climate Change: Emission Reduction and Uncertainties") explores solutions and future directions. Chapter 15 examines global emissions reduction commitments in the context of achieving a 1.5°C climate scenario, outlining the path forward. The book concludes with Chapter 16, a case study on measuring the environmental impact and uncertainties associated with cement production in South Africa.

This book transcends the boundaries of traditional scientific literature. It encompasses the scientific, social, and economic dimensions of climate change, providing a well-rounded perspective on this critical issue. Whether you are a seasoned researcher, a concerned citizen, or simply curious about our planet's future, *Global Warming – A Concerning Component of Climate Change* offers valuable insights and a call to action.

I would like to express my heartfelt gratitude to Publishing Process Manager Ms. Nina Miocevic at IntechOpen, whose exceptional efficiency and dedication were pivotal in bringing this book to fruition. Nina's meticulous editorial management and proactive approach ensured the timely completion of this project, making invaluable contributions to its quality and relevance. Thank you, Nina, for your outstanding professionalism and unwavering support throughout the publishing process.

Sincerely,

Vinay Kumar, Ph.D.
Department of Atmospheric Science, Environmental Science, and Physics,
University of the Incarnate Word,
San Antonio, TX, USA

Section 1

Climate Change: Land Use and Human Influence

Chapter 1

Perspective Chapter: Responses of the Water Balance Components under Land Use/Land Cover and Climate Change Using Geospatial and Hydrologic Modeling in the Dhidhessa Sub-Basin, Ethiopia

Damtew Degefe Merga

Abstract

The future water balance estimates are based on climatology and LULC reference period, as well as two future periods, 2030–2059 and 2070–2099, with monthly time-steps determined by bias modifications for both rainfall and temperature for every RCP. As depend on simulated results, the quantity of water stability element as in study area sub-basin would than decline in the coming centuries, with an annual mean river basin decrease of 10% throughout 2044 and 6.3% throughout 2084. Potential evaporation and evapotranspiration risen by 15.9 and 6.5, including both by percent, in the immediate term (2044), while rainfall, percolation of water, surface runoff, transmission losses, lateral runoff, groundwater flow, and water production declined by 5.6, 42.6, 44, 2.1, 30.8, 47.8, and 9.2, in all by percent. Long-term 2084 rainfall, percolation of water, surface runoff, transmission losses, lateral runoff, groundwater flow, and water production decreased by percent 1.6, 42.7, 43.1, 3.4, 29.1, 47.3, and 5.7 while evaporation increased by 11.6 and potential evapotranspiration increased by 22.4. Overall, the above research found that any impact on the watershed that causes a reduction through water balance has a major impact also on area continued efficient water management as well as improvement of living standards.

Keywords: Dhidhessa, climate change, land use land cover, balance, component, element

1. Introduction

The multiple key causes of the hydrological cycle are LULC variability and climate change. The issues of LULC effect on water balances like soil water, evapotranspiration, percolation, base flow, discharge, and water yield through changing land environmental factors and altering soil as well as atmospheric limitations [1–3]. Climate change, and the other, can direct effect rainfall as well as temperature, causing shifts in watersheds and water resource distribution [4, 5]. Changes of intensity, amplitude, and duration of rainfall influence the amount and variation of river flow [6], which often exacerbates floods and droughts while also having a negative impact on local and regional water resources [7]. As a result, evaluating the effects of water balance may be vital for water policy and administration [8, 9]. The scientific community and policymakers have paid close attention to LULC research and evaluating the climate impacts [10].

Several studies have been conducted across the world to discuss a consequences of climate warming as well as LULC impact on water balance element [8, 11–15]. Moreover, several research findings claim that climate change is really the primary cause of hydrological fluctuation [16–19], whereas others claim that LULC change is the primary driving factor [3, 20, 21].Others, however, argue that hydrology directly impacts are caused by both climate cause and variations throughout the LULC [22–26]. Outcome, of the consequences of climate effect and LULC on hydrologic processes have been unknown and necessitate additional research.

Undoubtedly, some Ethiopian River Basins are now undergoing water scarcity as a result of hydrological changes [27]. Human-induced LULC changes inside the river basin, like illegal logging, rural expansion, and urban development, have impacted the hydrological cycle components [28, 29]. The expected rise in temperature [30] will cause more forceful issue in regional rain fall and ET rates [16, 17], as well as affecting water source and surface run, among other issue [16, 31, 32]. Ethiopia is ranked as the second-most densely populated nation due to unknown future sources of food and water. Water scarcity is undoubtedly occurring in some Ethiopian river basins as a result of hydrological changes [27]. Human-induced changes in LULC within the river basin, such as deforestation, cultivation, and urban development, have impacted hydrological cycle components [28, 29]. Temperature rises [30] are expected to cause significant changes in local rainfall and ET rates [16, 31], as well as affect water accessibility and surface runoff, among other issues [16, 31, 32]. Dhidhessa Basin in Ethiopia's Abbay Basin was chosen for the study due to its security vulnerability to climate challenge due to variation in rainfall distribution as well as a warmer climate [31]. Increased agricultural land, uncontrolled LULC changes, and growing populations all have an impact on the water environment [33]. Furthermore, the Dhidhessa basin is one of the basin's productive areas [33]. Despite their importance, Ethiopian government officials have paid less attention to the climate issue as well as LULC cause on watershed hydrological components. With these concerns in mind, the aim of this research is to determine the effects of climate impacts, LULC & combined LULC and climate impact on water balance element. These scientific papers include (1) investigating climate change and future LULC; (2) examining the various climate issue plus LULC issue on different hydrological elements; (3) measuring the composite climate with LULC trends on various hydrological elements. SWAT, the cellular simulation prediction model, and local climate concepts were used in this research to achieve the above objectives.

2. Materials and methods

2.1 Description of study area

Dhidhessa sub-basin is among Ethiopian most productive areas. The Dhidhessa Sub-basin in western Ethiopia is formed by southwestern region of the Abay Basin, as depicts in **Figure 1**. Research area is located between 08.0001 and 09.0002 north and 35.345 and 37.653 east [2]. The total covers of study area 1,962,983 ha. Evaluation has included a vital development of water resources research programs and locations of involvement, in addition to circulation through the Dhidhessa River's main stream and significant rivers [2]: Agaroo, Dabana, Wama, and Dhidhessa River.

The altitude of the area of research varies from 818 meters above sea level to 3146 meters above sea level, as indicated by **Figure 2** [2].

2.2 Data used and sources

From 1991 to 2020, the ENMSA was measured climate required on rainfall and max and min temperature for monitoring stations (**Table 1**).To complete the gap rainfall and temperature values, by normal ratio technique was used. In Dhidhessa basin, the duration of solar radiation, wind speed, and relative humidity statistics are restricted, or otherwise these data are not available for many stations. The Arc-SWAT CSFR World Weather Generator for restricted data as a result. The WGEN CFSR World Weather dataset, created with the assistance of the National Centers for Environmental Prediction Climate Reanalysis System, incorporates SWAT inputs (Arc-SWAT CSFR World Weather Generator) that fully cover long-term weather survey data for the entire world [34].

The CORDEX Africa data set contains climate data for the period 2010–2099. The CORDEX models (http://www.ccafs-climate.com/data-spatial-down-scaling/) have daily rainfall, minimum temperature, and maximum temperature, so we used this website for this study. For the surface water analysis, flow data were collected from MoWIEE GIS database (See **Table 2**).

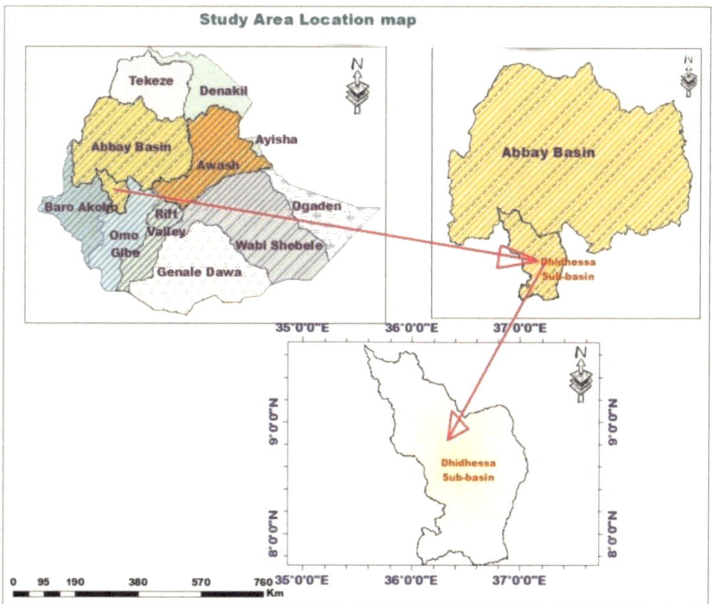

Figure 1.
Study area map.

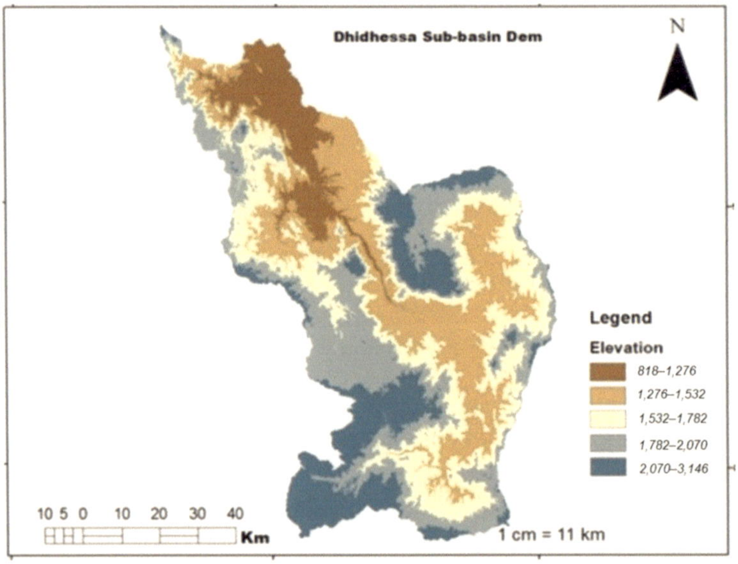

Figure 2.
Topography map.

Station Name	Zone	Station Elevation	Latitude	Longitude	Data coverage	% of missing	% of missing
Agaroo	Jimma	1666	7.85	36.6	1991–2020	22	22.5
Arjoo	East Wollega	2565	8.75	36.5	1991–2020	21	37.5
Bedele	Illuabba bor	2011	8.45	36.33	1991–2020	21	17.5
Dhidhessa	East Wollega	1310	9.38	36.1	1991–2020	20	46
Nekemte	East Wollega	2080	9.08	36.46	1991–2020	21	20.5

Table 1.
Records of weather stations.

Data type	description	Data source
Climate/meteorological data	Daily estimated Data: Rainfall, Max and Min Temperature from 1991 to 2020	NMAE
Hydrological data	Flow/Discharge data	MWIEE
DEM	30×30 m resolution	USGS
Landsat image	30 m resolution	USGS
Regional climate model	CORDEX Africa dataset 0.5×0.5 m	From http://www.ccafs-climate.org/data_spatial_down scaling/

Table 2.
Generate a data input description for the Dhidhessa watershed SWAT model.

2.3 Methodology

The method entails predicting future LULC changes, defining climate change impacts, compiling a watershed analysis model (SWAT) with calibration and

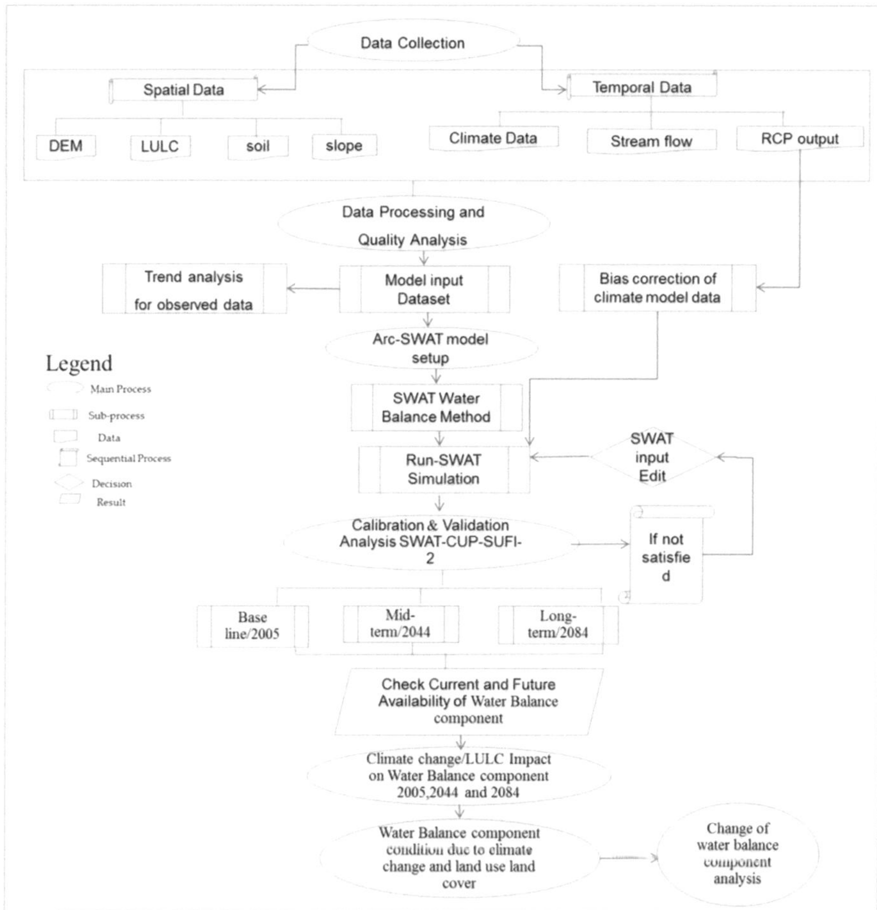

Figure 3.
Flow chart of the study.

validation, analyzing effects of LULC impact on water balance elements, examine climate impact variability on water balance elements, as well as evaluating the composite for LULC as well as climate impact variability on water balance elements, as demonstrated by simulation (see **Figure 3**).

2.4 Climate change background and model selection

Most every model may have benefits and drawbacks [35, 36]. This model used during research was linked with observations to achieve this specific goal. In addition, the models' ensemble performance was calculated relative to the observed models achievement. In this data analysis, climate source by high-resolution RCM from the CORDEX-Africa data model were used [37, 38]. Previous Dhidhessa Sub-basin research papers [36, 39] are used to select the driving model, or GCM. As a result, the driving models chosen were those from the Danish Hydrologic Institute (DMI-HIRHAM5), the Regional Atmospheric Climate Model Version 2.2 (KNMI-RACMO22T), and the Swedish Meteorological and Hydrological Institute (SMHI-RCA4), with the most recent regional climate model (RCM/RCP) gridded at 0.5° to

0.5° spatial resolution. Most previous research on the UBN has used coarse-resolution GCMs rather than regional climate models (RCMs). GCMs have been usually run at horizontal resolutions that tend to range from 250 to 600 km, which are inadequate for analyzing impacts on the environment. Furthermore, previous research depended on SRES climates, applying newly developed RCP. RCP lead the analysis by temperature and precipitation feedbacks to anthropogenic an activity, which includes data on a wide range of GHGs, including radioactive gases and vapor emission levels, land use, and socioeconomic conditions. In addition, commonly used statistical indicators from the World Meteorological Organization (WMO) were used to compare model and observational results. Bias, root mean square error, and coefficient of variation are examples of these survey results [40].

$$Bias = \frac{1}{N}\sum_{i=1}^{N}(P_i - Obs_i) \tag{1}$$

$$RMSE = \sqrt{\frac{1}{N}\sum_{i=1}^{N}(P_i - Obs_i)^2} \tag{2}$$

$$CV = \frac{\sum_{i=1}^{N}(P_i - \overline{P})\left(Obs_i - \overline{Obs}\right)}{\left(\sqrt{\frac{1}{N}\sum_{i=1}^{N}\left(Obs_i - \overline{Obs}\right)^2}\right)\left(\sqrt{\frac{1}{N}\sum_{i=1}^{N}\left(P_i - \overline{P}\right)^2}\right)} \tag{3}$$

P and Obs are the modeled and estimated values, and i and N are the modeled and estimated pairs.

2.4.1 Bias correction

RCM models frequently [41] have large bias during compared to estimated data, actually after segmentation to high resolution. Because these precipitation and temperature biases were found to be spatially variable [42], bias adjustment was done separately for every sub-basin. To bias-correct [43] linear transformation concepts and min and max-temperature scales (Eq. (4)) were used.

$$T_{Corrected} = T_{mean\ Obs} + \frac{\delta(T_{Obs})}{\delta(T_{RCM})}(T_{RCM} - T_{mean\ Obs}) + (T_{mean\ Obs} - T_{mean\ RCM}) \tag{4}$$

where: $T_{corrected}$ denotes the adjusted per day temperature, $T_{mean\ Obs}$ denotes the average mean temperature, and $T_{mean\ RCM}$ denotes the avg. modeled temperature. T_{RCM} is the RCM model's uncorrected per day temperature. T_{Obs} represent daily observed temperatures; $T_{mean\ Obs}$ represents the observed aveg temperature; and $T_{mean\ RCM}$ represents the modeled avg. temperature. This refinement was carried out on data for every 12-month period separately [43]. Coefficient of variation and the mean were both adjusted using a power transformation [44]. The following nonlinear correction is used to convert all daily rainfall quantities P to corrected P* (Eq. (5)):

$$\mathrm{P}^* = \mathrm{ap}^{\mathrm{b}} \tag{5}$$

Where; the adjusted variable value (rainfall), the scaling exponent "b" appears repeatedly, and the CV by the RCM rainfall during correspond to that of the observed rainfall. Then, to ensure t the average of the interpreted rainfall results equals the estimated mean, a constant "a" is determined [2].

$$b = \frac{\delta_{P\ obs_{day}}}{\mu_{P_{obs\ day}}} - \frac{\delta_{PRCM_{day}}}{\mu_{PRCM_{day}}},\ P^{+} = \left(P_{RCM\ day}\right)^{b},$$
$$a = \frac{\mu_{P\ obs_{day}}}{\mu_{P^{+}{}_{RCM\ day}}},\ and\ P^{*} = \left(aP^{+}\right)^{b} \tag{6}$$

2.5 SWAT model in hydrological characteristics

The SWAT model in the hydrological element is dependent on the water balance. SWAT's hydrological model only takes into account flow rates from the unsaturated zone [2]. Since water entering the shallow wells is presumed to make a contribution to the watershed through area outside the basin, groundwater from the shallow aquifer is not considered. Surface seepage, lateral seepage from the soil layer, and return seepage or base seepage from the deep ground water all make a contribution to the water in the flow, as per [2, 45]. Water discharged to shallow ground waters is regarded as lost from the drainage basin system and is not considered for that in water analysis [2, 46]. A basin soil moisture level (Eq. (7)) is calculated as follows:

$$S_{Wt} = S_{Wo} + \sum_{i=1}^{t}\left\{R_{day} - Q_{Surf} - E_a - W_{Seep} - Q_{gw}\right\}i \tag{7}$$

S_{Ws} denote the final soil moisture content i(mm), S_{Wo} is the initial soil moisture content (mm), and t is time (days), R_{day} i(mm) the quantity of rainfall on day, $Q_{surf}\,i$ (mm) the amount of surface runoff on day, the amount of evaporation per day is denoted by $E_a\,i$(mm). W_{seep} is the quantity of water approaching the vadose zone from the soil layer per day i (mm), and Q_{gw} is the quantity of return flow on per day i (mm). The water balance element in the SWAT modeled intended using temporal and spatial data from the hydrological cycle as well as the SWAT input time series, as shown above Eq. (7).

2.5.1 Surface runoff

SWAT-model generates storm water quantities and peak flow using per day or sub-daily rainfall quantities for every HRU. SWAT techniques for determining storm water volumes include the SCS-curve counting technique and the Green and Ampt infiltration technique. Although the latter technique is much more precise in estimating flow quantity, it requires sub-daily interval data, is not available. The SCS-curve totaling technique remained employed. SCS-curve numbering method is generalized (Eq. (8)):

$$Q_{Surf} = \frac{\left(R_{day} - I_a\right)^2}{\left(R_{day} - I_a + S\right)} \tag{8}$$

where Q_{surf} is rainfall excess (mm), R-day is the rainfall depth for the day (mm of water), I_a is the initial abstraction (mm of water), and S is the retention parameter (mm of water). S, the retention parameter, can be computed using Eq. (9).

$$S = 25.4\left(\frac{1000}{CN} - 10\right) \tag{9}$$

where CN is the in-day curve number as well as its value is affected by land cover, soil permeability, and soil hydrological group. The initial abstraction, Ia, is commonly estimated at 0.2 S, resulting in (Eq. (10)):

$$Q_{Surf} = \frac{\left(R_{day} - 0.2S\right)^2}{R_{day} + 0.8S} \tag{10}$$

2.5.2 Peak flow

A maximum volume discharge velocity beyond to a point during a rainfall incident is referred to as peak rainfall or surface maximum flow rate. Maximum flow rate is a measure of storm discharge power and is recycled toward determining soil injury and water excellence. SWAT computes peak flow rates using an obtainable rational method (see Eq. (11)) [47].

$$Q_{peak} = \frac{CiA_{total}}{3.6} \tag{11}$$

where:
Q_{peak} is peak flow rate (m^3/s),
C is the runoff coefficient,
i is the rainfall intensity (mm/hr.),
Sub-basin area (km^2) and 3.6 is conversion factor.

2.5.3 Water yield

The residual total volume of water gives every HRU for sub-basin and needs to enter the main line flow rate once the water balance scheme is filled. SWAT uses the following equation to calculate how much total water yield adds to flow circulation as freshwater through the HRU:

$$W_{YID} = SURQ + LATQ + GWQ - T_{LOSS} \tag{12}$$

2.6 SWAT model performance appraisal

Performance data were used in different concepts to determine if calibration and validation times, spatial and temporal restrictions, and particular performance evaluations were mandatory criteria [47]. Calibration is a composite of manual and automated processes that analyze model limits that cannot be easily observed. The validation determines whether the model is functioning properly and serves as a foundation for systematic model calibration and validation. A prior report's author describes how to calibrate model [47].The SWAT was used for 30 years, from 1991 to 2020, and watershed measurements were recorded from 1994 to 2011 for calibration. The validation of watersheds was also completed for the years 2012–2020. To be considered for the chosen group, a model must meet all three of the requirements [47, 48]. In this research, three conditions using: the coefficient of determination (R^2), the Nash-Sutcliff Index (NSE), and the percent bias (P_{BIAS}).

2.6.1 Coefficient of determination

The determination coefficient attempts to quantify a level of variability in the obtained data encountered by model. As shown in Eq. (13), the coefficient of decision (R^2) is a parameter that is widely used to evaluate the quality of a model.

$$R^2 = \left\{ \frac{\sum_{i=1}^{n} (O_i - O_{mean})(P_i - P_{mean})}{\sqrt{\sum_{i=1}^{n} (O_i - O_{mean})^2} \sqrt{\sum_{i=1}^{n} (P_i - P_{mean})^2}} \right\} \tag{13}$$

where:
O_i is the estimated flow at time i.
O_{mean} is an abbreviation for mean estimated flow discharge.
P_i represents the modeled flow at time i.
P_{mean} denotes mean modeled flow, and n is the amount of flow data documented. The R^2 quality performance criteria identified by [48] calibration and validation. The strength of the connection between measured and simulated attributes is indicated by R^2. It has a significance range of 0 to 1. A value near 0 indicates a very weak correlation, while a value near 1 indicates a higher connection among measured and modeled flow.

2.6.2 NSE stands for Nash Sutcliffe simulation coefficient

Nash-Sutcliffe modeling coefficient (NES) is indeed a plot of the measured data that shows how well the modeled data matches a1:1 line. The NSE calculation equation is shown in Eq. (14).

$$NSE = 1 - \frac{\sum_{i=1}^{n} (O_i - P_i)^2}{\sum_{i=1}^{n} (O_i - O_{mean})^2} \tag{14}$$

This classification causes the NSE performance level of such a model to be assessed on a variety of scales as indicated by [47, 48]. The NSE calculates the amplitude of the remaining variables in relation to observed value. Its values range start $-\infty$ to 1, with 1 representing a great model and principle and lower than 0 implying the time series average value would be a good indicator than that of the concept (**Table 2**).

2.6.3 P_{BIAS}

The P_{BIAS} calculates if the estimated model is greater or less than the measured data [47, 48]. The P_{BIAS} outcomes model simulation is within acceptable bounds. Positive numbers indicate that the simulation measures the bias, even though lower value/negative numbers imply that the model tends to overestimate the bias [49, 50]. Eq. (15) shows the method for calculating the P_{BIAS}:

$$P_{BIAS} = \frac{\sum_{i=1}^{n} (O_i - P_i)}{\sum_{i=1}^{n} O_i} 100 \tag{15}$$

Ref. [51] ENS, P_{BIAS}, and R^2 values were assigned to the models for assessment. It calculates the gap between the modeled and estimated amounts, with a best value of 0.

A high value signifies that the modeled value is lower value, while a negative value indicates that the model is exceeded.

3. Results

3.1 SWAT model

Using river flow data from Dhidhessa watershed gauge sites, model calibrated and validated. In past studies, around 18 most sensitive parameters affecting surface flow, lateral flow, groundwater flow, and evapotranspiration in the survey sub-basin were assessed [2]. The accurate values of the 18 stressed parameters shown in **Table 3**. During calibration and validation, hydrologic observations of recorded as well as simulated monthly river outflows at the Dhidhessa watershed sites are depict in **Figure 4**.

We used Dhidhessa River data (1994 to 2011) for model calibration and validation (2012 to 2020). For model warming, 3 years of data (1991 to 1993) have been used. The model's status was assessed using R^2, NES, and P_{BIAS} approaches together manual and automated calibration. The numerical results obtained throughout quantification had an influence of 0.80 for R2, 0.76 for NES, and 2.41 for PBIAS. The measured and simulated concentrations agree well in terms of measurements. Validity is critical for increasing user confidence in advanced analytics models. Monthly observations from 2012 to 2020 are also used for model validation, yielding values of 0.80 for R^2, 0.77 for

Parameter Name	Fitted Value	Min_value	Max_value
1:A__CN2.mgt	−0.2	−0.3	0.3
2:V__ALPHA_BF.gw	0.1	0.0	1.0
3:V__GW_DELAY.gw	6.3	1.0	150.0
4:V__GWQMN.gw	135.7	0.0	200.0
5:V__ESCO.hru	0.8	0.0	1.0
6:V__GW_REVAP.gw	0.1	0.0	0.2
7:V__OV_N.hru	0.2	0.1	0.3
8:V__SFTMP.bsn	2.5	−5.0	5.0
9:A__SLSUBBSN.hru	19.8	10.0	35.0
10:A__SOL_AWC(..).sol	0.0	0.0	0.0
11:A__SOL_K(..).sol	0.0	0.0	0.0
12:V__SURLAG.bsn	3.8	1.0	12.0
13:V__RCHRG_DP.gw	0.3	0.0	1.0
14:R__LAT_TTIME.hru	1.8	0.0	10.0
15:R__CH_N2.rte	0.0	0.0	0.1
16:R__CANMX.hru	1.0	0.0	1.0
17:R__RFINC(..).sub	0.5	0.1	0.5
18:R__CNCOEF.bsn	1.2	1.0	2.0

Table 3.
Parameters that are sensitive.

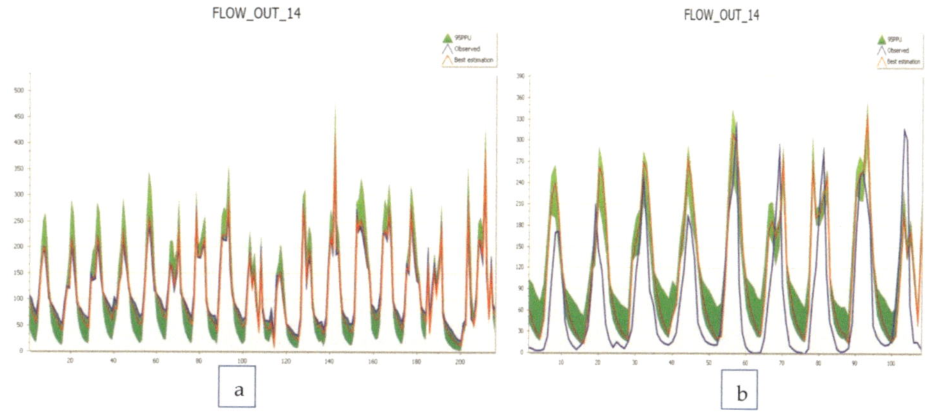

Figure 4.
(a) and (b) calibration and validation of sub-basin.

ENS, and 9.2% P_{Bias} for no correction to the fitted value that occurred during the measurement period.

3.2 Climate impact on water balance

Impact climate is expected to reduce impact of water balance element such as discharge, base flow, water yield, percolation, and evapotranspiration by 12.31% as in medium-term and 18.43% as in long term. Rising temperatures, on the other hand, may rise of the rate of water loss shown in **Table 4**.

LULC Class	LULC 2005	LULC 2044	LULC 2084
AGRL	34.46	38.23	38.52
URBN	3.86	5.52	6.97
AGRR	33.64	34.8	36.62
BANA	2.8	1.56	1.33
COFF	5.75	4.08	4.67
FRSE	2.43	2.72	2.64
FRSD	10.23	7.25	5.33
FRST	1.21	1.94	2.02
PAST	1.68	1.43	1.29
RNGE	2.25	1.42	2.51
WETF	0.17	0.19	2.06
WETL	1.2	0.75	0.61
WETN	1.32	1.11	0.89

Table 4.
LULC as in the Dhidhessa basin in 2005, 2044, and 2084.

3.3 Potential water balance elements

Composite percentage impact for every water balance element modeled from a selected-climate model at annual mean time steps (i.e., 2044 (2030 to 2059) and 2084 (2070 to 2099) in study area from the projected future climate. The aveg yearly water balance such as evapotranspiration and potential evapotranspiration rise marginally by 15.9 and 6.5, respectively in percent, whereas rainfall, runoff, surface runoff, transmission, lateral flow, groundwater flow, and water production dropped substantially by 5.6, 42.6, 44, 2.1, 30.8, 47.8, and 9.2, respectively in percent, over the short period (2044). The other side, the adjustment in flow back and reduction in water flow are largely caused by a reduction in the percent change in water flow and an overdose of the basic system. Through lateral and return flows, this water seepage returned the total water output from the Dhidhessa to the flow of the main stream (groundwater). This decrease in the percentage of total water yield over the sub-basin is issue concern for farm growth as well as aquifer growth in this watershed (**Figure 5**). Apart from potential evapotranspiration and evapotranspiration, which rose by insignificant volumes of 22.4 percent and 11.6 percent from the reference era, respectively, all water balance elements showed an insignificant reduction in future percent change in the long term 2084, which is identical to the short term 2044, percent change in mean annual water balance during the long-term period (2084) was significantly reduced, with 1.6, 42.7, 43.1, 3.4, 29.1, 47.3, and 5.7, respectively in percent, compared to the reference era (2084).

It's significantly reduced per variation throughout available water will jeopardize crop yields, and household tap water will be limited throughout this period, directly affecting welfare in study area. **Figure 5** depicts that potential evapotranspiration increased as a outcome of knowingly complex trends in average yearly temperature (2070 to 2099). All water balance elements, with the exception of potential evapotranspiration and evapotranspiration, showed a lower variance in 2044 in the RCP4.5. In the longer term of 2084, the study area suffered significantly, through all significantly changing upcoming water balance elements reducing since the reference point except potential evapotranspiration and evapotranspiration, which both decreased firmly with percent change. The reduced percent change in all elements over the long run (2084) in this scenario was caused by an increase in potential evapotranspiration and evapotranspiration. Rainfall was reduced by 5.6 and 1.6, respectively in percent, in both era (i.e.,

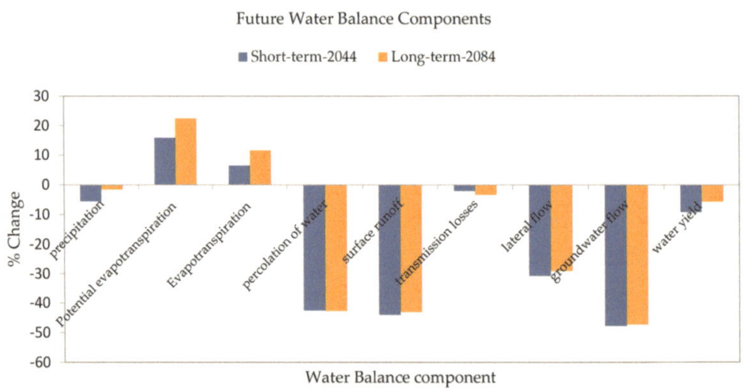

Figure 5.
Water balance element.

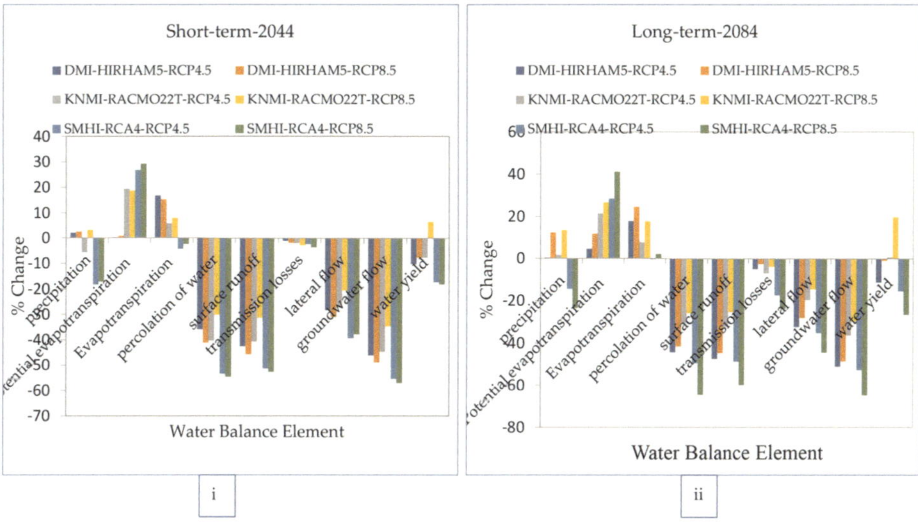

Figure 6.
Mean yearly water balance element under two RCP: (i) 2044 & (ii) 2084.

short and long), as a percent shifts of the reference era in this scenario **Figure 6**. Under the high-emission scenario (RCP8.5), potential evapotranspiration rose by 15.9 and evapotranspiration continued to increase by 22.4 from reference era (**Figure 6**).

3.4 LULC and climate influence combined on water balance

The consolidated effects of climate and LULC remain then measured against the impact of climate as well as LULC separately. It looked at the 2005 LULC map, projected 2044 and 2084 LULC maps see **Figure 7**, climate data from 1991 to 2020, and short-term (2044) (2030–2059) and long-term (2084) (2070 to 2099) future climate projections. The selected models influence the goodness of fit of the

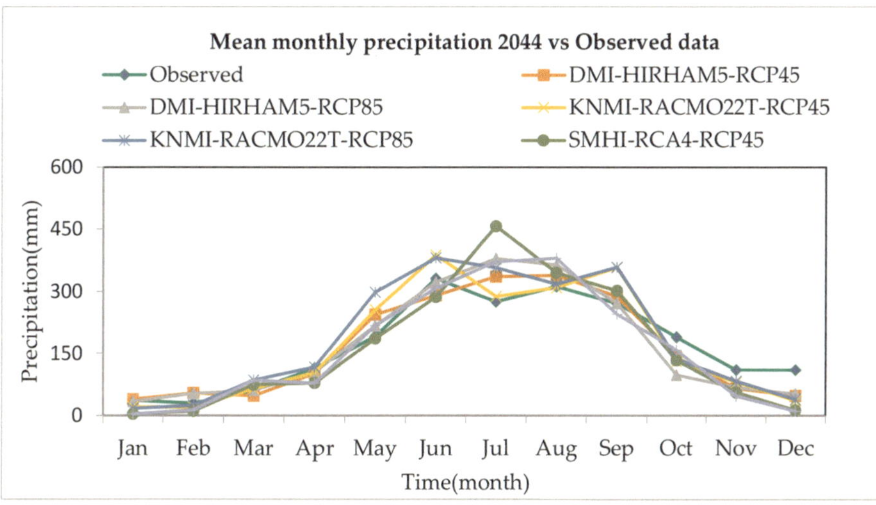

Figure 7.
Monthly mean rainfall short term with estimated data (1991–2020).

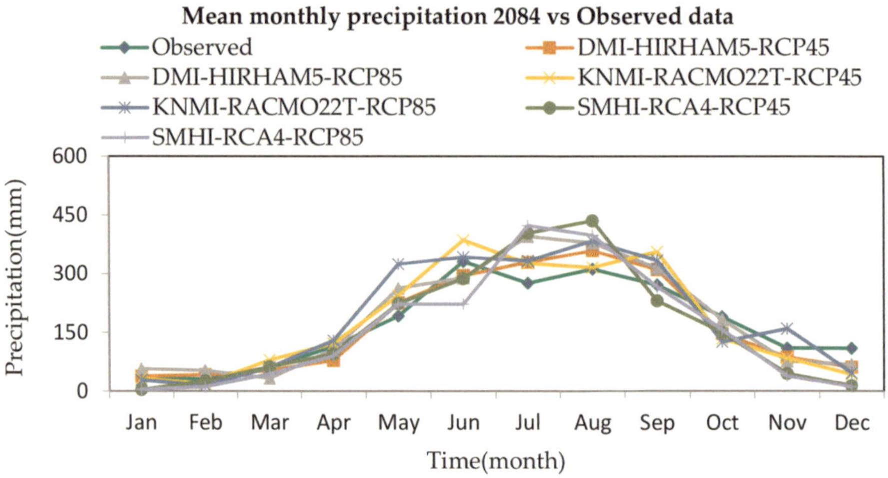

Figure 8.
Monthly mean rainfall long term with estimated data (1991–2020).

CORDEX-AFRICA RCP simulation results in the Dhidhessa basin (**Figures 7–10**) (**Table 5**).

Figure 11 depicts the significant amount of hydrological unpredictability in the Dhidhessa watershed future generated water balance characteristics. The cumulative discharge is a part of the total generated water balance; therefore, it is set for a short period of time (2030–2059) among the dual limits, since January through June as well as September through November, implying the impact of climate plus LULC is negligible throughout this period (flux changes are the cause of confirmation for other prevalent losses). Hydrological variation was not predicted for the long-term (2070–1999) scenario flow from September to December since the outflow occurred outside the boundary, implying that flow changes were caused by climate change, as implied by LULC. As a result, changes in climate and LULC were significant. In summary, the generated frequency was placed outside both boundaries; climate change and the impact of LULC in this sub-basin are causing more uncertainty about future flow.

4. Discussion

4.1 Implication of weather vs. LULC

The impact of separation, climate as well as future LULC on various hydrological components of the Dhidhessa sub-basin is investigated in this study. The SWAT simulated the elements of the future period water balance under three different scenarios (climate, LULC, and climate plus LULC). The results of each scenario type were compared to the base periods of 1991–2020 climate data and the 2005 LULC. Discussion point: Climate models predict reduce in rainfall and a rise in mean temperature in the mid-term (2030–2059) and long-term (2070–1999) era in RCP. The study of LULC changes predicted increases in AGRL, AGRR, and urban areas from 2005 to 2099 using this model, at the expense of evergreen forests and grasslands. Changes in LULC plus climate change scenarios revealed that in the short term (2044), evapotranspiration and potential evapotranspiration rise marginally by 15.9

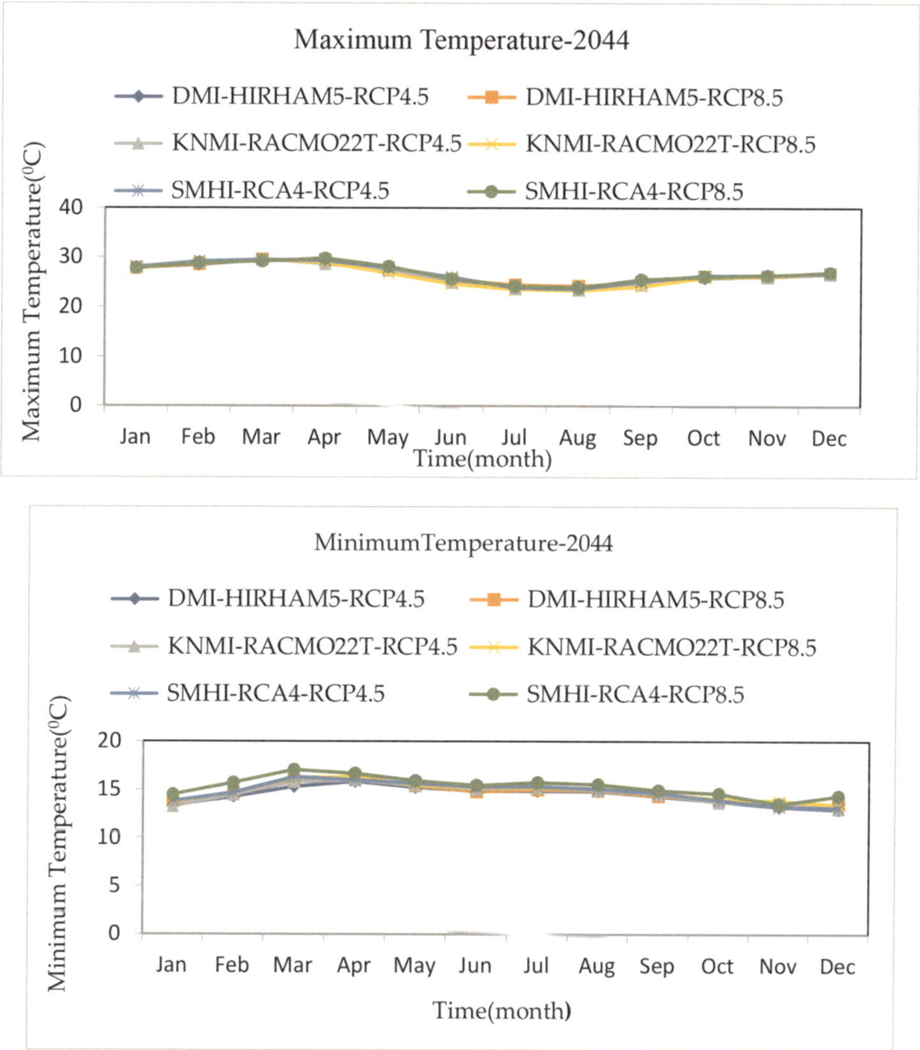

Figure 9.
Monthly mean temperature (max & min) short term.

and 6.5, respectively in percent, while rainfall, runoff, surface runoff, transmission losses, lateral flow, groundwater flow, and water production declined substantially by 5.6, 42.6, 44, 2.1, 30.8, 47.8, and 9.2, correspondingly in percent. The rate changes in mean annual water balances output declined effectively over time (2084) as a shown **Figure 10**, except for PET and ET, which rise marginally by 22.4 percent and 11.6 percent, respectively, from the reference period. In all cases, the composite impacts of climate and LULC result in decreases in discharge, base flow, water yield, percolation, and evapotranspiration. The decrease rates for discharge and water yield are lower in the composite influence analysis of climate change and LULC with single of climate change because by comparison effects of LULC. Even so, when compared to the specific issue of climate or LULC changes, the collective effects of BF, PC, as well as ET contribute to significant reductions. As a result, climate impacts and LULC are two important concurrent and constant impacts on a region's hydrological cycle that

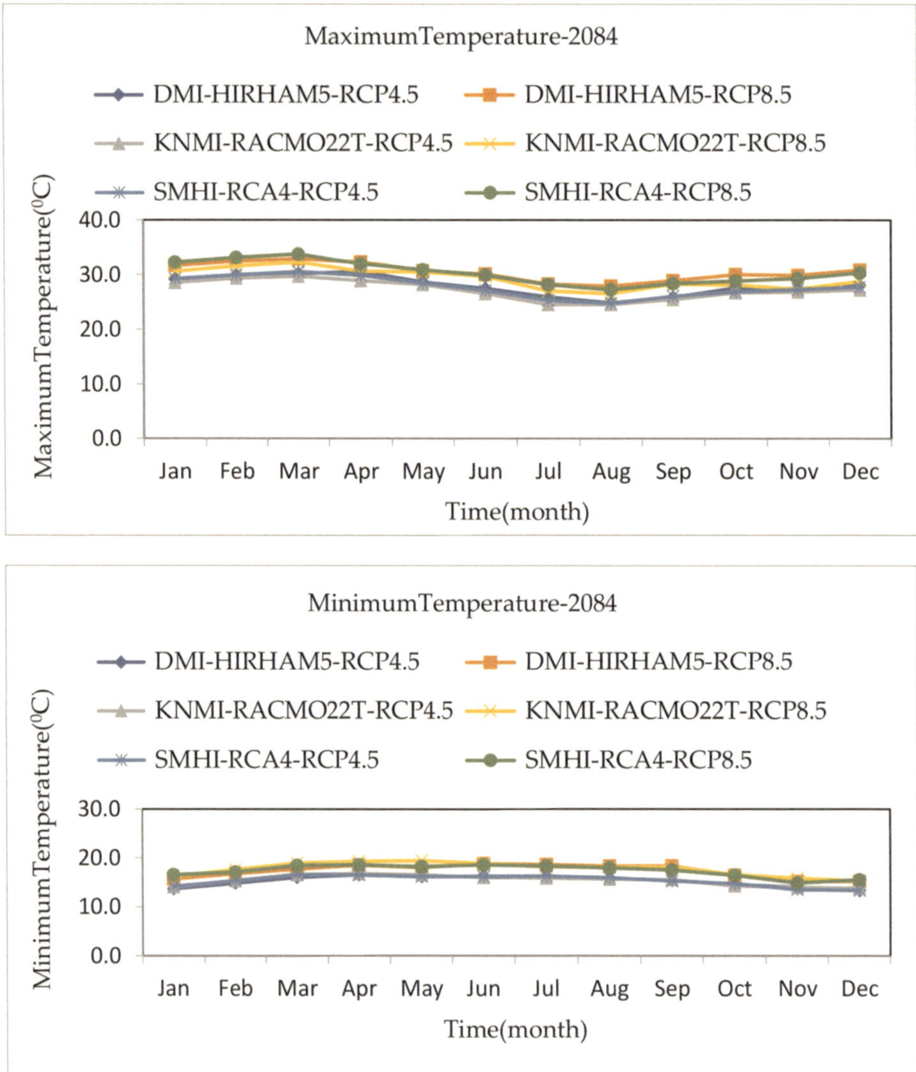

Figure 10.
Monthly mean temperature (max & min) long term.

require an understanding of the components of their effects. Finally, the study's findings may be applied to develop water security and options for adaptation (such as water harvesting and ability) as well as water reuse methodologies to remedial action of climate change and LULC on source, farming, and other areas.

5. Conclusions

According to the study, the Dhidhessa water balance component will decline in the coming hundred years, with an average annual river flow decline of 10 percent through 2044 and a 6.3 percent decline through 2084. Evapotranspiration and

	Statistical criteria		
Deriving RCP	R^2	P_{Bias}	Selection
DMI-HIRHAM5-RCP4.5	0.88	−15.62	✓
DMI-HIRHAM5-RCP8.5	0.84	−17.82	✓
KNMI-RACMO22T-RCP4.5	0.79	−10.55	✓
KNMI-RACMO22T-RCP8.5	0.74	−8.72	✓
SMHI-RCA4-RCP4.5	0.69	−22.94	✗
SMHI-RCA4-RCP8.5	0.71	−13.71	✓

Table 5.
Summary of goodness of fit of the CORDEX-AFRICA RCP modeled outputs in the Dhidhessa watershed.

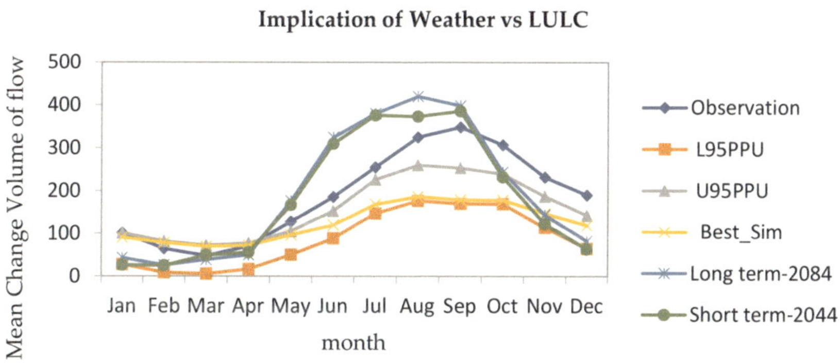

Figure 11.
Limitations with reference era and future flow rate.

potential evapotranspiration increased by 15.9 percent and 6.5 percent, respectively, while precipitation, runoff, surface runoff, transmission losses, lateral runoff, groundwater runoff, and water production significantly decreased. Potential Evapotranspiration and evapotranspiration increased by 22.4 percent and 11.6 percent in the long term, respectively. The water balance element distribution is even across years but uneven across years, indicating a high chances of water shortages. This has significant implications for the area's water security as well as ongoing social progress.

Acknowledgements

The authors would like to thank the University of Rostock for their willingness to fund this publication under the Open Access Publication funding program.

Author contributions

Conceptualization, D.D; methodology, D.D.; software, D.D.; validation, D.D., M. S., M.K and M.C.; formal analysis, D.D.; investigation, D.D.; resources, D.D.; Data curation, D.D.; writing—original draft preparation, D.D.; writing—review and editing, D.D. M.S., M.K, and M.C; visualization, D.D.; supervision, D.D.; project administration, D.D.; funding acquisition, M.K. All authors have read and agreed to the published version of the manuscript.

Funding

This study received no external funding.

Conflicts of interest

The authors declare no conflicts of interest or financial conflicts to disclose.

Data availability statement

The data used in this study can be obtained from the authors upon reasonable request.

Author details

Damtew Degefe Merga
Research and Technology Park, Wollega University, Nekemte, Ethiopia

*Address all correspondence to: hwre2020@gmail.com

IntechOpen

© 2023 The Author(s). Licensee IntechOpen. This chapter is distributed under the terms of the Creative Commons Attribution License (http://creativecommons.org/licenses/by/4.0), which permits unrestricted use, distribution, and reproduction in any medium, provided the original work is properly cited.

References

[1] Woldesenbet TA, Elagib NA, Ribbe L, Heinrich J. Hydrological responses to land use/cover changes in the source region of the upper Blue Nile Basin, Ethiopia. Science of the Total Environment. 2017;**575**:724-741. DOI: 10.1016/j.scitotenv.2016.09.124

[2] Merga DD, Adeba D, Regasa MS, Leta MK. Evaluation of surface water resource availability under the impact of climate Change in the Dhidhessa Sub-Basin, Ethiopia. Atmosphere (Basel). 2022;**13**(8):1296. DOI: 10.3390/atmos13081296

[3] Gashaw T, Tulu T, Argaw M, Worqlul AW. Modeling the hydrological impacts of land use/land cover changes in the Andassa watershed, Blue Nile Basin, Ethiopia. Science of The Total Environment. 2018;**619–620**:1394-1408. DOI: 10.1016/j.scitotenv.2017.11.191

[4] Kim BS, Kim BK, Kwon HH. Assessment of the impact of climate change on the flow regime of the Han River basin using indicators of hydrologic alteration. Hydrological Processes. 2011;**25**(5):691-704. DOI: 10.1002/hyp.7856

[5] Kis A, Pongrácz R, Bartholy J, Szabó JA. Projection of runoff characteristics as a response to regional climate change in a central/eastern European catchment. Hydrological Sciences Journal. 2020;**65**(13):2256-2273. DOI: 10.1080/02626667.2020.1798008

[6] Musie M, Sen S, Srivastava P. Application of CORDEX-AFRICA and NEX-GDDP datasets for hydrologic projections under climate change in Lake Ziway sub-basin, Ethiopia. Journal of Hydrology: Regional Studies. 2020;**31** (August):100721. DOI: 10.1016/j.ejrh.2020.100721

[7] Abdo KS. Assessment of climate change impacts on the hydrology of Gilgel Abay catchment in Lake tana basin, Ethiopia. 2010;**2274**:2267-2274. DOI: 10.1002/hyp

[8] Babar S, Ramesh H. Streamflow response to land use–land cover Change over the Nethravathi River basin, India. Journal of Hydrologic Engineering. 2015; **20**(10). Pp. 1, 4, 13. DOI: 10.1061/(asce)he.1943-5584.0001177

[9] Liu D et al. Assessing the effects of adaptation measures on optimal water resources allocation under varied water availability conditions. Journal of Hydrology. 2018;**556**:759-774. DOI: 10.1016/j.jhydrol.2017.12.002

[10] Chanapathi T, Thatikonda S, Raghavan S. Analysis of rainfall extremes and water yield of Krishna river basin under future climate scenarios. Journal of Hydrology: Regional Studies. 2018;**19**:287-306. DOI: 10.1016/j.ejrh.2018.10.004

[11] Liu J, Zhang C, Kou L, Zhou Q. Effects of climate and land use changes on water resources in the Taoer River. Advances in Meteorology. 2017;**2017**. Pp. 2, 6, 7, 11-16. DOI: 10.1155/2017/1031854

[12] Prasanchum H, Kangrang A, Hormwichian R, Compliew S. Impact of climate and rapid land use change on runoff quantities in Lower-Lampao river basin an improvement of irrigation planing view project impact of climate and rapid land use change on runoff quantities in Lower-Lampao River Basin, no. May 2014. 2013 [Online]. Available from: https://www.researchgate.net/publication/257942854

[13] Prasanchum H. Analyses of climate and land use changes impact on runoff

characteristics for multi- purpose reservoir system. 2017

[14] Dibaba WT, Miegel K, Demissie TA. Evaluation of the CORDEX regional climate models performance in simulating climate conditions of two catchments in upper Blue Nile Basin. Dynamics of Atmospheres and Oceans. 2019;**87**:101104. DOI: 10.1016/j.dynatmoce.2019.101104

[15] Barnes D. Climate change and water Inter-governmental Panel on Climate Change IPCC Technical Paper VI. 2008;**60** (8). DOI: 10.1525/9780520943933-008

[16] Melesse AM. Nile River Basin Ecohydrological Challenges Climate Change and Hydropolitics. Cham Heidelberg, New York, Dordrecht, London: Springer; 2014. ISBN 978-3-319-02719-7; ISBN 978-3-319-02720-3 (eBook). DOI: 10.1007/978-3-319-02720-3

[17] Gizaw MS, Biftu GF, Gan TY, Moges SA. Potential impact of climate change on streamflow of major Ethiopian rivers. August 2017;**143**(3): 371-383. DOI: 10.1007/s10584-017-2021-1

[18] Fentaw F, Hailu D, Nigussie A, Melesse AM. Climate change impact on the hydrology of Tekeze Basin, Ethiopia: Projection of rainfall-runoff for future water resources planning. Water Conservation Science and Engineering. 2018;**3**(4):267-278. DOI: 10.1007/s41101-018-0057-3

[19] Musie M, Sen S, Chaubey I. Hydrologic responses to climate variability and human activities in Lake Ziway Basin, Ethiopia. 2020;**12**(1):164. DOI: 10.3390/w12010164

[20] Zhou F, Xu Y, Chen Y, Xu C, Gao Y, Du J. Hydrological response to urbanization at different spatio-temporal scales simulated by coupling of CLUE-S and the SWAT model in the Yangtze River Delta region. 2013;**485**:113-125. DOI: 10.1016/j.jhydrol.2012.12.040

[21] Shawul AA, Chakma S, Melesse AM. The response of water balance components to land cover change based on hydrologic modeling and partial least squares regression (PLSR) analysis in the upper Awash Basin. Journal of Hydrology: Regional Studies. 2019;**26**: 100640. DOI: 10.1016/j.ejrh.2019.100640

[22] Chawla I, Mujumdar PP. Isolating the impacts of land use and climate change on streamflow. 2015. pp. 3633–3651. DOI: 10.5194/hess-19-3633-2015.

[23] Kundu S, Khare D, Mondal A. Individual and combined impacts of future climate and land use changes on the water balance. Ecological Engineering. 2017;**105**:42-57. DOI: 10.1016/j.ecoleng.2017.04.061

[24] Tirupathi C, Shashidhar T. Investigating the impact of climate and land-use land cover changes on hydrological predictions over the Krishna river basin under present and future scenarios. Science of the Total Environment. 2020;**721**(10):137736. DOI: 10.1016/j.scitotenv.2020.137736

[25] Wang Q et al. Individual and combined impacts of future land-use and climate conditions on extreme hydrological events in a representative basin of the Yangtze River Delta , China. 2020;**236**(163). Pp. 1-3, 4, 6-8. DOI: 10.1016/j.atmosres.2019.104805

[26] Dibaba WT, Demissie TA, Miegel K. Watershed Hydrological Response to Combined Land Use/Land Cover and Climate Change in Highland Ethiopia: Finchaa Catchment. Water. 2020;**12**(6): 2073-4441

[27] Melesse AM. Nile River Basin Ecohydrological Challenges, Climate Change and Hydropolitics. NewYork Dordrecht London: Springer Cham Heidelberg; 2014

[28] Gebremicael TG, Mohamed YA, Van Der Zaag P. Science of the Total environment attributing the hydrological impact of different land use types and their long-term dynamics through combining parsimonious hydrological modelling , alteration analysis and PLSR analysis. Science of The Total Environment. 2019;**660**: 1155-1167. DOI: 10.1016/j.scitotenv.2019.01.085

[29] Regasa MS. A Review on Land Use and Land Cover Change in Ethiopian Basins. 2021;**10**(6):585. DOI: 10.3390/land10060585

[30] Pirani A, Chen Y, Lonnoy E, Leitzell K, Connors SL, Berger S, et al. In: Proceedings of 2021 IPCC, The Physical Science Basis. Contribution of Working Group I to the Sixth Assessment Report of the Intergovernmental Panel on Climate Change. 2021. DOI: 10.1017/9781009157896

[31] Belihu M, Abate B, Tekleab S, Bewket W. Hydro-meteorological trends in the Gidabo catchment of the Rift Valley Lakes Basin of Ethiopia. Physics and Chemistry of the Earth. 2017;**104**. DOI: 10.1016/j.pce.2017.10.002. Available from: https://scholar.google.com/scholar?hl=en&as_sdt=0%2C5&q=+Hydro-meteorologi-cal+trends++in+the+Gidabo+catchment+of+the+Rift++Valley+Lakes+Basin+of+Ethiopia.+Physics++and+Chemistry+of+the+Earth&btnG=

[32] Hussen B, Mekonnen A, Murlidhar S. Integrated water resources management under climate change scenarios in the sub-basin of Abaya-Chamo, Ethiopia, model. Modeling Earth Systems and Environment. 2018;**4**(1):221-240. DOI: 10.1007/s40808-018-0438-9

[33] Aragaw HM, Goel MK, Kumar S. Hydrological responses to human-induced land use / land cover changes in the Gidabo River basin. Hydrological Sciences Journal. 2021;**66**(4):640-655. DOI: 10.1080/02626667.2021.1890328

[34] Odusanya AE, Mehdi B, Schürz C, Oke AO, Awokola OS. Multi-site calibration and validation of SWAT with satellite-based evapotranspiration in a data-sparse catchment in southwestern Nigeria. 4 February, 2019;**23**(2). pp. 1113–1144

[35] Endris HS et al. Assessment of the performance of CORDEX regional climate models in simulating east African rainfall. Journal of Climate. 2013;**26**(21):8453-8475. DOI: 10.1175/JCLI-D-12-00708.1

[36] Edamo ML, Bushira KM, Ukumo TY, Ayele MA, Alaro MA, Borko HB. Effect of climate change on water availability in Bilate catchment, southern Ethiopia. Water Cycle. 2022;**3**: 86-99. DOI: 10.1016/j.watcyc.2022.06.001

[37] Kanoma MS, Abdulkadir M. The impact of future climate change on water availability in Gusau, Zamfara state, Nigeria. Dutse Journal of Pure and Applied Sciences. 2022;7(4a):144-154. DOI: 10.4314/dujopas.v7i4a.15

[38] Hailemariam K. Impact of climate change on the water resources of Awash River basin, Ethiopia. Climate Research. 1999;**12**(2–3 SPEC. ISS. 6):91-96. DOI: 10.3354/cr012091

[39] Bekele WT. Implication of Representative Concentration Pathway's on Arjo-Didessa Catchment, Upper Blue

Nile Basin, Using Multiple Climate Models. Arba Minch University; 2017. Available from: https://scholar.google.com/scholar?hl=en&as_sdt=0%2C5&q=Implication+of+Representa-tive+Concentration+Pathway%E2%80%99s+on+Arjo-Didessa+Catchment%2C&btnG=

[40] Daba MH. Modelling the impacts of climate Change on surface runoff in Finchaa sub-basin, Ethiopia. International Journal of Food Science and Agriculture. 2018;**2**(1):14-29. DOI: 10.26855/jsfa.2018.01.002

[41] Kefeni K, Mokonnen B, Roba N. Evaluation the performance of regional climate models in simulating rainfall characteristics over upper awash Sub-Basin, Ethiopia. 2020;**5**(1):134-138

[42] Leander R, Buishand TA. Resampling of regional climate model output for the simulation of extreme river flows. Journal of Hydrology. 2007;**332**(3–4):487-496. DOI: 10.1016/j.jhydrol.2006.08.006

[43] Bourqui M, Mathevet T, Gailhard J, Hendrickx F. Hydrological validation of statistical downscaling methods applied to climate model projections. IAHS-AISH Publication. 2011;**344**:32-38

[44] Terink W, Hurkmans RTWL, Torfs PJJF, Uijlenhoet R. Evaluation of a bias correction method applied to downscaled precipitation and temperature reanalysis data for the Rhine basin. Hydrology and Earth System Sciences. 2010;**14**(4):687-703. DOI: 10.5194/hess-14-687-2010

[45] Setegn SG, Rayner D, Melesse AM, Dargahi B, Srinivasan R. Impact of climate change on the hydroclimatology of Lake Tana Basin, Ethiopia. Water Resources Research. 2011;**47**(4):1-13. DOI: 10.1029/2010WR009248

[46] Neitsch S, Arnold J, Kiniry J, Williams J. Soil & water assessment tool theoretical documentation version 2009, Texas water resources institute. 7 July, 2015;**5**(3):1-647. DOI: 10.1016/j.scitotenv.2015.11.063

[47] Moriasi DN, Gitau MW, Pai N, Daggupati P. Hydrologic and water quality models: Performance measures and evaluation criteria. Transactions of the ASABE. 2015;**58**(6):1763-1785. DOI: 10.13031/trans.58.10715

[48] Xue-song Z, Fang-hua HAO, Hong-guang C, Dao-feng LI. Application of Swat model In the upstream watershed of the Luohe River. 2003;**13**(4):334-339

[49] Mazengo M, Kifanyi GE, Mutayoba E, Chilagane N. Modeling surface water availability for irrigation development in Mbarali River sub-catchment Mbeya, Tanzania. Journal of Geoscience and Environment Protection. 2022;**10**(04):1-14. DOI: 10.4236/gep.2022.104001

[50] Liersch S et al. Are we using the right fuel to drive hydrological models? A climate impact study in the upper Blue Nile. Hydrology and Earth System Sciences. 2018;**22**(4):2163-2185. DOI: 10.5194/hess-22-2163-2018b

[51] Gupta HV, Sorooshian S, Yapo PO. Status of automatic calibration for hydrologic models: Comparison with multilevel expert calibration. Journal of Hydrologic Engineering. 1999;**4**(2):135-143. DOI: 10.1061/(asce)1084-0699(1999)4:2(135)

Chapter 2

Perspective Chapter: Quantifying the Carbon Footprint of a High School – A Three-Part Study

Pınar Rüya Uludağ and Başak Kiliç Taşeli

Abstract

Lokman Hekim Vocational and Technical Anatolian High School, which is located in Bulancak-Giresun in the Black Sea Region of Turkey, was selected to quantify its carbon footprint after previously conducting an awareness seminar. This study is divided into three parts. Firstly, pre-training was given to both students and teachers. Next, an awareness questionnaire was administered to determine their understanding of carbon footprints, climate change and global warming. The results showed that while the students were knowledgeable about climate change and global warming, they lacked sufficient understanding of carbon footprints. Secondly, a face-to-face measurement study was conducted with 295 students to calculate the annual average amount of CO_2 emitted by them based on various categories. The results revealed that the students emitted an average of 7.43 ton of CO_2e per year, with the highest amount of emissions (3 ton CO_2e/year) being attributed to household factors, and 2.135 ton CO_2e/year resulting from lifestyle habits. Finally, the institutional emission amounts from Scope 1, Scope 2, and Scope 3 were determined to be 85.4k, 16.7k, and 18.9k ton CO_2e/year, respectively. Consequently, the total institutional carbon footprint of the high school was calculated to be 121k ton of CO_2e/year.

Keywords: carbon footprint, institutional carbon footprint, awareness, global warming, climate change, CO_2 emissions, scope 1–2-3

1. Introduction

The primary priority in reducing the impact of human activities on the environment is awareness, and it is important to measure the direct and indirect effects of human activities on the environment in order to increase awareness and create behavior change. Environmental education aims to raise awareness of individuals at all levels about their position and role in the environment, to provide them with as much information as possible about all factors affecting the environment, and to make them conscious through an educational process. Environmental education involves directing students toward projects where they can learn by experiencing and observing, and where they can think and produce solutions in a student-centered approach, rather than through superficial explanations [1, 2]. Inadequate environmental education negatively affects students' attitudes toward the environment [3, 4].

Numerous studies have shown that carbon footprint is an effective tool in raising awareness about the impact of individual consumption habits on the environment [5–8]. Conway [9] calculated the ecological footprint of the University of Toronto and investigated the changes in the ecological footprint on campus by developing different scenarios (such as generating their own electricity, using public transportation, and using recycled office materials). By presenting the numerical data they obtained from the study to university students, they created awareness about the ecological footprint and encouraged students to reduce their own ecological footprints.

A study was conducted at Akdeniz University Health Services Vocational School to calculate the carbon emissions resulting from personnel transportation and electricity consumption, which totaled 98,307 kg CO_2e (CO_2 equivalent). The study was further conducted to assess the carbon footprints of gasoline, diesel, and LPG vehicles by changing fuel types, with calculated values of 102,214 kg, 95,749 kg, and 100,427 kg, respectively. The observed differences in these values are attributed to the fuel types' calorific values, chemical structures, and the engine types in which they are used. The study revealed that electricity consumption accounts for the largest share of the school's carbon emissions, and reducing electricity consumption directly reduces emissions. The study emphasized that the school's electricity consumption results from heating, cooling, lighting, office vehicles, and elevators. To reduce electricity consumption from heating and cooling, the study suggests applying thermal insulation to the building, replacing current air conditioning units with highly energy-efficient ones, and conducting awareness-raising training for staff and students. The study also proposes using photovoltaic solar panels and wind turbines to generate energy in a region that receives solar radiation, such as Antalya [10].

The primary carbon footprint of Erzincan Binali Yıldırım University was assessed using two different methods, namely Intergovernmental Panel on Climate Change (IPCC) Scope 1 and Department for Environment, Food and Rural Affairs (DEFRA). The IPCC Scope 1 approach yielded carbon footprints of 2753.2 tCO_2e (ton CO_2 equivalent) for 2019 and 2383.74 tCO_2e for 2020, with a reduction of 13.42% compared to 2019. The DEFRA conversion factor approach produced carbon footprints of 2314.53 tCO_2e for 2019 and 1826.53 tCO_2e for 2020, with a decrease of 21.08% compared to 2019. The decrease was attributed to the adoption of distance learning due to the Covid-19 pandemic and the warming weather resulting from global warming. The largest impact on the carbon footprint, following electricity-based emissions, was from natural gas consumption for heating, while the least impact was from gasoline consumption, according to the IPCC Scope 1 approach [11]. In contrast, the DEFRA conversion factor approach identified electricity and coal consumption as the highest and second-highest sources of carbon emissions, respectively, while gasoline consumption had the least impact, similar to the IPCC Scope 1 approach [12].

In another carbon footprint analysis conducted at Manisa Celal Bayar University, the annual carbon emissions of the university were calculated to be 8,953,906 ton of CO_2e. The majority of carbon emissions came from electricity consumption, which accounted for 7,865,721 ton of CO_2e, followed by coal with 613,701 ton of CO_2e, and the least amount of emissions were from gasoline consumption, which was 8223 ton of CO_2e. To reduce the carbon footprint of the university, several measures have been proposed, such as planting 4700 saplings on campus, launching the solar power plant project with renewable energy, demonstrating the senior management's determination and support to reduce the university's carbon footprint and promote sustainable resource use, conducting educational activities and projects to raise awareness in the fight against pollution, offering an elective course on environmental issues, nature conservation, global climate change, and sustainable use of natural resources in all departments of the university, and carrying out studies to save energy in electricity, heating, transportation, and water consumption [13].

In the study conducted by [14] at the Madrid University Forestry Faculty, which has 1150 students and 235 staff members, the total carbon footprint of the faculty was calculated as 2147 ton of CO_2e per year, with Scope 1 emissions of 169 ton CO_2e per year, Scope 2 emissions of 703 ton CO_2e per year, and Scope 3 emissions of 1275 ton CO_2e per year.

In another study conducted by [15] at the Mehmet Akif Ersoy University Bucak Health Vocational School, which has 357 students and 23 staff members, the total CO_2e emissions from natural gas, electricity, and student and staff transportation were reported as 217,503 kg CO_2e per year. Additionally, it was noted that heating from natural gas consumption had the biggest impact on carbon emissions, followed by emissions from electricity consumption, diesel, gasoline, and LPG vehicles, respectively.

In our study, we selected Lokman Hekim Vocational and Technical Anatolian High School located in Bulancak-Giresun, Turkey as the site for conducting awareness, individual carbon footprint, and institutional footprint studies. This was done after the school had previously conducted an awareness seminar. The study was divided into three main parts. The first part involved providing preliminary training to 304 students to determine their awareness of ecological and carbon footprints. The second part of the study involved face-to-face measurement studies with 295 students to calculate the amount of CO_2 generated annually by students and the category-based emission amounts. In the third part of the study, the emissions released institutionally in the Lokman Hekim Vocational and Technical Anatolian High School buildings were calculated. This included emissions from Scope 1 sources such as coal and generators, Scope 2 sources such as electricity, and Scope 3 sources including transportation, water consumption, paper consumption, kitchen cylinders, fire cylinders, air conditioning, and refrigerators.

2. Materials and methods

2.1 Awareness studies

A preliminary training was conducted for the school's 304 students and 29 teachers. Following the training, a 13-question awareness questionnaire on ecological and carbon footprint was administered to determine the level of awareness among students and teachers.

2.2 Individual carbon emission calculations

Following the analysis of the survey results, face-on-face meetings were held with 295 students to calculate their carbon footprints using a program based on DEFRA (Department for Environment, Food and Rural Affairs) measurement standards. This program offers a carbon footprint calculator that helps individuals measure their impact on environment. The calculator takes into account various factors such as energy consumption, transportation, waste management, and lifestyle choices to provide an estimate of carbon footprint of individuals.

2.3 Institutional carbon emission calculations

The calculation of institutional carbon emissions involves the application of the standards set forth by the Intergovernmental Panel on Climate Change (IPCC). The IPCC has established three distinct methods to determine CO_2 emissions: Scope 1, Scope 2, and Scope 3 [11].

Scope 1 pertains to the measurement of direct emissions resulting from an organization's activities. This encompasses all activities carried out within the organization, as well as any transportation activities conducted using the organization's own vehicles.

Scope 2 pertains to emissions resulting from sources such as electricity, water, heat, and other utilities that an institution purchases from external sources, rather than producing internally.

Scope 3 includes other emissions that are not encompassed in Scope 2, but are indirectly caused by an organization's corporate activities. This includes emissions stemming from packaging used for products, advertising brochures, subcontracting activities outsourced by institutions, rental vehicles, and the fuel utilized by these vehicles. Additionally, Scope 3 includes all emissions related to work-related travel by employees, including air, sea, and land transportation.

Table 1 presents the sources, units, and monitoring frequencies of data collected for emission sources in order to calculate institutional carbon emissions. These data were obtained from information and documents provided by the administrative staff of high school. The institutional emission sources of the high school include Scope 1 emissions from fuel use, Scope 2 emissions from purchased electricity, and Scope 3 emissions from paper consumption, water consumption, school bus transportation, public transportation, personal vehicle arrival and departure, energy use, air conditioning, kitchen cylinder, fire extinguisher, and refrigerator.

Scope	Emission source	Unit	Data source
Scope 1	School fuel consumption	L	School administration
Scope 2	Purchased electricity	kWh	School administration
Scope 3	Paper consumption	kg	School administration
	Water consumption	ton	School administration
	School bus	km	Service company
	Public transport	km	Questionnaire
	Personal vehicle arrival and departure	km	Questionnaire
	Energy	V	School administration
	Air conditioning	number, type	School administration
	Kitchen tube	number	School administration
	Fire extinguisher	number, type	School administration
	Freezer	number, type	School administration

Table 1.
Lokman Hekim vocational and technical Anatolian high school emission sources.

The literature commonly employs the ISO 14064-1 series guide and Greenhouse Gas Protocol's calculation groups to determine institutional carbon footprints, utilizing data collection, calculation, reporting, and reference value tables outlined in the IPCC guidelines. Carbon footprint calculations for products and services during their life cycle also make use of PAS 2050 and ISO 14067 standards, developed and published by the British Standards Institution (BSI) [16].

The high school has 304 students and 29 personnel, who collectively own 12 vehicles. Transportation for students is facilitated by 10 shuttles, 8 of which are public and 2 private. In order to determine the institutional carbon footprint of the high school, various factors contributing to emissions, including coal consumption for heating, electricity usage, and the annual transportation of students and staff, were examined. To gather the necessary data, information from 2021 were utilized because it covers all the invoices related to emission sources. These invoices were provided by the high school administration, and the calculations were based on the information contained within them and given in **Table 2**.

2.3.1 Emission calculations associated with electricity usage

The emission calculations associated with electricity usage employ the formula provided in Eq. (1) [17].

$$\mathrm{EtCO_2e} = \left(\left(\left(\mathrm{FV}_{\frac{\mathrm{kWh}}{\mathrm{year}}} * \mathrm{EF}_{\frac{\mathrm{kgCO_2}}{\mathrm{kWh}}}\right) * \mathrm{T\&}DL\right) + \mathrm{FV}_{\frac{\mathrm{kWh}}{\mathrm{year}}} \times \mathrm{EF}_{\frac{\mathrm{kgCO_2}}{\mathrm{kWh}}}\right) * 10^{-3} \tag{1}$$

where,
E tCO_2e = Emission amount,
FV = Total amount of electricity consumed annually (kWh/year),
EF = Emission factor ($kgCO_2e$/kWh), (for Turkey 0,437 $kgCO_2e$/kWh),
T&DL = Transmission and distribution loss value, (for Turkey %13.3).

2.3.2 Emission calculations due to coal use

The formula given in Eq. (2) was used in the calculations of emissions from public transportation [17].

$$\sum_{k=1}^{n} \mathrm{AD} * \mathrm{EFk} * (1 - \mathrm{BR})\mathrm{k} * \mathrm{CF} * 3.664 \tag{2}$$

Number of students	**304**
Staff number	29
Staff vehicle number	12
School fuel consumption	coal
Student transport distance of the shuttle (Total)	350 km/day - 63,000 km/year
Number of shuttle	8 + 2
Shuttle fuel type	Diesel

Table 2.
The data associated with Lokman Hekim vocational and technical Anatolian high school.

where,
AD = Activity data,
EF = Emission factor,
BO = Biomass rate,
CF = Conversion factor,
3.664 = Indicates the conversion to tCO_2e with the input value conversion factor of 3.664.

2.3.3 Emission calculations for gasoline usage

The formula given in Eq. (3) was used in the calculations of emissions caused by gasoline use [17].

$$EtCO_2e = \left(\left(\left(GAD * GA\right) * 10^{-3}\right) * \left(EF * 10^{-3}\right)\right) * \left(FO * OF\right) * 10^{-3} \quad (3)$$

where,
E tCO_2e = Total gasoline emission,
GAD = Gasoline average density (kg/m^3),
GA = Yearly gasoline amount (L/year),
EF = Emission factor ($kgCO_2/TJ$),
FO = Factor of oxidation (TJ),
OF = Oxidation factor (accepted as 1 in IPCC guidelines).

2.3.4 Emission calculations for diesel usage and generators

The formula given in Eq. (4) was used in the calculations of diesel fuel emissions [17].

$$EtCO_2e = \left(\left(\left(DAD * DA\right) * 10^{-3}\right) * \left(EF * 10^{-3}\right)\right) * \left(FO * OF\right) * 10^{-3} \quad (4)$$

where,
E tCO_2e = Total diesel emission,
DAD = Diesel average density (kg/m^3),
DA = Yearly diesel amount (L/year),
EFk = Emission factor, ($kgCO_2e/TJ$),
FO = Factor of oxidation, (TJ).
OF = Oxidation factor (accepted as 1 in IPCC guidelines).

2.3.5 Emission calculations for air conditioner, refrigerator, and fire protection systems

The formula given in Eq. (5) was used in the emission calculations from fire protection systems [17].

$$EtCO_2e = \left(\left(THFC * 10^{-3}\right) * TC * \right)\left(GWP * \left(\left(0,5 * R - 32\right) + \left(0,5 * R - 125\right)\right.\right. \quad (5)$$

where,
E tCO_2e = Total HFC-227ea (FM200) gas emission carbon dioxide equivalent,
THFC = Total HFC-227ea gas (kg),
GWP = Global warming potential ($tCO_2e/tHFC\text{-}227ea$),

TC = The total amount of charging made to the system due to emission (the ratio of the total HFC-227ea gas loaded to the system from the % value—an annual emission of around 1% occurs).

2.3.6 Emission calculations from public transportation

The formula given in Eqs. 2 and 6 was used in the calculations of emissions from public transportation [17].

$$DU = (KMY * 0,25)/SN \tag{6}$$

where.
KMY = km per year,
0,25 = Average diesel burned by the bus per km (L),
SN = Staff number,
DU = Diesel usage (L/year.person).

3. Results and discussion

3.1 Awareness survey

Upon evaluation of the data from the ecological footprint awareness survey, data reveals that out of the 295 students who took part in the survey, 66% were female and 34% were male. Additionally, the survey found that 40.2% of participants were in 10th grade, 27.8% were in 9th grade, 18.6% were in 11th grade, and 13.4% were in 12th grade.

A 13-question awareness questionnaire was then administered to determine their understanding of carbon footprints. Results are presented in **Table 3**.

Questions	Answers		
	Yes (%)	No (%)	Partially (%)
1. Do you have information about global warming?	53.6	43.3	3.1
2. Do you have information about climate change?	64.9	30.9	4.2
3. Do you have information about the causes of climate change?	46.4	42.3	11.3
4. What does the future hold in terms of global and regional climate change?	46	41	13
5. Are you aware of the climate changes that have been observed in Turkey?	38.1	52.6	9.3
6. Do you have information about the ecological footprint?	15.5	63.9	20.6
7. Do you have information about the carbon footprint?	17.5	58.8	23.7
8. Do you have information about what should be done to mitigate the carbon footprint?	9.3	66	24.7
9. Do you think you know enough about this subject?	11.3	37.2	51.5

Questions	Answers		
	Yes (%)	No (%)	Partially (%)
10. Do you think that the course content on these subjects is sufficient?	25.8	46.4	27.8
11. Do you read the news in the newspapers about environmental problems?	21.6	37.1	41.2
12. Do you want environmental problems to be covered in the lessons?	62.9	14.4	22.7
13. Does your family inform you about environmental pollution?	26.8	32	41.2

Table 3.
Awareness questionnaire for determining student's understanding of carbon footprints, global warming, and climate change.

According to the results of the survey, it was understood that the majority of the students had only knowledge about climate change and global warming, and whether they had read the articles in publications such as newspapers and magazines, and the majority of the answers were no, and it was understood that they obtained this information only from sources such as media and television. It has been determined that they do not have enough knowledge on issues such as ecological footprint and carbon footprint and they want these issues to be more on the agenda and to be covered in their curriculum.

3.2 Individual carbon emissions

With the one-to-one measurement study conducted with 295 students, the annual average amount of CO_2 created by the students and the amount of emissions on a category basis were calculated. The carbon footprint emissions created by students with the program based on the measurement standards of DEFRA (England Department of Environment, Food and Agricultural Affairs) are calculated according to the categories. This program offers a carbon footprint calculator that helps individuals measure their impact on the environment. The calculator takes into account various factors such as energy consumption, transportation, waste management, and lifestyle choices to provide an estimate of an individual's carbon footprint. The calculator asks questions listed in **Table 4** to obtain the necessary information.

The annual average amount of CO_2 created by the students and the amount of emissions on a category basis were calculated and results are given in **Table 5** and schematized in **Figure 1**. As can be seen, the annual CO_2 amount created by the students is 7.43 ton CO_2e/year. The emissions by category are from household factors with 3 ton of CO_2e/year, from lifestyle habits with 2.135 ton of CO_2e/year, from flights with 1.169 ton of CO_2e/year, from car use with 0.427 ton of CO_2e/year, from motorcycle use with 0.104 ton of CO_2e/year. It was determined that it was caused by bus-train travel with 0.315 ton of CO_2e/year. Annual CO_2 amount created by students (7.43 ton of CO_2e/year) is much higher than the Turkey average of 4.1 ton CO_2e/year, and the main reason for this is the use of coal for heating in Bulancak District and the limited public transport. In order to balance this emission, the amount of trees that students need to plant annually was found to be 11 on average.

A study conducted by [14] at the Faculty of Forestry of the University of Madrid reported 1.87 ton of CO_2e/year emissions per student. This difference is due to the

Home
What is your country of residence?
What is your household size?
How much electricity do you consume on a monthly basis?
What type of heating do you use in your home?
Transportation
How many km do you travel by car each year?
How many km do you travel by bus each year?
How many km do you travel by train each year?
How many km do you travel by plane each year?
How many km do you travel by motorbike each year?
Lifestyle questions
How much waste do you generate on a monthly basis?
How much water do you consume on a monthly basis?
How much meat do you consume on a weekly basis?
How much dairy do you consume on a weekly basis?
How much seafood do you consume on a weekly basis?

Table 4.
Computer program input questions.

Categories	Average generated CO_2 (ton CO_2e /year)
Home	3
Flight	1.169
Car	0.427
Motorbike	0.104
Bus-train	0.315
Life style	2.135
Total	7.432
Number of trees need to be planted per person	10.53

Table 5.
Average carbon footprint of the students.

widespread use of natural gas in the faculty where the study was conducted, the fact that coal is not used as a fuel and public transportation is common in transportation.

Upon analyzing the major contributors to household emissions, it can be concluded that fuel and electricity consumption used within the household is the most significant source. In terms of lifestyle factors, personal preferences such as food and drink choices, entertainment, and shopping habits play a significant role. Taking into account the age, lifestyle, and shopping preferences of students, it can be observed

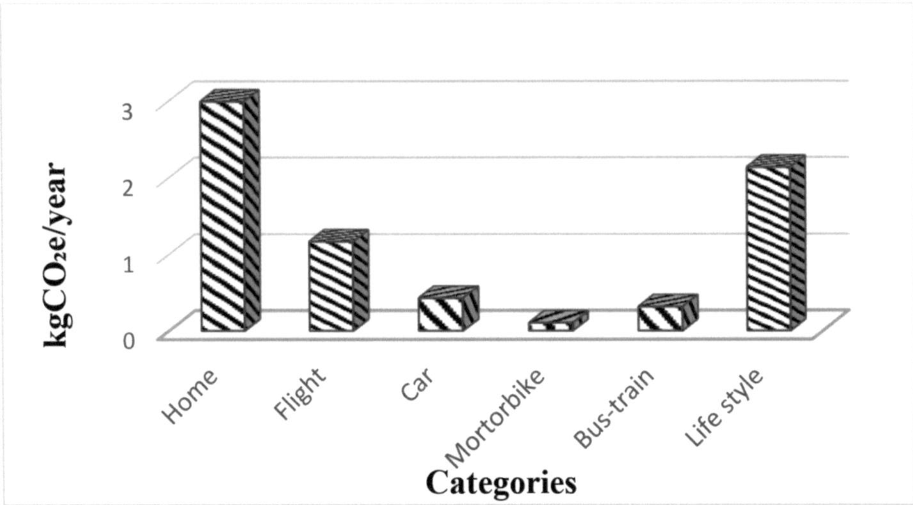

Figure 1.
CO_2e averages created by students according to categories.

that their limited knowledge of alternative social activities, eating and drinking habits, organic and local products may lead to higher emissions through their lifestyle choices. Therefore, the impact of lifestyle factors on carbon footprint emissions should not be overlooked.

3.3 Institutional carbon emissions

Table 6 provides information on the emission sources utilized to calculate the carbon footprint of the school. The table includes the emission calculations for each source and the total amount of emissions generated.

Emission amounts from Scope 1 (coal, generator), Scope 2 (electricity), and Scope 3 (transportation, water consumption, paper consumption, kitchen tube, fire extinguisher, air conditioner, and refrigerator) are calculated as 85,368 ton CO_2e/year, 16,716 ton CO_2e/year, and 18,916 ton CO_2e/year, respectively. As a result of the second part of the study, the total institutional footprint of the high school is 120,984 ton CO_2e/year.

Similarly, the largest share of the total institutional carbon footprint of the Madrid University Forestry Faculty was originated from Scope 1 (coal, generator) and followed by Scope 3 (transportation, water consumption, paper consumption, kitchen tube, fire extinguisher, air conditioner, and refrigerator), and by Scope 2 (electricity) [14].

In addition, the study conducted at Mehmet Akif Ersoy University Bucak Health Vocational School noted that heating from natural gas consumption had the biggest impact on carbon emissions, followed by emissions from electricity consumption, diesel, gasoline, and LPG vehicles, respectively [15].

It was also reported that the largest impact on the carbon footprint, following electricity-based emissions, was due to natural gas consumption for heating, while the least impact was due to gasoline consumption, according to the Scope 1 approach [11].

Scope	Emission source	Emission amount (ton CO_2e/year)	Total emission (ton CO_2e/year)
Scope 1	Generator	0.18	85,368
	Coal	85,188	
Scope 2	Electric	16,716	16,716
Scope 3	Paper consumption	0.23	18,916
	Water consumption	0.296	
	School shuttle	11.97	
	Arrival and departure by personal vehicle	3.088 diesel 0.958 gasoline 0.77 LPG	
	Fire extinguisher	0.15	
	Air conditioner, refrigerator	1.27	
	Kitchen tube	0.18	
Total			120,984

Table 6.
Total emission amount for high school.

4. Conclusion

Throughout their lives, individuals leave their mark on the world through the consumption and production activities they engage in. When considering the food, clothing, consumer goods, resources used for heating and transportation, and the resulting waste generated throughout their lives, it is clear that the footprint left behind is not insignificant. The top priority in reducing the impact of human activities on the environment is awareness. It is crucial to be able to measure the direct and indirect effects of human activity on the environ4ment to increase awareness and create behavioral change. One widely used method of measurement in recent times is the carbon footprint.

During the 27th Conference of the Parties to the United Nations Framework Convention on Climate Change, also known as COP 27, emphasis was placed on the fact that the young generation will be the most affected by the climate crisis. As a result, it was recommended that young people should participate in future climate crisis negotiations and that countries should establish targets that prioritize the urgent reduction of greenhouse gases. To achieve this goal, it is crucial to raise awareness among young people about carbon footprint, ecological footprint, and climate change.

The survey results revealed that most students possessed only a basic understanding of climate change and global warming. Furthermore, when asked if they had read articles on the topic in publications like newspapers and magazines, the majority responded negatively. Instead, they relied on sources such as media and television for information. Additionally, the survey indicated that students lacked sufficient knowledge of concepts such as ecological footprint and carbon footprint. They expressed a desire for these topics to receive more attention and be incorporated into their curriculum.

Upon analyzing the major contributors to household emissions (3 ton of CO_2e/year), it can be concluded that fuel and electricity consumption used within the household is the most significant source. In terms of lifestyle factors (2.135 ton of CO_2e/year), personal preferences such as food and drink choices, entertainment, and shopping habits play a significant role. Taking into account the age, lifestyle, and shopping preferences of students, it can be observed that their limited knowledge of alternative social activities, eating and drinking habits, organic and local products may lead to higher emissions through their lifestyle choices. Therefore, the impact of lifestyle factors on carbon footprint emissions should not be overlooked. Annual CO_2 amount created by the students is 7.43 ton CO_2e/year, and is much higher than the Turkey average of 4.1 ton CO_2e/year, and the main reason for this is the use of coal for heating in Bulancak District and the limited public transport. In order to balance this emission, the amount of trees that students need to plant annually was found to be 11 on average.

The total institutional carbon footprint of the high school was found to be 120,984 ton CO_2e/year, with the largest share originating from Scope 1 (coal, generator) with 85,368 ton CO_2e/year, followed by Scope 3 (transportation, water consumption, paper consumption, kitchen tube, fire extinguisher, air conditioner, and refrigerator) with 18,916 ton CO_2e/year and by Scope 2 (electricity) with 16,716 ton CO_2e/year.

Increasing the number of studies on calculating and reducing carbon footprints in schools is crucial to raising awareness about this issue. It is also important to organize training seminars that educate students and teachers about global climate change, as well as to develop projects that address this topic. Furthermore, incorporating subjects such as climate change, global warming, efficient resource utilization, carbon emission reduction, sustainability, and sustainable production and consumption into the education curriculum can help reduce carbon footprint in schools. Regular afforestation activities can also help reduce carbon emissions, as well as saving fuel, waste, electricity, heating, and transportation, which are major sources of carbon footprint.

In addition, it is essential to encourage school personnel to adopt electric and hybrid vehicle technologies, which are being implemented worldwide, including Turkey, in order to reduce emissions from transportation, the most significant source of CO_2 emissions.

Acknowledgements

The authors gratefully thank Mr. Cem Mutlu Türkseven, the Assisstant Director of Lokman Hekim Vocational and Technical Anatolian High School, for providing data support.

Conflict of interest

The authors declare no conflict of interest.

Declarations

The authors have no relevant financial or non-financial interests to disclose. The authors have no competing interests to declare that are relevant to the content of this paper.

Author details

Pınar Rüya Uludağ[1] and Başak Kiliç Taşeli[2]*

1 Gürgentepe Municipality, Ordu, Turkey

2 Department of Environmental Engineering, Giresun University, Giresun, Turkey

*Address all correspondence to: basak.taseli@giresun.edu.tr

IntechOpen

© 2023 The Author(s). Licensee IntechOpen. This chapter is distributed under the terms of the Creative Commons Attribution License (http://creativecommons.org/licenses/by/4.0), which permits unrestricted use, distribution, and reproduction in any medium, provided the original work is properly cited.

References

[1] Yalvaç GH. The Effect of Cooperative Learning Approach on the Mental Structures of Teacher Candidates Regarding the Environment. Master Thesis. Turkey: Abant İzzet Baysal Univesity; 2008

[2] Yücel A. Morgil F: Investigation of environmental phenomenon in higher education. Journal of Hacettepe University Faculty of Education. 1998; **14**:84-91

[3] Atasoy E, Ertürk H. A field study on primary school students' environmental attitudes and environmental knowledge. Journal of Erzincan Faculty of Education. 2008;**10**(1):105-122

[4] Keleş Ö, Naim U, Özsoy S. Calculation and evaluation of pre-service teachers' ecological footprints. Aegean Education Journal. 2008;**9**(2):1-15

[5] Schaller D. Our footprints - they're all over the place. Newsletter of the Utah Society for Enviromental Education. 1999;**9**(4):1-10

[6] Dawe FMG, Vetter A, Martin S. An overview of ecological footprinting and other tools and their application to the development of sustainability process: Audit and methodology at Holme Lacy College, UK. International Journal of Sustainability in Higher Education. 2004;**5**(4):340-371

[7] Janis AJ. Quantifying the Ecological Footprint of the Ohio State University. Dissertation. USA: The Ohio State University; 2007

[8] Wada Y, Izumi K, Mashiba T. Development of a web-based personal ecological footprint calculator for the Japanese. In: International Ecological Footprint Conference. Cardiff: Cardiff University; 2007

[9] Conway TM, Dalton C, Loo J, Benakoun L. Developing ecological footprint scenarios on university campus: A case study of the University of Toronto at Mississauga. International Journal of Sustainability in Higher Education. 2008;**9**(1):4-20

[10] Yaka F, Koçer A, Güngör A. Detection of the carbon footprint of Akdeniz University vocational school of health services. Electronic Journal of Machine Technologies. 2015;**12**(3):37-45

[11] IPCC. Guidelines for National Greenhouse Gas Inventories. 2006. https://www.ipcc-nggip.iges.or.jp/public/2006gl/ [Accessed: August 13, 2022]

[12] Seyhan AK, Çerçi M. Determination of carbon footprint with IPCC tier 1 and DEFRA methods: Fuel and electricity consumption example of Erzincan Binali Yıldırım university. Journal of Suleyman Demirel University Institute of Science. 2022;**26**(3):386-397

[13] Binboğa G, Ünal A. A research on calculating the carbon footprint of Manisa Celal Bayar University on the axis of sustainability. International Journal of Economic and Administrative Studies. 2018;**21**:187-202

[14] Alvarez S, Blanquer M, Rubio A. Carbon footprint using the compound method based on financial accounts: The case study of the School of Forestry Engineering. Journal of Cleaner Production. 2014;**66**:224-232

[15] Kumaç K, Akyüz A, Zaman M, Güngör A. Carbon footprint determination for a sustainable

environment: Maku Bucak health college example. Al-Cezerî Journal of Science and Engineering. 2019;**6**(1):108-117

[16] The Guide to PAS 2050:2011 How to Carbon Footprint Your Products, Identify Hotspots and Reduce Emissions in Your Supply Chain. https://www.bsigroup.com/globalassets/localfiles/en-th/carbon-footprint/pas-2050-2011-guide.pdf [Accessed: January 15, 2023]

[17] Üreden A, Özden S. How to calculate corporate carbon footprint: A theoretical study. Anatolian Journal of Forest Studies. 2018;**4**(2):98-108

Chapter 3

Dual Nature of Land Ocean Thermal Contrast during Pre-Monsoon and Onset Phase of Indian Summer Monsoon

Jayanti Pal and Sayoni Sarkar

Abstract

Under faster Indian Ocean (IO) warming, several thermodynamical properties of the atmosphere over the Indian sub-continent change abruptly. The present study has evaluated the temperature field using ERA5 and IMDAA reanalysis. From the climatological evolution, it has been observed that before monsoon onset over Kerala (MOK) not only does meridional tropospheric temperature gradient reverses from negative to positive but the surface LOTC also decline very sharply. Interannual variation of LOTC shows that there is no significant trend, however, warming since 1980 may lead to an increase in variability. The reason behind having no trend in LOTC may be attributed to land ocean warming ratio (WR). Composite analysis depicts that except for early MOK, surface LOTC decreases sharply before MOK while deep tropospheric LOTC or meridional tropospheric temperature increases. The climatological average of pre-monsoonal average is about -5.52 K which has been found slightly higher (-4.9 K) during the early MOK years and found slightly lower (-5.7 K) during the delayed MOK years. Hence deep tropospheric LOTC is mostly used to identify the onset of MOK while surface LOTC can be utilized to predict MOK. However, to make a more precise MOK prediction, the interaction between three-dimensional temperature field with large-scale flow needs to explore.

Keywords: land ocean thermal contrast, surface, deep tropospheric, pre-monsoon, MOK

1. Introduction

India being home to a large section of the agriculture-dependent population, thrives largely on the Indian Summer Monsoon Rainfall (ISMR). Besides deciding the agricultural output of a year, the onset of ISMR plays a pivotal role in the transition of the weather from the scorching summer season to a rainy/monsoon season. More than a billion people in South Asia are impacted by ISMR, which accounts for approximately 80% of the total annual rainfall from June to September. The onset of ISMR is marked by the monsoon onset over Kerala

(MOK) in early June followed by a northward propagation of the monsoon winds and thereby covering the whole country by mid-July. The onset and advancement of the monsoon are primarily attributed to the differential heating of land and ocean that results in the formation of a gradient between the land surface temperature and ocean surface temperature [1–4], often known as the Land Ocean Thermal Contrast (LOTC). LOTC is usually expressed by an amplification factor or a ratio factor, which is defined as the relationship between the change in land temperature to the change in ocean temperature. From various observations and simulations of the general circulation model (GCM), it has been seen that the global amplification factor has a value of 1.5 approximately [5–7]. However, the factor also undergoes variation depending on the latitude. In the multi-model average, the local minimum in the tropics is 1.2, and the maximum in the subtropics is 1.6 [8].

The role behind the land ocean thermal contrast being the primary driver of monsoon was well explained by Sir Edmund Halley in the seventeenth century. During the summer season, the land gets heated up faster than the oceans because of which a low-pressure region is created over the land mass followed by the formation of a high-pressure region over the oceans. Also, the polar shift of the Inter Tropical Convergence Zone (ITCZ) to the position over South Asia strengthens the center of the thermally induced low-pressure center. This creates a pressure gradient from the ocean to the land, causing warm and humid air to flow from the ocean to the land. The situation is exactly the opposite during the winter season. The landmass cools faster than the oceans during the winter season. So, a high-pressure area is formed over the land and a low-pressure region is established over the ocean. This results in the establishment of the land-to-ocean pressure gradient followed by the outflow of cold and dry air from the land to the ocean. The advancement of the monsoon winds northward implies a shift in the deep convection from the equatorial to continental regions [3, 9, 10].

As stated earlier, the monsoon onset is mostly thought to be due to the LOTC. The classic theory of LOTC only takes account of the surface temperature and can only lead to shallow circulation. However, the meridional gradient in the upper troposphere is directly proportional to the meridional gradient of the deep troposphere heating, which can accelerate the deep tropospheric circulation [11]. The upper levels show an appreciable difference between the temperatures of the pre-monsoon and the monsoon months [12]. He et al. [13] through their work suggested that a considerable amount of tropospheric warming starts in May over the areas stretching from the Eastern Tibetan Plateau (TP) to the South China Plain. As a result, the meridional temperature gradient on south of the eastern Plateau reverses and the low-level south-westerlies start blowing over the Bay of Bengal. Then, the onset of the Indian monsoon occurs in June with the increase of temperature over a wide area covering Saudi Arabia, Iran, Afghanistan and the western Plateau. The thermal effect of TP is quite prominent as the air over it, is always warmer than its surrounding areas [14]. Flohn [15, 16] proposed that seasonal warming of the upper surface of the TP and subsequent changes in meridional temperature gradient and pressure gradient to the south of latitude 35°N caused large-scale changes in Asian atmospheric circulation and outbreaks of monsoon over the Indian subcontinent. Li and Yanai [17] confirmed and concluded that the change in the meridional temperature gradient is due to the increase in temperature over the Tibetan Plateau, and there is no obvious temperature change on the Indian Ocean (IO), indicating that in the spring month of the year, TP acts as an independent heat source, separate from the more intense heat sources that are associated with the equatorial IO rain belt. They further noted that when, on one hand, the sensible heating of the surface leads to the Tibetan heat source, the release of latent heat of condensation contributes to the oceanic heat source and it is the sensible heating of the land surface that is responsible for the reversal of the meridional temperature gradient than the latent heat of condensation. This is because though the intensity of latent heat of condensation may be high, due to the adiabatic cooling, it is unable to produce any significant temperature change over the IO. Hence LOTC may get controlled by temperature change over land during onset. After the onset, ISMR has a northward propagation and moves from the southern part of the country to cover the whole country. The progress of ISMR over India is perpendicular to the mean winds blowing from the southwest from June to September [18]. After the onset of ISMR, there is a significant decrease in land temperature thereby resulting in a cooling of the land surface. However, the latent heat released from the convection phenomenon keeps the troposphere above the land warm thereby maintaining the LOTC and consequently helping in the advancement of the monsoon throughout the country [19]. Hence, deep tropospheric LOTC is capable of driving winds that carry a massive amount of moisture from the warm oceans and can impact large-scale changes. The above discussion clears that surface LOTC and deep tropospheric LOTC have different aspects and effects during onset and advancement. Under the global warming scenario, it has been observed that the Land surface temperature (LST) increases more rapidly than the ocean surface temperature (OST) in the northern hemisphere under increasing greenhouse gasses. Although through literature review may provide insight about the dependency of LOTC on land temperature, it is not clear how does LOTC modify in response to change in land and ocean temperature. To understand, how does thermal contrast impact ISM, studies are mostly focused on surface temperature rather than deep tropospheric temperature. Jin and Wang [20] showed that ISM has been started reviving since 2002 under favorable surface LOTC while the theory of association of surface LOTC with surface LOTC has been proved to be null and void by Gadgil et al. [21]. However, how the deep tropospheric LOTC evolves and effect Indian monsoon onset and advancement is not well explored.

In this present study, LOTC at the surface and for the deep troposphere has been evaluated and explored in detail. A brief climatology of surface temperature and deep tropospheric temperature along with monthly change and variation at an inter-annual scale has been assessed to understand the spatial pattern of LOTC evolution during ISM onset and advancement. The present study has also explored the sensitivity of LOTC towards the surface and deep tropospheric temperature for both land and ocean. Afterward, the study has tried to explore the effect of LOTC during different types of ISM onsets. Section 2 describes about data and methodology used in the present study. Section 3 includes the detailed observed analysis followed by Section 4 which summarizes and provides a brief conclusion.

2. Data and methodology

2.1 Data

To study the impact of LOTC on the onset and advancement of ISMR the analysis has been carried out for five months - March, April, May, June and July using long-term data. Parameters considered for the purpose of study are - Temperature at 2 m, Temperature at pressure levels, Precipitation. All meteorological data except precipitation used for evaluation have been taken from the fifth-generation reanalysis datasets of the European Centre for Medium-Range Weather forecasts (ERA5;[22]) and the Indian Monsoon Data Assimilation and Analysis (IMDAA; [23]) reanalysis. From both sources, data has been collected at monthly and hourly time scales for all available time period. The evaluation has been carried out using two datasets (IMDAA and ERA5) so as to check the robustness of the data and draw a stronger conclusion from the results obtained. ERA5 provides all data at 25 km spatial resolution from 1959 to 2021 and IMDAA provides data at a horizontal resolution of 0.12° × 0.12° from 1979 to 2021. For each case, pressure level temperature has been taken for 600 hPa, 500 hPa, 400 hPa, 300 hPa and 200 hPa. Hourly data from both sources have been converted into daily by averaging. Tropospheric Temperature has been calculated by averaging temperature from 600 hPa to 200 hPa as per [11].

2.2 Methodology

The climatological and monthly tendency has been analyzed to understand the change in the temperature field and development of LOTC. Through this analysis, the study has obtained a domain over both land and ocean to study LOTC. A spatial linear trend analysis has been performed. Since the linear trend does not have any statistical significance, a Man-Kendall trend analysis has been implemented along with linear trend analysis for area-averaged data. To explore the variability of LOTC, two parameters have been evaluated. First is Land Ocean thermal contrast which is the standardized difference between the land temperature and the ocean temperature. The positive value of LOTC signifies higher land temperature. Another parameter is land ocean warming ratio (WR) which is the ratio of change in land temperature and change in ocean temperature. WR provides insight about into the direction in which land and ocean change simultaneously. The positive value of WR signifies that land and ocean temperature changes are in phase i.e. either two are increasing or decreasing. The negative value of WR depicts out-phase change in both temperature fields. For further analysis, monsoon onset over Kerala (MOK) has been classified as early, delay and normal MOK considering mean MOK as 1st June with a standard deviation of 8 days. Thereafter a composite analysis has been performed to analyze LOTC field during different MOK.

3. Results

3.1 Climatological analysis

Pre-monsoon and pre-onset climatology of the temperature field has been analyzed using all available data from both data sources. The climatological evolution of the Surface temperature at 2 m (**Figure 1**) from both sources - IMDAA and ERA5 show similar results. The graphs show a relatively low temperature over northern India and the Tibetan Plateau (TP) region compared to the rest of India for all the months from March to July. However, it can be observed that as the season changes from the pre-monsoon period to the monsoon period the temperature over the TP increases but remains low than the rest of the Indian Subcontinent. Also, the

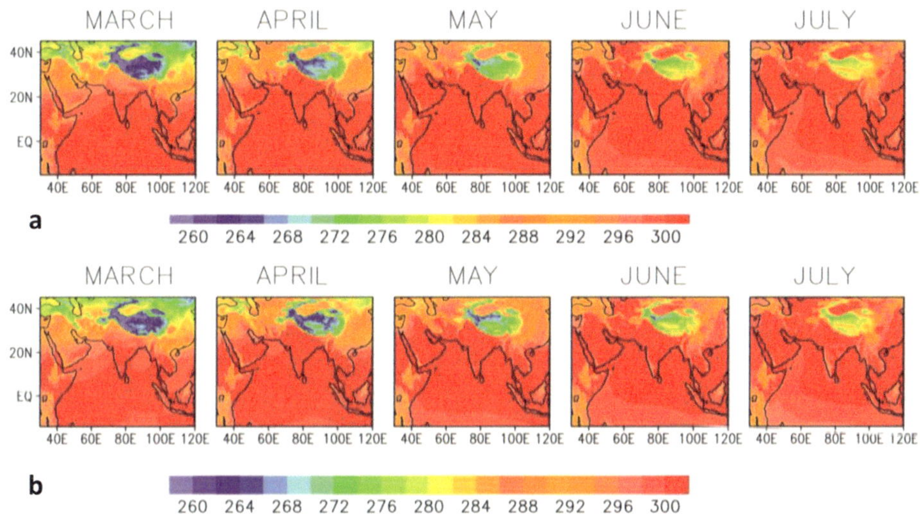

Figure 1.
Climatology of surface temperature at 2 m during March, April, May, June and July from (a) IMDAA reanalysis and (b) ERA5 reanalysis.

temperature over the Indian landmass is seen to increase by very small amounts from March to July. The surface temperature over the Indian Ocean remains warm from March to July only varying by small amounts. Tropospheric temperature (TT) over the northern part of India and the Tibetan Plateau shows an increase from March to July in **Figure 2**. The increase is small during March, April and May which intensify in the month of June and July. In July, the tropospheric temperature is high in the Tibetan plateau region thereby helping the monsoon advancement. The IMDAA data

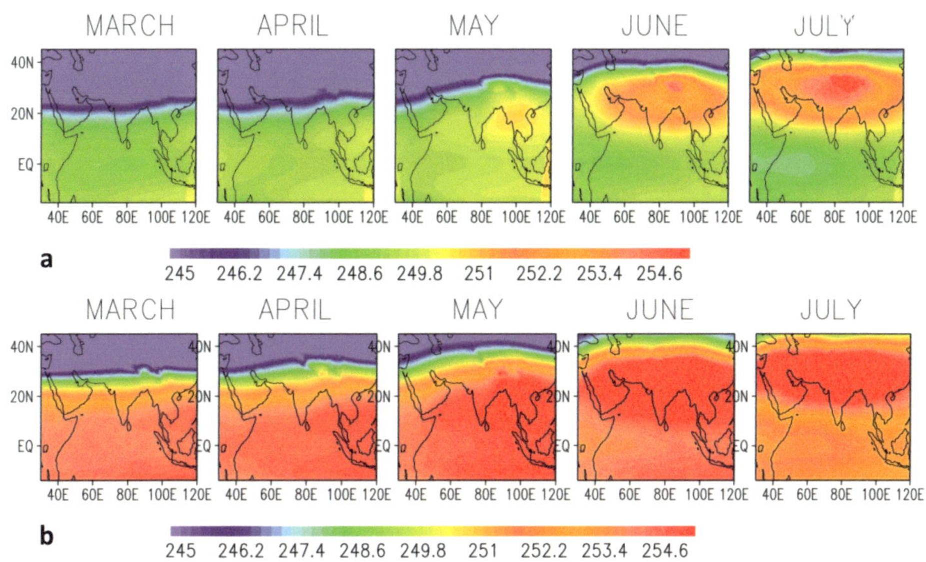

Figure 2.
Climatology of tropospheric temperature during March, April, May, June and July from (a) IMDAA reanalysis and (b) ERA5 reanalysis.

shows a less amount of warming over the rest of India compared to the ERA 5 data. The TT over the Indian Ocean is also seen to increase from March to July and similar to the land temperature IMDAA shows a lesser temperature than that shown by the ERA 5 data.

The monthly tendency plots (**Figure 3**) show that the intensity of warming up of the land surface is increasing during the pre-monsoon period. The land surface of the Indian subcontinent and Tibetan plateau region is warmer in April and May compared to March and April, respectively. However, a decrease in land surface temperature is seen in Southern India in June compared to May. This decrease in land temperature shifts from Southern India to the Central part of India and so a decline of surface temperature over land is seen in this part from July to June. A similar pattern of tendency is seen in the Tropical Indian Ocean. The warming of the ocean surface temperature increases in April and May compared to March and April, respectively. On the other hand, there is a decrease in the temperature during June and July compared to May and June, respectively. From the graphs, it can be concluded that the warming intensity and the change in temperature are more on land than that in the ocean. The monthly tendency plots of the TT (**Figure 4**) show that over the land the tropospheric temperature warming intensity increases by a small amount from March to April, and increases significantly over the Northern India and Tibetan Plateau region during the onset and advancement of the ISMR. But, the amount of increase of the warming decreases during the advancement than during the onset phase. On the other hand, the tropospheric temperature over the Indian ocean shows a positive tendency in April and May compared to March and April, respectively. However, the temperature over the TIO starts decreasing in June and July when compared to the Temperature in May and June, respectively. Both sources

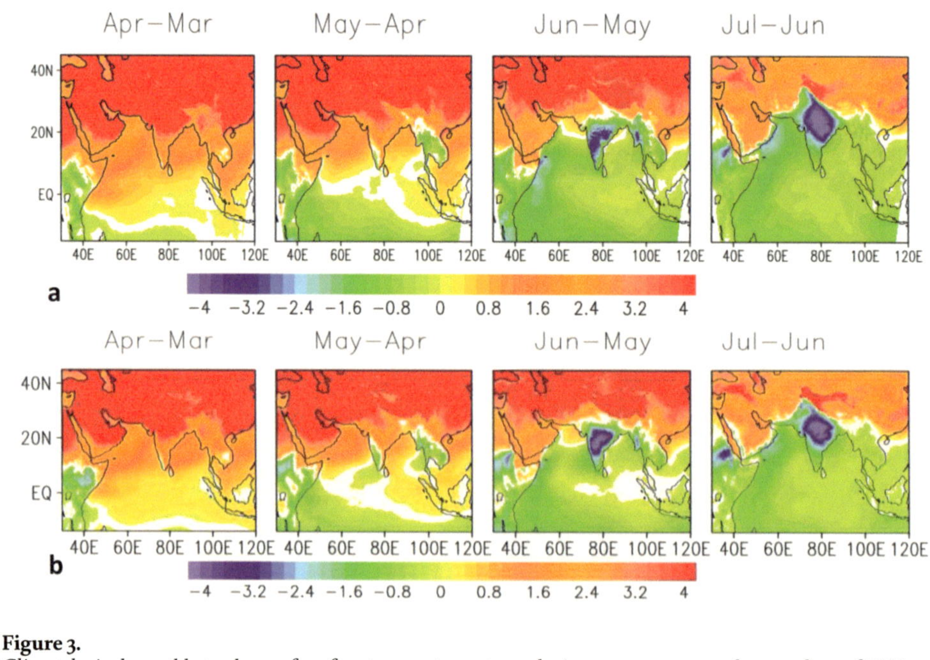

Figure 3.
Climatological monthly tendency of surface temperature at 2 m during pre-monsoon and onset phase of ISM evaluated from (a) IMDAA reanalysis and (b) ERA5 reanalysis.

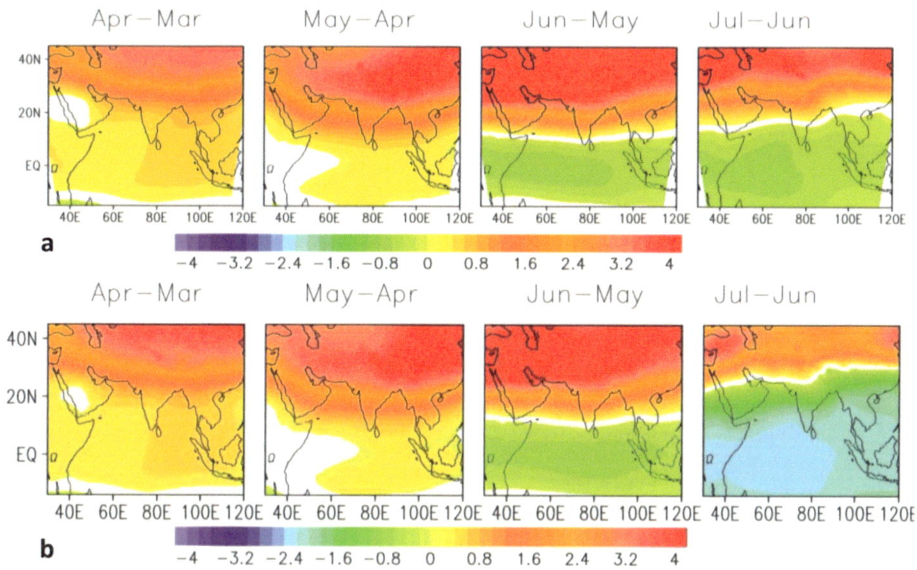

Figure 4.
Climatological monthly tendency of tropospheric temperature during pre-monsoon and onset phase of ISM evaluated from (a) IMDAA reanalysis and (b) ERA5 reanalysis.

(IMDAA and ERA5) show nearly the same results but the tendency of tropospheric temperature over the Indian Ocean is slightly different for the July–June plot where the ERA5 data shows a larger decrease in the tropospheric temperature of the Indian Ocean than that in IMDAA data.

Based on the above analysis, it has been observed that thermal contrast regions are not the same for the surface and deep troposphere. Surface temperature mostly changes due to the radiative effect and due to the contrast of heat capacity between land and ocean whereas the temperature of the deep troposphere is mostly controlled by latent heating. The above analysis shows that before the onset of the monsoon, a small region of the southern peninsular develops LOTC near the Kerala region where deep tropospheric LOTC becomes prominent and a significant structure during the onset of monsoon. Hence for further land analysis of LOTC, land and ocean areas have been identified separately for surface and deep troposphere. For surface land and ocean, area is bounded by 8-18 N;78-80E and 10-20 N;60-75E, respectively, and for the deep troposphere, land and ocean area are bounded by 25-35 N;60-100E and 10-5 N; 50-95E, respectively (**Figure 5**). **Figure 6** shows the climatological evolution of LOTC from March to July using the IMDAA reanalysis data and ERA5 reanalysis data. The y-axis on the left and right sides shows the surface LOTC plotted in black and deep tropospheric LOTC plotted in green respectively. From the graphs of both sources, it is evident that during the pre-monsoon period, surface LOTC decreases and increases with the onset of the monsoon. While the deep tropospheric LOTC shows almost a linear increase during the pre-monsoon and the monsoon periods. The 0 line is plotted for the deep tropospheric LOTC as it shows a negative value during the pre-monsoon period, however, the surface LOTC has a positive value through the 5 monthsperiod. It has been observed that not only meridional tropospheric temperature gradient becomes positive but also the surface land ocean thermal contrast shows a sharp decline just before onset (**Figures** 7 and **8**).

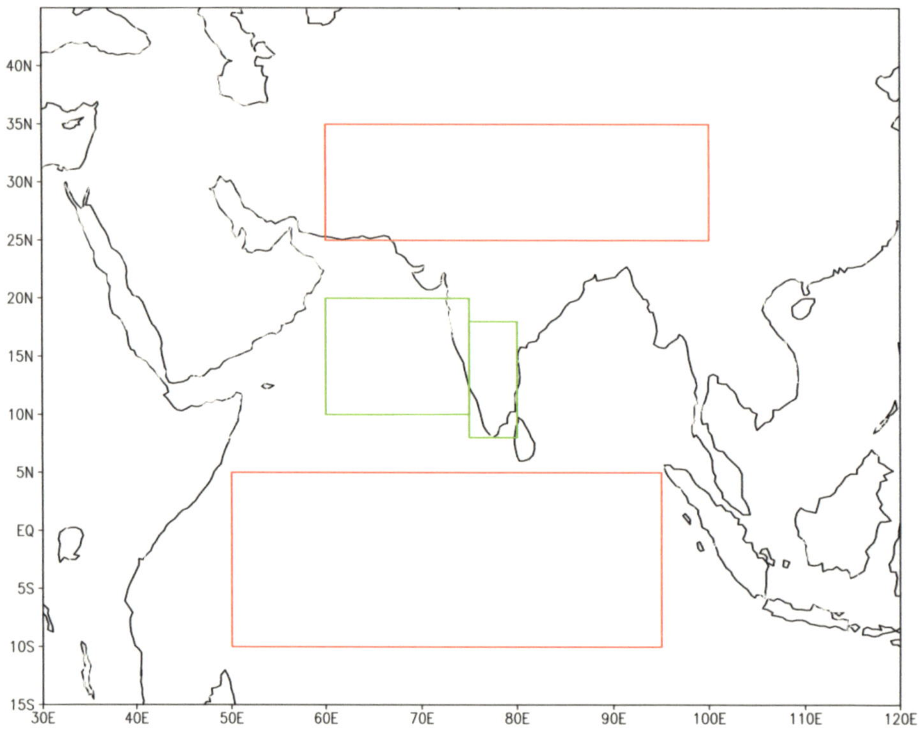

Figure 5.
Land Ocean area for surface temperature (green box) and tropospheric temperature (red box).

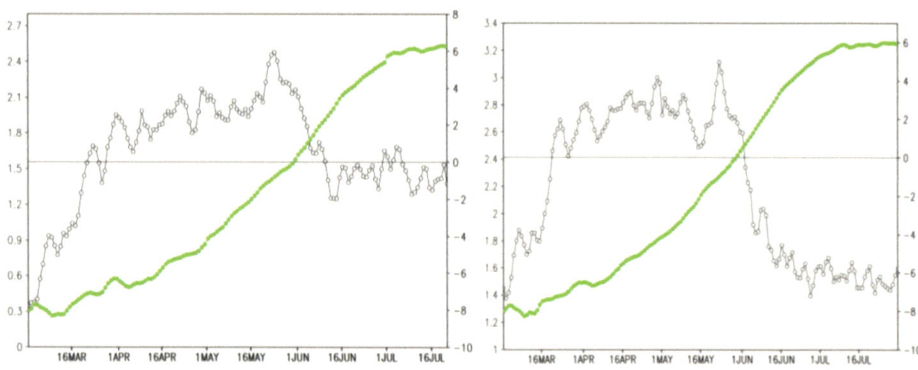

Figure 6.
Climatological evolution of Land Ocean thermal contrast starting from March to July evaluated from (a) IMDAA reanalysis and (b) ERA5 reanalysis. Left side y-axis signify surface LOTC (black line) and right-side y-axis signify deep tropospheric LOTC (green line). "0" line is marked against deep tropospheric LOTC.

3.2 Variability analysis

Interannual variation of LOTC at the surface level and in the deep troposphere has been attributed through trend analysis. It has been observed that there is no

DOI: http://dx.doi.org/10.5772/intechopen.110536

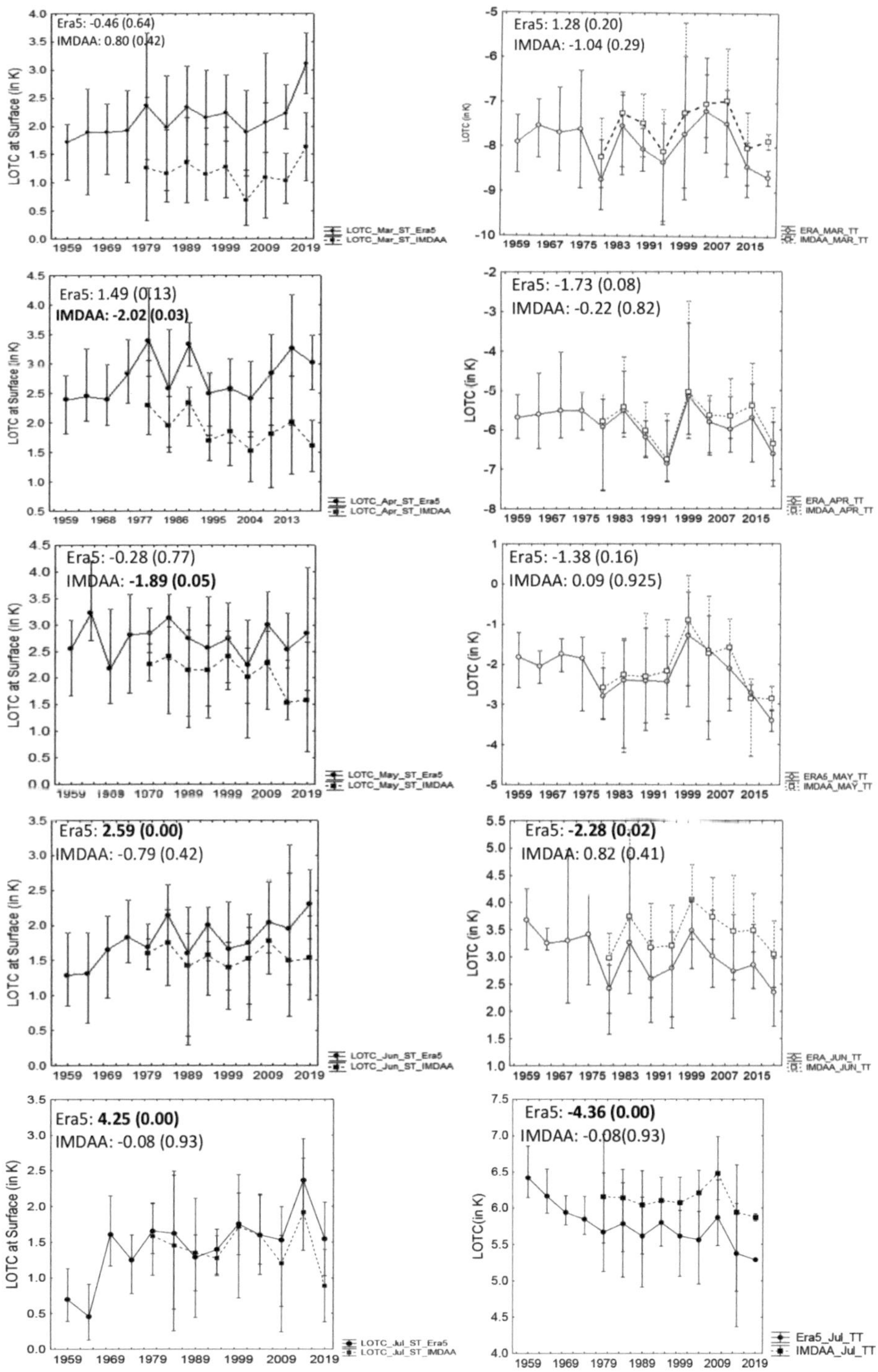

Figure 7.
Interannual variation of LOTC at the surface (left panel) and in the deep troposphere (right panel). Man-Kendall trend analysis has been marked with its statistical significance. Trend value significant at 95% level has been marked in bold.

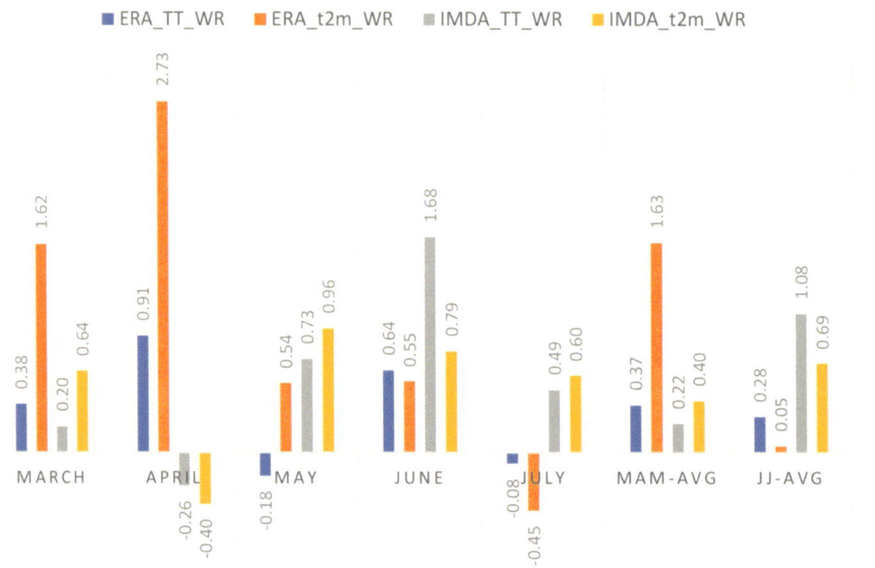

Figure 8.
Monthly climatology of Land Ocean warming ratio.

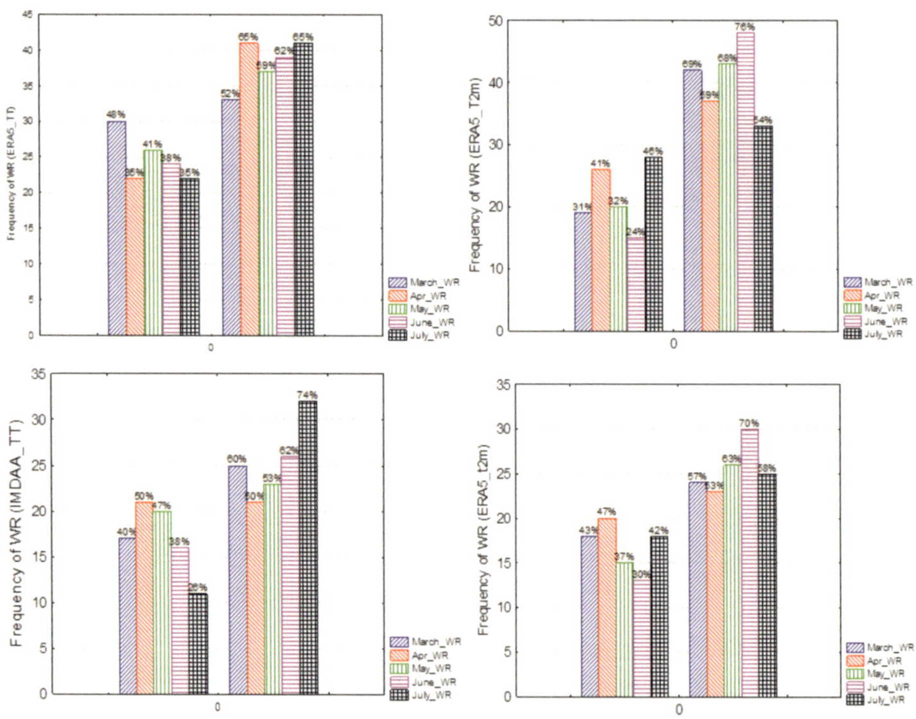

Figure 9.
Frequency analysis of Land Ocean warming ratio.

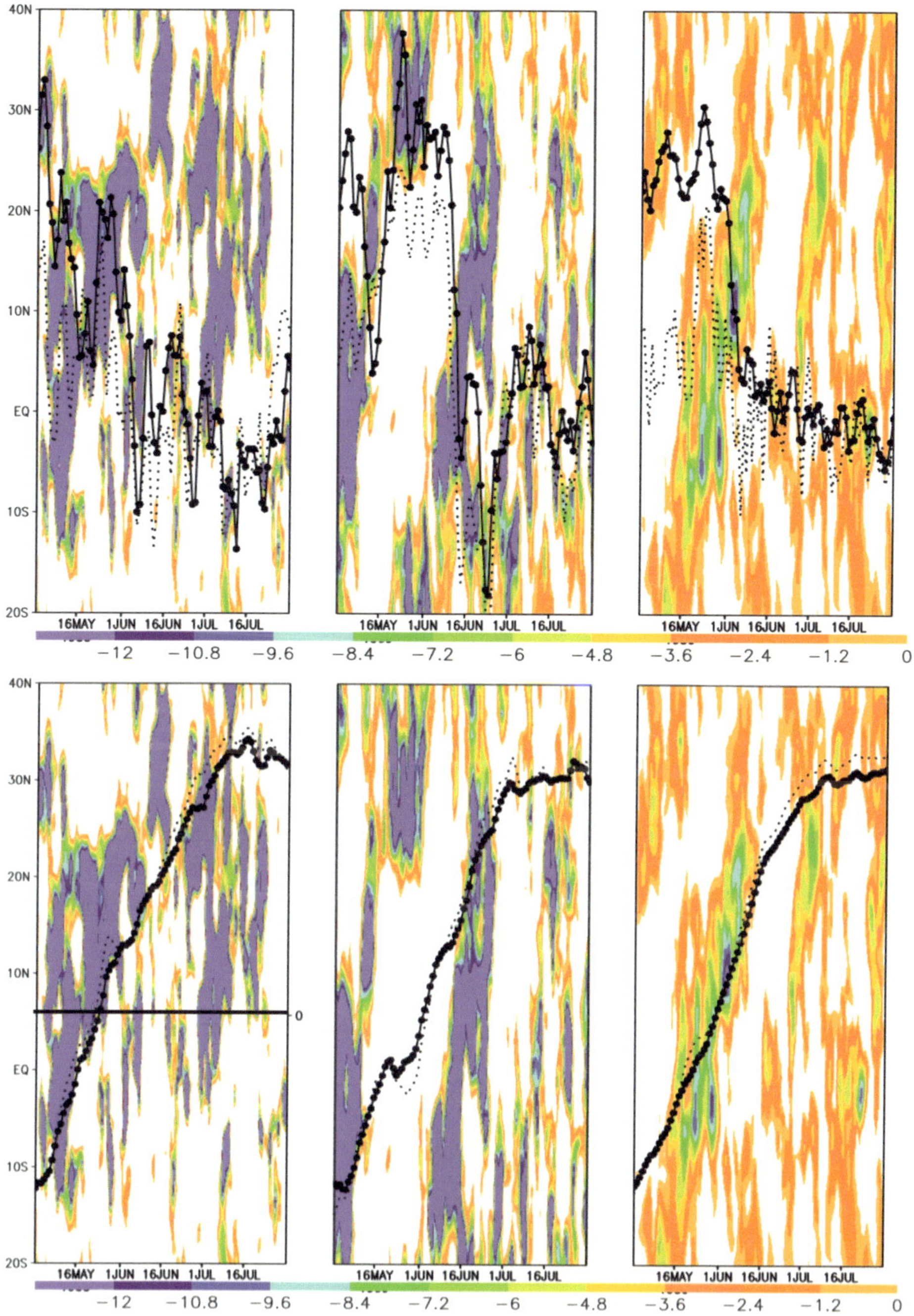

Figure 10.

Composite analysis of northward propagation of convective activity averaged over 70E-80E and evolution of surface and deep tropospheric LOTC during early (left panel), delay (middle panel) and normal (right panel) MOK years. Left y-axis defines latitude (in degree N) and right y-axis defines LOTC (in degree K). Solid black color for ERA5 and dotted line for IMDAA reanalysis. The upper panel is for surface LOTC and the bottom panel for deep tropospheric LOTC. In both cases shaded color define OLR anomaly.

Early	Delay	Normal
1960,1961,1962,1969, 1985,1990,1999,2009,2013	1967,1972,1979,1983,1986,1 995,1996,1997,2002,2003	1959,1963,1964,1965,1966,1968,1970, 1971,1973,1974,1975,1976,1977,1978, 1980,1981,1982,1984,1987,1988,1989, 1991,1992,1993,1994,1998,2000,2001, 2004,2005,2006,2007,2008,2010,2011, 2012,2014,2015,2016,2017,2018,2019, 2020,2021

Table 1.
Classification of early, delay and Normal monsoon onset over Kerala based on MOK mean date 1st June.

significant visual linear trend in LOTC at both levels. Man-Kendall trend analysis depicts that the surface LOTC shows a significant decreasing trend (IMDAA reanalysis) during May and a significant increasing trend (ERA5 reanalysis) during June and July. The deep troposphere LOTC interannual variation, on the other hand, shows a significant declining trend (ERA5 reanalysis) during June and July.

Change in LOTC may be attributed through changes in land and ocean temperature which can be understood through the warming ratio which provides insight into the direction of change. In the month of March positive value of WR for all data sources signifies that the direction of change is the same over both land and ocean. In the month of April, there is some inconsistency in the outcome. ERA5 data signifies that land and ocean temperature changes in the same direction while IMDAA signifies the opposite. In the month of May, except for the deep tropospheric, both data source signifies surface temperature changes over land and ocean in the same direction. For the month of June, all data depicts the change in temperature in the same way. However, if seasonal averages are considered, it may be stated that land and ocean both manifest changes in temperature in the same direction. Frequency analysis (**Figure 9**) suggests that more than 60% cases of land and ocean temperature changes are in phase.

3.3 Impact of Land-Ocean thermal contrast during onset and advancement of Indian summer monsoon

To understand the impact of LOTC on monsoon onset and advancement, the present study has mainly focused on the movement of convective activity. Northward propagation of convection attributed through an anomaly in OLR, is a significant trigger during ISM onset. The present study has analyzed and compared the LOTC field along with northward propagation for early, delay and normal MOK years. It has been observed that during early MOK, surface LOTC decreases gradually while during delay and normal onset, the declination LOTC is sharp. Also, Deep tropospheric LOTC evolves according to convective activity. Hence it is clear that to trigger MOK, surface LOTC is significant rather than deep LOTC. Deep tropospheric LOTC evolution is the manifestation of convective activity (**Figure 10** and **Table 1**). It has been observed that during early MOK deep tropospheric LOTC is higher than normal during pre-monsoon season while during delay, it has been noted to be lower than normal. For surface LOTC, the variation in surface LOTC during early, delay and normal is not same (**Figure 11**).

Figure 11.
Variation of land ocean thermal contrast during normal MOK, early MOK and delay MOK for (a) deep tropospheric LOTC and (b) surface LOTC.

4. Conclusions

Land ocean thermal contrast has been the most debated topic in the field of Indian summer Monsoon studies. One group of scientists denied the theory of LOTC behind the Indian monsoon and another group provides justification of keeping the LOTC theory. The present study aims to understand the contradiction and explore a completely new insight about LOTC. Most of the study pertaining to ISM circulation and rainfall has been associated with surface LOTC while deep tropospheric LOTC has been associated with MOK.

The present study evaluated the temperature field at the surface and for the deep troposphere using the two most current reanalyzes i.e., ERA5 and IMDAA reanalysis. Although these two data have been at two different spatial resolutions, these two data can explicitly capture the climatological behavior. While comparing these two data for the temperature field, it has been observed that it has range differences with a similar spatial pattern. With both data sets, trend analysis has been performed which shows the disparity in results unless the trend is very prominent in both data in the same way. It has been assumed that IMDAA reanalysis resolved the Indian feature much more than ERA5 because of high resolution as IMDAA reanalysis has been only generated by focusing on the Indian sub-continent.

A detailed climatological analysis has been performed with the temperature field at monthly, daily and interannual scales. Spatial climatology and monthly tendency analysis explored that surface LOTC and deep tropospheric LOTC has different domain over land and ocean which has been identified and further analysis of LOTC over those areas has been evaluated. From the climatological evolution of surface and deep tropospheric LOTC, it has been observed that before MOK not only meridional tropospheric temperature gradient reverses from negative to positive but surface LOTC also decline very sharply. Interannual variation of LOTC shows that there is no significant trend, however, warming since 1980 may lead to an increase in variability. The reason behind having no trend in LOTC may be attributed through WR ratio which depicts that in more than 60% cases land and ocean changes are in phase.

The evolution of LOTC analysis depicts that except for early MOK, surface LOTC decreases sharply before MOK while deep tropospheric LOTC or meridional tropospheric temperature increase may be a manifestation of increasing convective activity. Hence deep tropospheric LOTC is mostly used to identify the onset of MOK while surface LOTC can be utilized to predict MOK.

Acknowledgements

The authors are also grateful to IntechOpen for giving the opportunity to contribute to this book and also thankful to the editor for accepting the proposal.

Conflict of interest

"The authors declare no conflict of interest."

Author details

Jayanti Pal* and Sayoni Sarkar
Department of Atmospheric Science, Central University of Rajasthan, India

*Address all correspondence to: jayanti.pal@curaj.ac.in

IntechOpen

© 2023 The Author(s). Licensee IntechOpen. This chapter is distributed under the terms of the Creative Commons Attribution License (http://creativecommons.org/licenses/by/4.0), which permits unrestricted use, distribution, and reproduction in any medium, provided the original work is properly cited.

References

[1] Lau K, Li M. The monsoon of East Asia and its global associations—A survey. Bulletin of the American Meteorological Society. 1984;**65**(2):114-125. DOI: 10.1175/1520-0477(1984)065<0114:TMOEAA>2.0.CO;2

[2] Webster PJ. The elementary monsoon. In Fein JS, Stephens PL, editors. New York: Monsoons, Wiley Co.; 1987. p. 3-32

[3] Webster PJ, Magaña VO, Palmer TN, Shukla J, Tomas RA, Yanai M, et al. Monsoons: Processes, predictability, and the prospects for prediction. Journal of Geophysical Research. 1998;**103**(C7):14451-14510. DOI: 10.1029/97JC02719

[4] James IN. Introduction to Circulating Atmospheres. Cambridge Atmospheric and Space Series (9). Cambridge University Press; 1994. p. 444

[5] Sutton RT, Dong B, Gregory JM. Land/sea warming ratio in response to climate change: IPCC AR4 model results and comparison with observations. Geophysical Research Letters. 2007;**34**:L02701. DOI: 10.1029/2006GL028164

[6] Drost F, Karoly D, Braganza K. Communicating global climate change using simple indices: An update. Climate Dynamics. 2012;**39**:989-999. DOI: 10.1007/s00382-011-1227-6

[7] Joshi MM, Lambert FH, Webb MJ. An explanation for the difference between twentieth and twenty-first century land–sea warming ratio in climate models. Climate Dynamics. 2013;**41**:1853-1869. DOI: 10.1007/s00382-013-1664-5

[8] Byrne MP, O'Gorman PA. Land–Ocean warming contrast over a wide range of climates: Convectiv e quasi-equilibrium theory and idealized simulations. Journal of Climate. 2012;**26**:4000-4016. DOI: 10.1175/JCLI-D-12-00262.1

[9] Rao YP. Southwest Monsoon. New Delhi: India Meteorological Department; 1976

[10] Sikka DR, Gadgil S. On the maximum cloud zone and the ITCZ over Indian longitudes during the southwest monsoon. Monthly Weather Review. 1980;*108*(11):1840-1853. DOI: 10.1175/1520-0493(1980)108<1840:OTMCZA>2.0.CO;2

[11] Xavier PK, Marzin C, Goswami BN. An objective definition of the Indian summer monsoon season and a new perspective on the ENSO–monsoon relationship. Meteorological Society. 2007;**133**:749-764. DOI: 10.1002/qj.45

[12] Kothawale DR, Rupa Kumar K. Tropospheric temperature variation over India and links with the Indian summer monsoon 1971-2000. Mausam. 2002;**53**:289-308

[13] He H, McGinnis JW, Song Z, Yanai M. Onset of the Asian summer monsoon in 1979 and the effect of the Tibetan plateau. Monthly Weather Review. 1987;*115*(9):1966-1995. DOI: 10.1175/1520-0493(1987)115<1966:OOTASM>2.0.CO;2

[14] Yanai M, Li C, Song Z. Seasonal heating of the Tibetan plateau and its effects on the evolution of the Asian summer monsoon. Journal of the Meteorological Society of Japan Ser.

II. 1992;**70**(1B):319-351. DOI: 10.2151/jmsj1965.70.1B_319

[15] Flohn H. Large-scale aspects of the “summer monsoon” in south and East Asia. Journal of the Meteorological Society of Japan Series II. 1957;**35A**:180-186. DOI: 10.2151/jmsj1923.35A.0_180

[16] Flohn H. Recent Investigations on the Mechanism of the “Summer Monsoon” of Southern and Eastern Asia: Monsoons of the World. Civil Lines, Delhi, India: The Manager of Publications; 1960. pp. 75-88

[17] Li C, Yanai M. The onset and interannual variability of the Asian Summer Monsoon in relation to Land–Sea thermal contrast. Journal of Climate. 1996;**9**(2):358-375. DOI: 10.1175/1520-0442(1996)009<0358:TOAIVO>2.0.CO;2

[18] Menon A, Turner AG, Volonté A, Taylor CM, Webster S, Martin G. The role of mid-tropospheric moistening and land-surface wetting in the progression of the 2016 Indian monsoon. Quarterly Journal of the Royal Meteorological Society. 2022;**148**(747):3033-3055. DOI: 10.1002/qj.4183

[19] Roxy M, Ritika K, Terray P, et al. Drying of Indian subcontinent by rapid Indian Ocean warming and a weakening land-sea thermal gradient. Nature Communications. 2015;**6**:7423. DOI: 10.1038/ncomms8423

[20] Jin Q, Wang C. A revival of Indian summer monsoon rainfall since 2002. Nature Climate Change. 2017;7:587-594. DOI: 10.1038/nclimate3348

[21] Gadgil S. The monsoon system: Land–sea breeze or the ITCZ? Journal of Earth System Science. 2018;**127**:1-29

[22] Hersbach H, Bell B, Berrisford P, Hirahara S, Horányi A, Muñoz-Sabater J, et al. The ERA5 global reanalysis. Q.J.R. Meteorological Society. 2020;**146**:1999-2049. DOI: 10.1002/qj.3803

[23] Indira RS, Arulalan T, George JP, Rajagopal EN, Renshaw R, Maycock A, et al. IMDAA: High resolution satellite-era reanalysis for the Indian monsoon region. Journal of Climate. 2021;**34**:5109-5133. DOI: 10.1175/JCLI-D-20-0412.1

Chapter 4

Delving into Recent Changes in Precipitation Patterns over the Western Himalayas in a Global Warming Era

Rohtash Saini, Nischal Sharma and Raju Attada

Abstract

Western Himalayas (WH) have experienced a two-fold temperature increase compared to the Indian sub-continent post-2000, strongly linked to global warming with significant implications for precipitation patterns. Using ERA5 reanalysis, we examine seasonal precipitation changes in the WH between recent (2001–2020) and past decades (1961–2000). Mean summer precipitation has increased over foothills but declined at higher elevations, while winter precipitation has increased region-wide except in certain parts of Jammu-Kashmir (JK), Uttarakhand (UK), and Punjab. In summer, light precipitation has increased in JK, while moderate precipitation has decreased over foothills but enhanced at higher altitudes. Moreover, extreme precipitation has significantly increased in the UK and Himachal Pradesh. During winter, light and extreme precipitation has increased, while moderate and heavy precipitation declined. Maximum one and five-day precipitation extremes (Rx1day, Rx5day) have increased in the foothills with more consecutive wet days. Winter extremes have increased in the northern region, while consecutive dry and wet days have declined, except for specific areas in eastern Ladakh and JK. Furthermore, rising sea surface temperatures, enhanced moisture transport, increased precipitable water and cloud cover in WH are associated with increasing mean and extreme precipitation, emphasizing the impacts of global warming on temperature and precipitation transitions in the region.

Keywords: climate change, western Himalayas, summer monsoon, winter precipitation, global warming

1. Introduction

The Western Himalayas (WH), comprising the hilly terrains of Jammu and Kashmir (JK), Himachal Pradesh (HP), Uttarakhand (UK), and Ladakh is known for its unique landscape, biodiversity, and cultural significance [1–6]. The region is not only a critical water source for local communities but also plays a vital role in maintaining the climate of the entire northern belt of the Indian subcontinent [5–9]. However, the diverse topography, varying elevations, soil compositions, and vegetation patterns in the mountain ranges lead to distinct and contrasting weather patterns over short distances. In recent years, concerns have been raised about the potential impact of global warming on precipitation patterns over the WH [10, 11]. The sixth assessment report of the Intergovernmental Panel on Climate Change revealed that the climate of WH might alter significantly due to increasing trends of temperature at regional scales [12]. The report underscores elevation-dependent warming in the Himalayas, indicating that higher altitudes undergo more significant temperature increases compared to lower regions, thereby impacting the hydroclimate system [13–16]. As a result, it can lead to accelerated changes in critical parameters such as the snowline, glacier equilibrium-line altitude, and the transition height between snow and precipitation. Such changes can also alter the precipitation phase and impact the distribution and timing of precipitation and snowfall events, with potential consequences for water availability, hydrological processes, and ecosystems in the region [1, 5, 17, 18].

WH observes two distinct precipitation seasons, the Indian summer monsoon (ISM) (June to September; JJAS) and winter precipitation (December to February; DJF). During summer monsoon, a substantial amount of precipitation (approximately 65–76%) [6] occurs over the WH, which is primarily attributed to orographic-induced mechanisms, as explained in various research studies [13, 19–25]. Moisture-laden winds from the Arabian Sea (AS) and the Bay of Bengal (BoB) converge along the WH foothills resulting in widespread precipitation throughout the region [26–28]. On the other hand, winter precipitation in the WH is characterized by the influence of western disturbances (WDs), which are extra-tropical cyclonic systems embedded in the subtropical westerly jet [5, 29]. This precipitation plays a critical role in meeting the ecological and socioeconomic needs of the region, sustaining regional glaciers and supporting agricultural activities, and also influences the glacial mass balance and river runoff [5, 28, 29]. The spatiotemporal distribution of precipitation in the WH exhibits high variability influenced by atmosphere-land surface exchange processes, diverse land surface characteristics, and orography [5, 30, 31].

In recent years, global warming and climate change have emerged as significant subjects of interest for both the scientific community and the media. The dynamic climate and atmospheric warming have profound implications for water resources, agriculture, and the national economy [32, 33]. While there is substantial evidence supporting global climate change, further investigation is required to understand its implications for regional climates. Regional-scale forcings influenced by local topographical features and land-use characteristics can influence climate change patterns, including weather and climate extremes, notably over complex terrains with distinctive geographical attributes such as WH [34]. Given the dense population in the WH and surrounding areas, they are highly susceptible to climate extremes [5, 28, 35, 36]. Moreover, elevation-dependent warming in the Himalayan region is a major concern, causing shifts in precipitation patterns, including a reduced snow season and increased liquid precipitation [37].

Research focussing on ISM-attributed precipitation over the WH indicates that the region is undergoing a complex array of changes in precipitation and exhibits heterogeneous patterns. An enhancement of ISM precipitation is evident in certain areas, while others observe decreased precipitation [3, 6, 19, 24, 38, 39]. Such spatiotemporal variability in precipitation patterns, in the context of global warming, can be attributed to several factors. Alterations in the intensity and frequency of ISM mean and extreme precipitation, shifts in dynamics such as moisture transport, sea surface temperature fluctuations, etc. can be some of these factors [9, 28, 29]. According to climate models, the region is likely to experience an increase in precipitation variability, with more extreme precipitation events (EPEs) and longer dry spells during the ISM [38, 40, 41]. Das and Meher [24] reported that station-based precipitation records in the WH indicate a reduction in the mean ISM precipitation, however, an increase in the intensity and frequency of EPEs is evident. Such changes can contribute to adverse consequences including increased risks of flash floods, landslides, and glacial lake outburst floods. The impact of climate change on the hydrology of the Hindu Kush Himalayan region during ISM was investigated in a few research studies [8, 39, 42, 43]. Their findings agree with the presence of heterogeneous precipitation patterns, implying an increase in precipitation at certain hotspots and decrease in other regions. Warming trends have also been observed during the winter season in a few research studies [1, 44]. Shekhar et al. [35] aimed to comprehend the influence of climate variability on winter precipitation trends in the WH, and observed a significant decline in snowfall days during Jan-Mar from 1984 to 2007. Observations by Dimri and Dash [45] highlighted decreasing precipitation at specific locations in the WH during the period from 1975 to 2006. Additionally, studies have indicated a decline in spring snow cover since 1993, accompanied by accelerated snowmelt from winter to spring, potentially linked to global warming [46–49]. Gusain et al. and Singh et al. [17, 50] observed a decline in seasonal snowfall at stations below 4000 m and 3250 m, respectively, while an increasing trend in seasonal rainfall was observed at eight stations.

However, a significant research gap exists in understanding the long-term precipitation trend changes in the WH before and after the twenty-first century, highlighting the need for a comprehensive investigation to address this knowledge gap. Since the WH is vulnerable to significant changes in seasonal precipitation patterns in the context of global warming, the current study aims to investigate changes in precipitation trends during the summer (JJAS) and winter (DJF) seasons over the WH region (27.5°N–37.5°N and 72°E–82°E) within the timeframe of 1961–2020.

2. Data and methods

This study aims to investigate precipitation changes over the WH by utilizing the ERA5 dataset, which is a comprehensive and high-resolution fifth-generation reanalysis dataset developed by the European Centre for Medium-Range Weather Forecasts (ECMWF). The ERA5 dataset offers a spatial resolution of 0.25° × 0.25° and derives precipitation data through an advanced 4D-Var assimilation system that integrates various observations, including surface observations, weather station measurements, satellite data, and weather radar information [51]. These observations are assimilated with the ECMWF's atmospheric model to generate a reliable representation of precipitation estimates. The reanalysis encompasses several decades of data and has gained widespread usage in climate studies due to its accuracy and consistency [46, 50, 51]. In this study, we specifically analyze the alterations in precipitation during the ISM and winter season over the WH region (27.5°N–37.5°N and 72°E–82°E) during 1961–2020.

Various factors such as precipitation categories, climate indices, and associated atmospheric variables have been investigated by considering two distinct time periods: a 40-year period representing the pre-2000 era (1961–2000; referred to as past decades) and a 20-year period representing the post-2000 era (2001–2020; referred to as recent decades). This division allows for a comparative assessment of precipitation changes and associated variables, as notable temperature shifts have been observed in the recent decades over the WH region. The study examines the precipitation climatology during the longer time period of 1961–2000 and compares the precipitation differences between the recent and past decades. The comparative analysis of various precipitation events is conducted through a percentile-based classification namely: light (below the 25th percentile), moderate (between the 25th and 90th percentiles), heavy (between the 90th and 95th percentiles), and extreme (above the 95th percentile) precipitation. Here, the percentile approach has been adapted, considering the complex nature of the WH and the high spatiotemporal variability of precipitation. We further examined four extreme climate indices, defined by the World Meteorological Organization's Expert Team on Climate Change Detection and Indices (ETCCDI) [52–54], to represent the magnitude and duration of EPEs over WH. The "Rx1day" index represents the maximum single-day precipitation value per year, indicating the high-intensity precipitation events. The "Rx5day" index identifies the maximum consecutive five-day precipitation, providing insights into prolonged precipitation events. Additionally, the study analyzes the count of continuous wet days (CWD) and consecutive dry days (CDD) to assess the duration of wet and dry periods, respectively. CWD represents a persistent wet period with precipitation exceeding a specified threshold, while CDD indicates a prolonged dry spell without any precipitation.

The study also explores the factors contributing to changes in precipitation, considering global warming and its impact on thermal, moisture, and moist circulation patterns. To achieve this, changes in the recent decades for key atmospheric variables such as sea surface temperature (SST), vertically integrated moisture transport (VIMT), precipitable water, and outgoing longwave radiation (OLR) have been analyzed. The available moisture in the atmosphere is calculated by integrating specific humidity from the surface level to the mid-tropospheric level. Additionally, the investigation of vertically integrated moisture transport (VIMT) is conducted using relevant equations given thus,

$$VIMT = \frac{1}{g}\int_{1000}^{300} (qVdp) \tag{1}$$

where q is specific humidity, V is the horizontal velocity, and dp denotes the vertical incremental change in pressure.

3. Results and discussion

The temperature anomalies over the WH and India over the long-term period 1961–2020 are shown in **Figure 1**. The analysis reveals that notable climatic variations have transpired within both regions toward the start of the twenty-first century. In the Indian subcontinent, a gradual and consistent warming trend has been observed, characterized by an average annual temperature increase of approximately 0.1°C per decade (**Figure 1a**). In contrast, the WH has experienced a more accelerated rate of warming at annual time scales, estimated to be around 0.2°C per decade. Examining the seasonal variations, it becomes evident that the summer surface temperature in the WH region is increasing at a faster rate of 0.23°C per decade compared to the Indian landmass, which is experiencing a rise of 0.17°C per decade (**Figure 1b**). Moreover, we find that winter surface temperatures also show a significantly

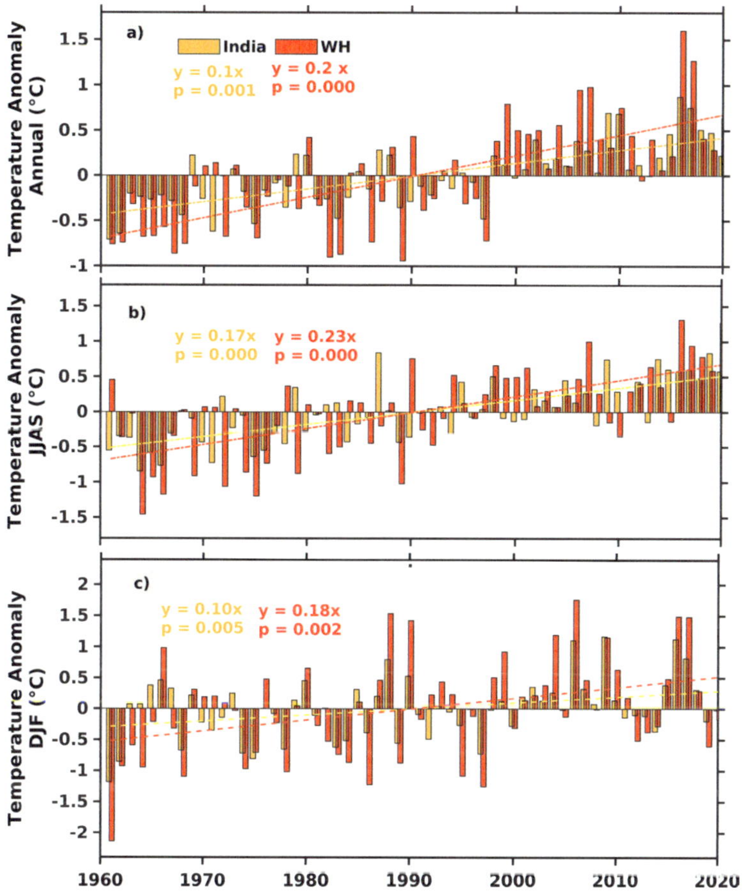

Figure 1.
Time series of temperature anomalies across the western Himalayas (Ladakh, Jammu and Kashmir, Himachal Pradesh, Uttarakhand, Punjab, and Haryana; red bars) and Indian landmass (yellow bars) from 1961 to 2020 for (a) annual, (b) summer and, (c) winter seasons using ERA5 data.

increasing trend of about 0.1°C per decade for the entire India, while those in the WH region are rising much more rapidly, approximately at a rate of 0.18°C per decade (**Figure 1c**). The findings suggest that the WH has exhibited more pronounced tendencies toward warming at both annual and seasonal timescales.

Overall, it becomes clear that the temperature anomalies have predominantly leaned toward the positive side rapidly over the WH since the beginning of the twenty-first century. To delve deeper into the influence of such alterations on the precipitation patterns across the region, we specifically examine the disparities for the summer and winter seasons using two distinct time frames: the pre-2000 era (1961–2000) and the post-2000 era (2001–2020). By focusing on these seasonal variations, we aim to gain a scientific understanding of the precipitation changes and their extreme characteristics in the complex terrains of the WH.

3.1 Mean precipitation

The precipitation climatology and the respective differences for the summer and winter seasons across the WH have been presented in **Figure 2**. The summer

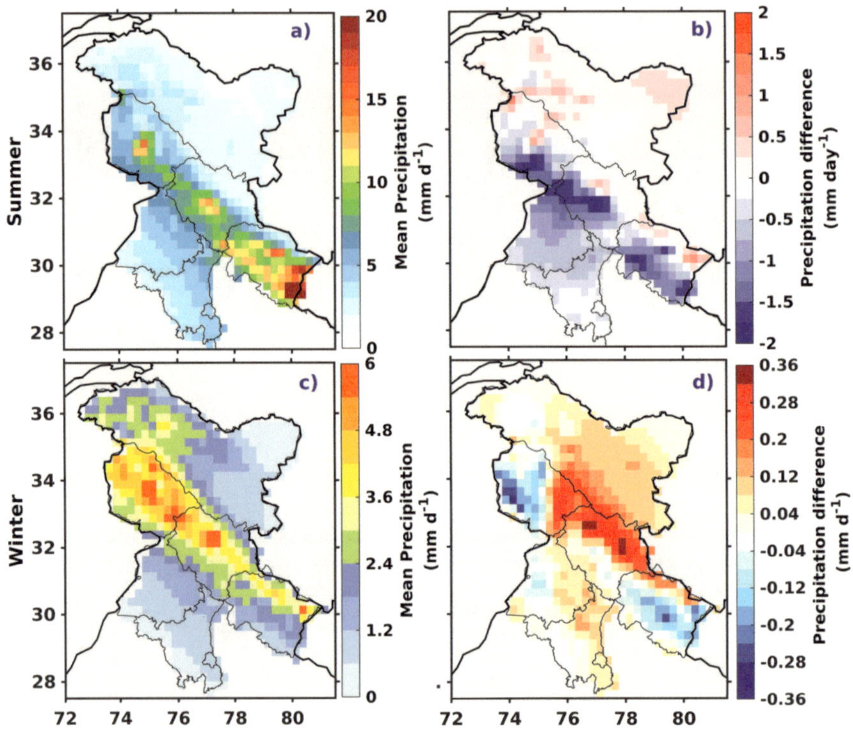

Figure 2.
Spatial distribution of precipitation climatology (1961–2020) and difference in precipitation between post-2000 era (2001–2020) and pre-2000 era (1961–2000) for summer (a and b) and winter (c and d) seasons, respectively based on ERA5 data.

precipitation climatology reveals that the WH foothill belt is a significant zone for precipitation (reaching up to 20 mm/day; **Figure 2a**). Moreover, this region experiences comparatively higher amounts of precipitation in comparison to other parts of the WH. The foothills serve as a transitional area between the plains and the mountainous terrains, where orographic effects play a crucial role in influencing precipitation patterns. **Figure 2b** provides evidence of decreasing mean ISM precipitation over the WH foothill regions during the recent decades. Declining summer precipitation during 2001–2015 in the WH foothills, and an increased snowmelt observed by Riley et al. [55] supports these findings. Our findings are suggestive of a shift in summer precipitation patterns in response to changing climate.

The precipitation climatology for winter season showcases that the Great and lower Himalayas and the foothills are the regions of the precipitation maxima (reaching up to 6 mm/d), with highest intensity observed in JK, followed by HP and UK (**Figure 2c**). The precipitation differences depict increasing intensities over most of the region in the recent decades (reaching up to 0.36 mm/day; **Figure 2d**). The increase in precipitation is relatively stronger in eastern JK and northern HP, whereas a decrease in precipitation is observed over western JK and parts of the UK. As the drivers of precipitation observed during summer and winter seasons over the WH differ significantly, these findings imply the seasonal dependency of precipitation changes over the complex regimes of WH under global warming scenarios. Such changes have potential implications for various sectors, including water resources, ecosystems, and agriculture in the region.

3.2 Precipitation categories

The spatial distribution of precipitation intensities in various categories (encompassing extremes) across the WH during the summer and winter, including the climatological precipitation patterns and the differences noted between recent and past decades is displayed in **Figure 3**. The distributions reveal that the southern slopes of the WH, specifically the foothills, experience substantial precipitation throughout the summer season (**Figure 3a–d**), however, the intensity and geographical locations of precipitation patterns vary across different categories. The precipitation amounts range from about 3 mm/day for light precipitation events (**Figure 3a**) to more than 100 mm/day in the case of EPEs (**Figure 3d**). The UK region, located in the foothills of the WH, receives the highest precipitation amounts for all types of events. Moderate (**Figure 3b**) and heavy (**Figure 3c**) precipitation categories exhibit a similar spatial pattern, with highest precipitation observed over the foothills. The differences in the climatology of light precipitation events (**Figure 3e**) indicate an increasing trend over JK and mixed trends in HP and UK. Moderate precipitation events (**Figure 3f**) are declining over the WH foothills in recent decades, while an increment is observed in the high-altitude regions. Although the heavy precipitation events (**Figure 3g**) exhibit a mixed trend throughout the region, increasing trends for EPEs (**Figure 3h**) have been observed in southeastern parts of HP and most of the UK.

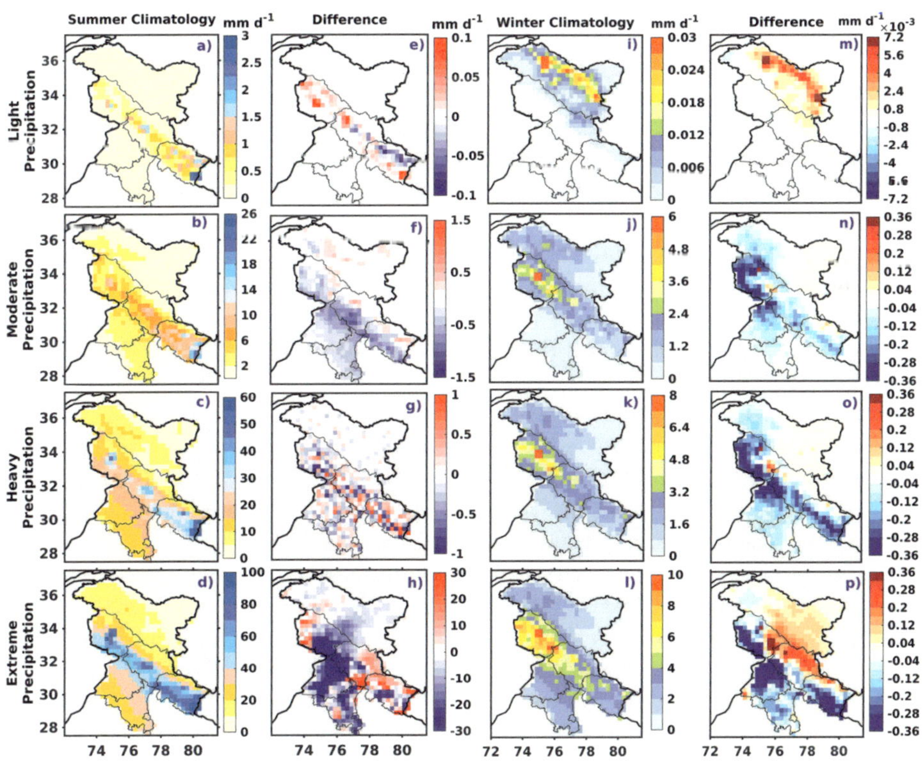

Figure 3.
Spatial distribution of mean precipitation intensity for light, moderate, heavy and extreme precipitation in summer (a–d), and winter seasons (i–l), respectively. (e–h) and (m–p) represents the difference in precipitation between post-2000 era (2001–2020) and pre-2000 era (1961–2000), for the summer and winter seasons, respectively.

Light precipitation events during the winter season primarily occur in the Ladakh region, with intensities averaging around 0.03 mm/day. The region with the highest occurrence of moderate (0–6 mm/day), heavy (0–8 mm/day), and extreme precipitation (10 mm/day or higher) is located in the lower Himalayas and foothills. The highest magnitudes of precipitation intensities are observed over JK, followed by HP and UK. Recent decades indicate notable changes in precipitation patterns, with a substantial increase in the intensity of light precipitation events but a region-wide decrease in moderate and heavy precipitation intensities. Conversely, extreme precipitation exhibits a more heterogeneous trend, characterized by an increase in the Ladakh region and the Great Himalayan range encompassing eastern JK and northern HP, while other areas experience a decline. These findings underscore the intricate nature of precipitation patterns within the WH, including variations in the types of precipitation events. The spatial distribution and disparities in precipitation have significant implications for water availability, flood risk, and ecosystem dynamics in the region.

3.3 Climate indices

The spatial distribution of extreme climate indices during the summer and winter seasons is illustrated in **Figure 4** to understand the altering characteristics of EPEs, wet spells, and dry spells over the WH, in response to warming patterns. **Figure 4a** displays the climatology of Rx1day, which represents the maximum one-day precipitation for each year over the WH during ISM. The WH foothill belt exhibits high Rx1day precipitation (>120 mm), while the downstream side and high-altitude regions of the WH represent comparatively drier areas with an average Rx1day below 30 mm. Similarly, the Rx5day precipitation is very high in the foothill regions (>240 mm), indicating a similar spatial distribution of extreme events. **Figure 4c** and **d** illustrate the climatology for CWD and CDD, which represents the duration of consecutive wet and dry days, respectively. The WH foothills exhibit the highest values of CWD (>70 days) and the lowest values for CDD (<12 days), indicating that the foothills are the main precipitation belt of WH during ISM. In contrast, other parts of WH exhibit comparatively shorter wet periods (<40 days) and longer dry periods. **Figure 4(e–h)** illustrates the climatological differences in EPEs in recent decades compared to past decades. The magnitude-based indices in **Figure 4(e** and **f)** indicates an increasing trend in EPEs over the northern regions of the UK and the southeastern region of HP. Conversely, a significant decreasing trend is observed over the majority of JK and western HP. **Figure 4(g** and **h)** shows an increasing trend for CWD and decreasing trend for CDD over the high-altitude regions of the WH, whereas comparatively mixed trends are evident across the foothills.

The wintertime climate indices illustrate magnitudes of Rx1day and Rx5day reaching up to 60 mm and 150 mm respectively, with maxima situated over zones of highest precipitation (**Figure 4i,j**). Interestingly, the intensity of Rx1day exhibits a notable disparity across the study area, with the eastern part of JK experiencing comparatively higher values compared to other regions. This discrepancy may be attributed to the influence of orographic effects and local-scale atmospheric processes on the distribution of precipitation extremes in the region. In the CWD climatology, Karakoram Himalayas exhibit the longest duration of wet days, exceeding 12 days, followed by the Great Himalayas with a duration of over 6 days (**Figure 4k**). In contrast, the foothills and surrounding plains exhibit dry durations of more than 25 days, while the eastern parts of Ladakh experience an extended period of CDDs, exceeding

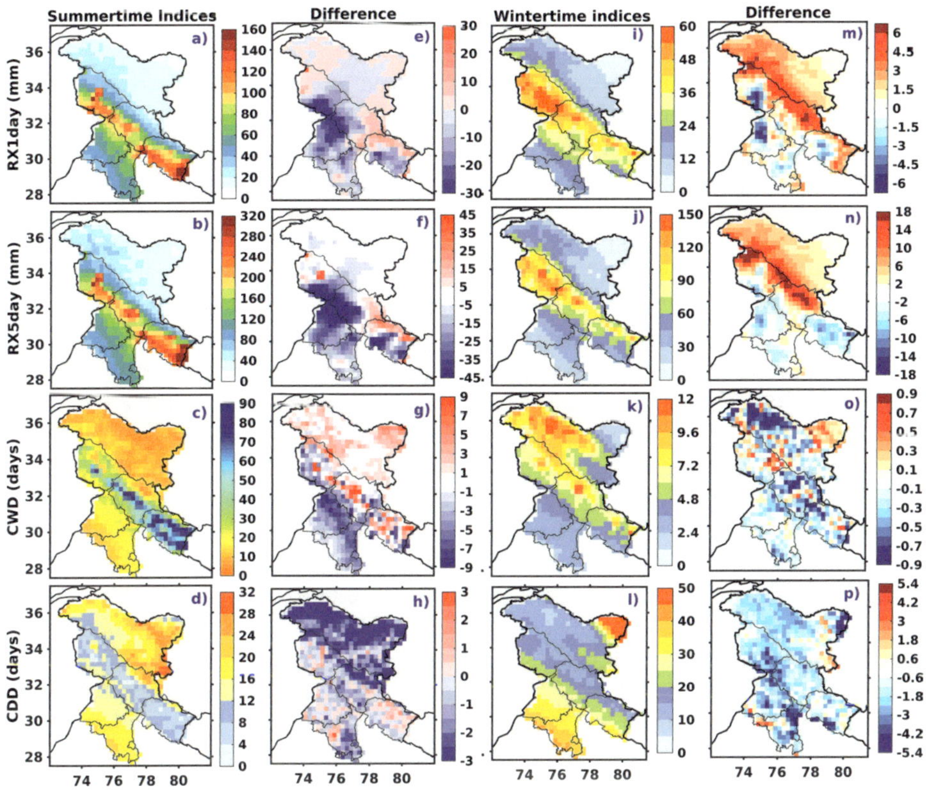

Figure 4.
Spatial distribution for the climatology of different extreme climate indices (Rx1day, Rx5day, CWD, and CDD) using ERA5 during summer (a–d) and winter (i–l) seasons, respectively. (e–h) and (m–p) present the differences in the climate indices between pre-2000 ERA (1961–2000), and post-2000 era (2001–2020) for summer and winter, respectively.

50 days (**Figure 4l**). These findings suggest a pronounced contrast in the precipitation characteristics between the various subregions within the study area. The Karakoram range, the Great Himalayas, and the foothills are the primary zones where winter precipitation occurs. In contrast, eastern Ladakh falls within the precipitation shadow zone, resulting in a notable duration of CDDs in this region during winter.

Examining the changes in these climate indices over the recent decades, a clear increasing trend is observed in the precipitation extremes (both Rx1day and Rx5day) across the majority of the region (**Figure 4m** and **n**). Notably, the Karakoram range and the Great Himalayas exhibit stronger trends in precipitation extremes compared to other areas. However, pockets of slight decrease in these indices are observed in certain parts of JK, UK, and Punjab. These localized variations may be attributed to complex interactions between large-scale climate patterns, local atmospheric processes, and orographic effects. Regarding the duration of consecutive wet and dry days, a heterogeneous pattern emerges for wet days, with a general decrease observed throughout the region (**Figure 4o**). However, certain pockets in eastern Ladakh and JK exhibit a contrary trend, indicating a localized persistence of wetter conditions. In contrast, the duration of CDDs displays a homogenous decrease across the entire region, indicating a general trend toward a decrease in the frequency of persistent dry periods during winter (**Figure 4p**).

3.4 Mechanisms

In light of the substantial modifications observed in precipitation patterns since the beginning of the twenty-first century, which coincide with the prevailing warming trends in the WH, our study further focuses on an in-depth investigation into the changes occurring in key atmospheric variables as a direct response to these warming trends. This investigation aims to discern the potential impacts of these changes on precipitation patterns within the region. **Figure 5(a** and **b)** exhibits the mean SST (1961–2020) and its disparities between recent and past decades during the ISM. Notably, significant SST increases have been observed in the adjacent oceans, namely the Indian Ocean, AS, and BoB (reaching up to 0.16°C). The warming rate appears to be more pronounced in the eastern and central regions of the Indian Ocean, followed by the AS and BoB. The findings indicate a regional warming effect over the surrounding oceans, which can impact the atmospheric moist circulation patterns, thereby influencing the mean summer precipitation over the WH. Moreover, during EPEs, such fluctuations in the monsoon circulation can facilitate an enhanced moisture transport toward the WH, thereby increasing the moist convective activity over the region, in context of atmospheric instability.

Furthermore, we have examined the moisture transport fields (VIMT) to ascertain whether they demonstrate comparable patterns, and influence the evolving precipitation patterns across the WH in recent decades. **Figure 5(c** and **d)** presents the VIMT climatology during the entire study period and the difference between recent and past decades. On a climatological basis, AS and BoB as the major moisture sources for summer precipitation across the WH. Interestingly, an increase of VIMT, though with low magnitudes, is observed over the higher elevations across the WH whereas a decreasing trend can be seen for moisture transport associated with the WH foothills and surrounding plains, which are the regions of reduced precipitation during the recent decades.

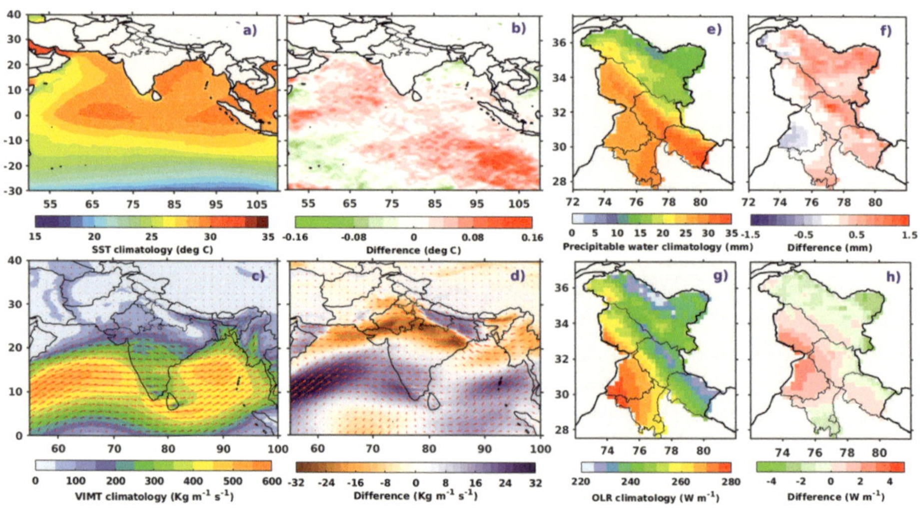

Figure 5.
Spatial distribution of climatological characteristics, and the differences between post-2000 period (2001–2020) and pre-2000 period (1961–2000), during ISM for (a and b) sea surface temperatures (SST), (c and d) vertically integrated moisture transport (VIMT), (e and f) precipitable water and (g and h) outgoing longwave radiation OLR, respectively.

Figure 5e illustrates the mean precipitable water over the WH, representing the available moisture content in the atmosphere, which is highest over the foothills and surrounding plains, followed by that in high altitudes. However, the differences associated with precipitable water between recent and past decades, show high levels of precipitable water in the elevated regimes of WH whereas comparatively much lower amounts are present over the foothills including southern parts of JK and HP as well as surrounding plains of Punjab and western Haryana (**Figure 5f**). These low magnitudes suggest that the long-term changes in moisture availability over these regions have remained relatively stable during the recent decades. The outgoing longwave radiation (OLR) serves as a valuable proxy for assessing deep convection and presents significant potential in unraveling large-scale precipitation variability in complex regions like the Himalayas [56]. A low OLR value is indicative of strong convection, representing an active precipitation process, while a high OLR value suggests cloud-free regions [57–59]. **Figure 5(g** and **h)** demonstrates the spatial distribution of mean OLR and the climatological difference of OLR between recent and past decades. The results indicate that during the ISM, lower values of OLR are prominent over the upper regions of WH foothills, specifically in the northern parts of HP and UK (**Figure 5g**), indicating the presence of more convective activity and cloud cover there. Comparatively higher OLR values are observed in lowermost parts of foothills and surrounding plains. Significant positive anomalies can be observed in the WH foothills during recent decades (**Figure 5f**), suggesting reduced cloud cover and decreased precipitation during the ISM. However, the higher elevations exhibit comparatively negative anomalies, suggesting increased convective activity and enhanced precipitation as suggested earlier.

Upon examining the responses of these variables during the winter season (**Figure 6**), a distinct and homogenous rise in SSTs becomes evident, reaching a maximum of 0.27°C (**Figure 6b**). Notably, this temperature rise appears to exhibit greater magnitudes than that observed during the summer season. Of particular interest, the Indian Ocean demonstrates the most pronounced rate of SST increase, followed by the AS and the BoB. The AS represents a pivotal moisture supplier for winter precipitation associated with WDs [29, 60], while Dimri et al. [61] have identified the BoB as a primary moisture contributor for non-WD-related winter precipitation over the WH. The rising SSTs, consequently, trigger enhanced evaporation processes, fostering the availability and transport of additional moisture, thereby substantially contributing to enhanced mean and extreme precipitation across the WH. Additionally, the Mediterranean, Caspian, Red Seas, and AS play a crucial role in transporting significant amounts of moisture eastward with propagating WDs toward the WH (**Figure 6c**), thereby contributing to moisture availability for winter precipitation [29, 60]. The changes in winter seasonal moisture supply over the WH in recent decades (**Figure 6d**) clearly indicate a noticeable increase in both the magnitude and transport of moisture over the region.

Significant amounts of total precipitable water content, primarily concentrated in areas of maximum precipitation, are observed during winter over the region (**Figure 6e**). The observed rise in temperatures across the WH in recent decades corresponds to an increase in moisture content, consistent with the Clausius-Clapeyron relationship. The presence of positive anomalies in precipitable water (up to 1.2 mm) recently (**Figure 6f**) indicates a greater availability of moisture, which enhances the efficiency of orographic precipitation and cloud formation in the WH's mountainous regions. This additional moisture can be contributed by sources such as the AS, BoB, and conventional moisture sources associated with eastward moving WDs. In conclusion, these changes in regional dynamic and thermodynamic variables provide

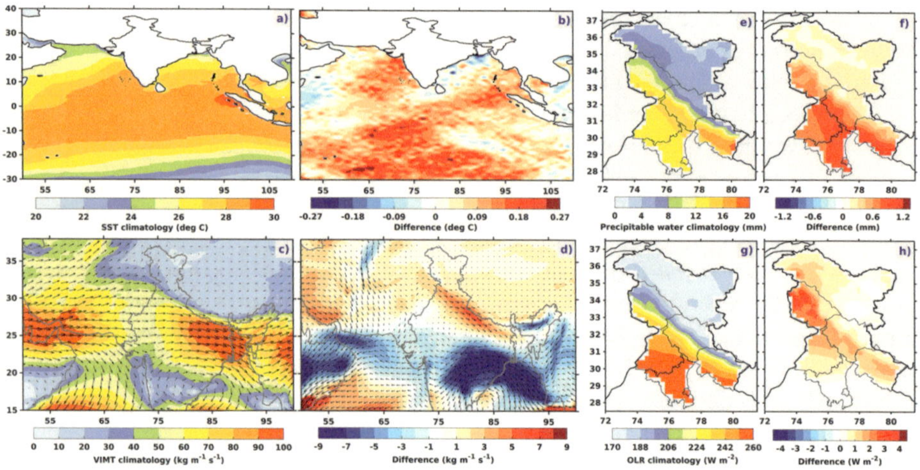

Figure 6.
Same as **Figure 5**, *but for the winter season.*

compelling evidence for the emergence of rising mean and extreme precipitation patterns over the WH in recent decades. OLR variability serves as a significant indicator for cloud formations associated with precipitation; however, it is important to note that changes in near-surface temperature resulting from winter snow cover over complex regions like the WH can also contribute to the variations in OLR [62]. Typically, WDs that bring precipitation contribute to overcast skies and increased cloud cover across the WH [63]. Recent decades reveal slight negative anomalies in OLR during winter (**Figure 6h**) over the regions where precipitation has shown an increase (as depicted in **Figure 1d**). This suggests an intensification of convective activity and cloud cover. Conversely, strong positive anomalies are evident over JK and the lower foothills, indicating reduced convective activity and subsequently decreased cloud cover, which aligns with the observed decline in precipitation over those particular regions.

4. Summary and conclusion

The changes in precipitation patterns over the WH in the context of global warming are complex and multifaceted, acknowledging the complexities of the region's topography and the high spatiotemporal variability of precipitation. The current study aims to examine the intricate and diverse changes in precipitation patterns during summer and winter seasons across the WH, driven by the significant warming trends observed in the area in recent decades due to global warming.

- Findings emphasize that the WH region has been more significantly impacted by warming trends compared to India as a whole, at both annual and seasonal scales. The accelerated warming in the WH has implications for the local climate, ecosystems, and various sectors dependent on this mountainous region.

- The analysis of mean precipitation patterns in the WH reveals a complex and varied picture. Our findings suggest a decline in mean ISM precipitation over the

WH foothills in recent decades, while an increment is observed in the high-altitude regions. Winter mean precipitation exhibits amplified precipitation mostly throughout the entire WH region (reaching up to 0.36 mm/d), except for specific areas in JK, UK, and Punjab, where the observed decline is relatively weaker compared to the overall increase in precipitation.

- During summer, an increasing trend for light precipitation events is observed over JK and mixed trends in HP and UK whereas the moderate events are declining in the foothills and increasing at high altitudes. Heavy events show a mixed trend, while EPEs are increasing in southeast HP and most of the UK. In winter, precipitation patterns in Ladakh exhibit a noteworthy increase in the occurrence of light precipitation events whereas the extreme precipitation portrays a heterogeneous trend, characterized by an upswing in the Ladakh region and the Great Himalayas encompassing eastern JK and northern HP, while witnessing a decline in other areas. Concurrently, moderate and heavy precipitation are experiencing a decline across the entire region.

- Changes in climate indices offer insights into changing precipitation patterns, extreme events, and climate variability, aiding in understanding risks associated with heavy precipitation, droughts, and wet/dry periods. Magnitude-based extreme indices indicate an increase in summertime precipitation extremes in the northern UK and southeastern HP and decrease in JK and western HP. At the same time, the duration of wet days is enhanced in the high-altitude regions of WH alongside a decrease in dry days. In the foothills, however, the trends for both CWD and CDD are mixed. Winter climate indices reveal a significant intensification of extreme precipitation (both Rx1day and Rx5day) primarily in the northern parts of WH (up to 7.5 mm and 18 mm, respectively). However, a minimal decrease in extreme precipitation is observed in the foothills. Analysis of consecutive dry and wet days demonstrates decreasing trends for both categories across the region. Nonetheless, specific localized areas, notably eastern Ladakh and J&K, exhibit pockets of increasing wet days.

- The examination of SST patterns in both summer and winter reveals positive anomalies for SSTs in the surrounding Indian Ocean, AS and BoB, with comparatively higher magnitudes of increase during winter than ISM. During summer, higher elevations in the WH show a slight increase in VIMT, while moisture transport decreases in WH foothills and plains with reduced precipitation. Moreover, higher precipitable water in WH elevations but lower amounts in foothills is also observed, suggesting stable moisture availability in the latter. However, the winter seasonal VIMT as well as precipitable water over the WH has experienced significant increases in recent decades.

- In summer, the WH foothills exhibit positive OLR anomalies, indicating reduced cloud cover and decreased precipitation during the ISM, while higher elevations display negative anomalies, suggesting increased convective activity and enhanced precipitation over the WH. The changes associated with winter OLR depict minute negative anomalies over the regions experiencing enhanced precipitation and positive anomalies in western JK and foothills where the precipitation patterns have been found to be decreasing, thus confirming increased and decreased cloud cover respectively.

In summary, the spatial and temporal variability in precipitation patterns over the WH can have significant implications for the region's environment, economy, and society. It is crucial to develop adaptation strategies that take into account the complex and uncertain nature of these changes, and to work toward mitigating the underlying causes of global warming to minimize the impact of these changes.

Acknowledgements

This research work was supported by the Prime Minister's Research Fellowship (PMRF) and the Science and Engineering Research Board, Department of Science and Technology, Government of India under the "Start-up Research Grant (SRG) scheme" (Grant SRG/2020/001857). Rohtash and Nischal Sharma gratefully acknowledge the financial assistance from the PMRF, Ministry of Education, Government of India.

Conflict of interest

The authors declare no conflict of interest.

Nomenclature

WH	Western Himalayas
ISM	Indian summer monsoon
WD	Western disturbances
JK	Jammu Kashmir
HP	Himachal Pradesh
UK	Uttarakhand
IPCC	Intergovernmental Panel on Climate Change
EPEs	extreme precipitation event
CWD	continuous wet days
CDD	consecutive dry days
SST	sea surface temperature
OLR	outgoing longwave radiation
VIMT	Vertically integrated moisture transport

Author details

Rohtash Saini, Nischal Sharma and Raju Attada*
Department of Earth and Environmental Sciences, Indian Institute of Science Education and Research Mohali, Punjab, India

*Address all correspondence to: rajuattada@iisermohali.ac.in

IntechOpen

© 2023 The Author(s). Licensee IntechOpen. This chapter is distributed under the terms of the Creative Commons Attribution License (http://creativecommons.org/licenses/by/4.0), which permits unrestricted use, distribution, and reproduction in any medium, provided the original work is properly cited.

References

[1] Bhutiyani MR, Kale VS, Pawar NJ. Climate change and the precipitation variations in the northwestern Himalaya. 1866-2006. International Journal of Climatology. 2010;**30**:535-548. DOI: 10.1002/joc.1920

[2] Rasmussen KL, Houze RAA. Flash-flooding storm at the steep edge of high terrain. Bulletin of the American Meteorological Society. 2012;**93**:1713-1724. DOI: 10.1175/BAMS-D-11-00236.1

[3] Bharti V, Singh C, Ettema J, Turkington TAR. Spatiotemporal characteristics of extreme rainfall events over the northwest Himalaya using satellite data. International Journal of Climatology. 2016;**36**:3949-3962. DOI: 10.1002/joc.4605

[4] Meher JK, Das L, Benestad RE, Mezghani A. Analysis of winter rainfall change statistics over the Western Himalaya: The influence of internal variability and topography. International Journal of Climatology. 2018;**38**:e475-e496. DOI: 10.1002/joc.5385

[5] Nischal AR, Hunt KMR. Evaluating winter precipitation over the Western Himalayas in a high-resolution Indian regional reanalysis using multisource climate datasets. Journal of Applied Meteorology and Climatology. 2022;**61**:1607-1627. DOI: 10.1175/JAMC-D-21-0172.1

[6] Saini R, Attada R. Analysis of Himalayan summer monsoon rainfall characteristics using Indian high-resolution regional reanalysis. International Journal of Climatology. 2023:1-22. DOI: 10.1002/joc.8087 [Online]

[7] Bharti V. Investigation of Extreme Rainfall Events Over the Northwest Himalaya Region Using Satellite Data [Master's thesis]. University of Twente; 2015. p. 72

[8] Krishnan R, Shrestha AB, Ren G, Rajbhandari R, Saeed S, Sanjay J, et al. Unravelling climate change in the Hindu Kush Himalaya: Rapid warming in the mountains and increasing extremes. In: Wester P, Mishra A, Mukherji A, Shrestha A, editors. The Hindu Kush Himalaya Assessment. Cham: Springer; 2019. DOI: 10.1007/978-3-319-92288-1_3

[9] Vellore RK, Bisht JS, Krishnan R, Uppara U, Capua GD, Coumou D. Sub-Synoptic Circulation Variability in the Himalayan Extreme Precipitation Event during June 2013. Vol. 132. Austria: Springer-Verlag GmbH; 2020

[10] Webster PJ, Toma VE, Kim HM. Were the 2010 Pakistan floods predictable? Geophysical Research Letters. 2011;**38**:1-5. DOI: 10.1029/2010GL046346

[11] Sabin TP, Krishnan R, Vellore R, Priya P, Borgaonkar HP, Singh BB, et al. Climate change over the Himalayas. In: Assessment of Climate Change over the Indian Region. Springer Singapore: Singapore; 2020. pp. 207-222

[12] IPCC, 2021: Climate Change 2021: The Physical Science Basis. In: Masson-Delmotte V, Zhai P, Pirani A, Connors SL, Péan C, Berger S, et al, editors. Contribution of Working Group I to the Sixth Assessment Report of the Intergovernmental Panel on Climate Change. Cambridge, United Kingdom and New York, NY, USA: Cambridge University Press. 2391 p. DOI: 10.1017/97810091578

[13] Giorgi F, Hurrell JW, Marinucci MR, Beniston M. Elevation dependency of

the surface climate change signal: A model study. Journal of Climate. 1997;**10**:288-296. DOI: 10.1175/1520-0442(1997)010<0288:EDOTSC>2.0.CO;2

[14] Beniston M. Climatic change in mountain regions: A review of possible impacts. In: Climate Variability and Change in High Elevation Regions: Past, Present & Future. 2003. pp. 5-31. DOI: 10.1007/978-94-015-1252-7_2

[15] Kang S, Xu Y, You Q, Flügel WA, Pepin N, Yao T. Review of climate and cryospheric change in the Tibetan plateau. Environmental Research Letters. 2010;**5**:8. DOI: 10.1088/1748-9326/5/1/015101

[16] You Q, Kang S, Pepin N, Flügel WA, Yan Y, Behrawan H, et al. Relationship between temperature trend magnitude, elevation and mean temperature in the Tibetan plateau from homogenized surface stations and reanalysis data. Global and Planetary Change. 2010;**71**:124-133. DOI: 10.1016/j.gloplacha.2010.01.020

[17] Gusain H, Mishara V, Bhutiyani M. Winter temperature and snowfall trends in the cryospheric region of north-west Himalaya. Mausam. 2014;**65**:425-432. DOI: 10.54302/mausam.v65i3.1053

[18] Hunt KMR, Curio J, Turner AG, Schiemann R. Subtropical westerly jet influence on occurrence of Western disturbances and Tibetan plateau vortices. Geophysical Research Letters. 2018;**45**:8629-8636. DOI: 10.1029/2018GL077734

[19] Singh P, Ramasastri KS, Kumar N. Topographical influence on precipitation distribution in different ranges of Western Himalayas. Nordic Hydrology. 1995;**26**:259-284. DOI: 10.2166/nh.1995.0015

[20] Ménégoz M, Gallée H, Jacobi HW. Precipitation and snow cover in the Himalaya: From reanalysis to regional climate simulations. Hydrology and Earth System Sciences. 2013;**17**:3921-3936. DOI: 10.5194/hess-17-3921-2013

[21] Houze RA. Mesoscale convective systems. Geophysical Journal International. 2014;**104**:237-286. DOI: 10.1016/B978-0-12-374266-7.00009-3

[22] Vellore RK, Krishnan R, Pendharkar J, Choudhury AD, Sabin TP. On the anomalous precipitation enhancement over the Himalayan foothills during monsoon breaks. Climate Dynamics. 2014;**43**:2009-2031. DOI: 10.1007/s00382-013-2024-1

[23] Kesarwani K, Dobhal DP, Durgapal A, Mehta M. High altitude meteorology and cloud cover conditions A study from the chorabari glacier catchment, Central Himalaya, India. Himalayan Geology. 2015;**36**(2):134-142

[24] Das L, Meher JK. Drivers of climate over the Western Himalayan region of India: A review. Earth-Science Reviews. 2019;**198**:102935. DOI: 10.1016/j.earscirev.2019.102935

[25] Singh D, Ghosh S, Roxy MK, McDermid S. Indian summer monsoon: Extreme events, historical changes, and role of anthropogenic forcings. Wiley Interdisciplinary Reviews: Climate Change. 2019;**10**:1-35. DOI: 10.1002/wcc.571

[26] Buckner CA, Lafrenie RM, Dénommée JA, Caswell JM, Want DA. Complementary and alternative medicine use in patients before and after a cancer diagnosis. Current Oncology. 2018;**25**:e275-e281. DOI: 10.3747/co.25.3884

[27] Aggarwal D, Attada R, Shukla KK, Chakraborty R, Kunchala RK. Monsoon

precipitation characteristics and extreme precipitation events over Northwest India using Indian high resolution regional reanalysis. Atmospheric Research. 2021;**267**:105993. DOI: 10.1016/j.atmosres.2021.105993

[28] Nischal S, Saini R, Pathaikara A, Punde P, Attada R. Hydrological extremes in Western Himalayas-trends and their physical factors. In: Natural Hazards – New Insights. London, UK: IntechOpen; 2023

[29] Hunt KMR, Turner AG, Shaffrey LC. The evolution, seasonality and impacts of Western disturbances. Quarterly Journal of the Royal Meteorological Society. 2018;**144**:278-290. DOI: 10.1002/qj.3200

[30] Dimri AP, Thayyen RJ, Kibler K, Stanton A, Jain SK, Tullos D, et al. A review of atmospheric and land surface processes with emphasis on flood generation in the southern Himalayan Rivers. Science of the Total Environment. 2016;**556**:98-115. DOI: 10.1016/j.scitotenv.2016.02.206

[31] Hunt KMR, Turner AG, Schiemann RKH, Norris J, Carvalho LMV, Jones C, et al. WRF simulations of two extreme snowfall events associated with contrasting extratropical cyclones over the Western and Central Himalaya. Journal of Geophysical Research: Atmospheres. 2015;**120**:3114-3138. DOI: 10.1002/qj.4275

[32] Dimri AP, Allen S, Huggel C, Mal S, Ballesteros-Cánovas JA, Rohrer M, et al. Climate change, cryosphere and impacts in the Indian Himalayan region. Current Science. 2021;**120**:1012-1012

[33] Mishra AK. Quantifying the impact of global warming on precipitation patterns in India. Meteorological Applications. 2019;**26**:153-160. DOI: 10.1002/met.1749

[34] Rao VB, Franchito SH, Gerólamo ROP, Giarolla E, Ramakrishna SSVS, Rao BRS, et al. Himalayan warming and climate change in India. American Journal of Climate Change. 2016;**05**:558-574. DOI: 10.4236/ajcc.2016.54038

[35] Shekhar MS, Chand H, Kumar S, Srinivasan K, Ganju A. Climate-change studies in the Western Himalaya. Annals of Glaciology. 2010;**51**:105-112. DOI: 10.3189/172756410791386508

[36] Cannon F, Carvalho LMV, Jones C, Hoell A, Norris J, Kiladis GN, et al. The influence of tropical forcing on extreme winter precipitation in the Western Himalaya. Climate Dynamics. 2017;**48**:1213-1232. DOI: 10.1007/s00382-016-3137-0

[37] Pepin N, Bradley RS, Diaz HF, Baraer M, Caceres EB, Forsythe N, et al. Elevation-dependent warming in mountain regions of the world. Nature Climate Change. 2015;**5**:424-430. DOI: 10.1038/nclimate2563

[38] Bookhagen B, Burbank DW. Toward a complete Himalayan hydrological budget: Spatiotemporal distribution of snowmelt and rainfall and their impact on river discharge. Journal of Geophysical Research - Earth Surface. 2010;**115**:1-25. DOI: 10.1029/2009JF001426

[39] Li H, Haugen JE, Xu CY. Precipitation pattern in the Western Himalayas revealed by four datasets. Hydrology and Earth System Sciences. 2018;**22**:5097-5110. DOI: 10.5194/hess-22-5097-2018

[40] Anders AM, Roe GH, Hallet B, Montgomery DR, Finnegan NJ, Putkonen J. Spatial patterns of precipitation and topography in the Himalaya. Special Paper of the Geological Society of America. 2006;**398**:39-53. DOI: 10.1130/2006.2398(03)

[41] Shrestha D, Singh P, Nakamura K. Spatiotemporal variation of rainfall over the central Himalayan region revealed by TRMM precipitation radar. Journal of Geophysical Research – Atmospheres. 2012;**117**:1-14. DOI: 10.1029/2012JD018140

[42] Bolch T, Shea JM, Liu S, Azam FM, Gao Y, Gruber S, et al. Status and Change of the Cryosphere in the Extended Hindu Kush Himalaya Region. Cham: Springer; 2019. DOI: 10.1007/978-3-319-92288-1_7

[43] Boschi R, Lucarini V. Water pathways for the Hindu-Kush-Himalaya and an analysis of three flood events. Atmosphere (Basel). 2019;**10**:1-32. DOI: 10.3390/atmos10090489

[44] Fowler HJ, Archer DR. Conflicting signals of climatic change in the upper Indus Basin. Journal of Climate. 2006;**19**:4276-4293. DOI: 10.1175/JCLI3860.1

[45] Dimri AP, Dash SK. Wintertime climatic trends in the Western Himalayas. Climatic Change. 2012;**111**:775-800. DOI: 10.1007/s10584-011-0201-y

[46] Kripalani RH, Kulkarni A, Sabade SS. Western Himalayan snow cover and Indian monsoon rainfall: A Re-examination with INSAT and NCEP/NCAR data. Theoretical and Applied Climatology. 2003;**74**:1-18. DOI: 10.1007/s00704-002-0699-z

[47] Baudouin JP, Herzog MA, Petrie C. Cross-validating precipitation datasets in the Indus River Basin. Hydrology and Earth System Sciences. 2020;**24**:427-450. DOI: 10.5194/hess-24-427-2020

[48] Lucas-Picher P, Christensen JH, Saeed F, Kumar P, Asharaf S, Ahrens B, et al. Can regional climate models represent the Indian monsoon? Journal of Hydrometeorology. 2011;**12**:849-868. DOI: 10.1175/2011JHM1327.1

[49] Wang S, Dieterich C, Döscher R, Höglund A, Hordoir R, Markus Meier HE, et al. Development and evaluation of a new regional coupled Atmosphere-Ocean model in the North Sea and Baltic Sea. Tellus A: Dynamic Meteorology and Oceanography. 2015;**67**(1):24284. DOI: 10.3402/tellusa.v67.24284

[50] Singh D, Sharma V, Juyal V. Observed linear trend in few surface weather elements over the Northwest Himalayas (NWH) during winter season. Journal of Earth System Science. 2015;**124**:553-565. DOI: 10.1007/s12040-015-0560-2

[51] Hersbach H, Dee D. ERA5 reanalysis is in production, ECMWF newsletter 147. European Centre for Medium-Range Weather Forecasts. 2016;**147**:7

[52] Sillmann J, Kharin VV, Zwiers FW, Zhang X, Bronaugh D. Climate extremes indices in the CMIP5 multimodel ensemble: Part 2. Future climate projections. Journal of Geophysical Research – Atmospheres. 2013;**118**:2473-2493. DOI: 10.1002/jgrd.50188

[53] Chervenkov H, Slavov K. ETCCDI climate indices for assessment of the recent climate over Southeast Europe. Studies in Computational Intelligence. 2021;**902 SCI**:398-412. DOI: 10.1007/978-3-030-55347-0_34

[54] Bhattacharyya S, Sreekesh S, King A. Characteristics of extreme rainfall in different gridded datasets over India during 1983-2015. Atmospheric Research. 2022;**267**:105930. DOI: 10.1016/j.atmosres.2021.105930

[55] Riley C, Rupper S, Steenburgh JW, Strong C, Kochanski AK, Wolvin S. Characteristics of historical precipitation in High Mountain Asia based on a 15-year high resolution dynamical downscaling.

Atmosphere (Basel). 2021;**12**(3):355. DOI: 10.3390/atmos12030355

[56] Haque MA, Lal M. Space and time variability analyses of the Indian monsoon rainfall as inferred from satellite-derived OLR data. Climate Research. 1991;**1**:187-197. DOI: 10.3354/cr001187

[57] Xie P, Arkin PA. Global monthly precipitation estimates from satellite-observed outgoing longwave radiation. Journal of Climate. 1998;**11**:137-164. DOI: 10.1175/1520-0442(1998)011<0137:GMPEFS>2.0.CO;2

[58] Kumar A, Sarthi PP, Kumari A, Sinha AK. Observed characteristics of rainfall indices and outgoing longwave radiation over the Gangetic plain of India. Pure and Applied Geophysics. 2021;**178**:619-631. DOI: 10.1007/s00024-021-02666-6

[59] Prasad KD, Bansod SD. Interannual variations of outgoing longwave radiation and Indian summer monsoon rainfall. Int. J. Climatol. 2000. A Journal of the Royal Meteorological Society. 2000 Dec;**20**(15):1955-1956

[60] Dimri AP, Niyogi D, Barros AP, Ridley J, Mohanty UC, Yasunari T, et al. Western disturbances: A review. Reviews of Geophysics. 2015;**53**(2):225-246

[61] Dimri AP, Pooja JG, Mohanty UC. Western disturbances vs Non-western disturbances days winter precipitation. Climate Dynamics. 2023:1-23 [Online]

[62] Dimri AP. Interannual variability of Indian winter monsoon over the Western Himalayas. Global and Planetary Change. 2013;**106**:39-50

[63] Rao VB, Rao ST. A theoretical and synoptic study of Western disturbances. Pure and Applied Geophysics. 1971;**90**(7):193-208

Chapter 5

Navigating the Frontlines of Climate Change: Resilience and Perspectives of Climate Champions

Camellia Moses Okpodu and Bernadette J. Holmes

Abstract

The frontline of climate change in the African diaspora has been championed by the efforts of women. Although visibility has been given to women on the continent, the heroic efforts of African American women have often been largely ignored. Just as in the developing world, women of color have disproportionally experienced the repercussions of the assaults of a changing climate—rising sea levels, changing weather patterns, and increasing pollution are threatening food security, increasing infant mortality, exacerbating poverty, and maternal death rates exposing other health inequalities. This chapter will explore the life's work of three climate champions at Historically Black College or Universities (HBCUs) via teaching the next generation of climate activists. We use a conceptual framework that situates their work in three broad areas: (1) risk reduction and emergency preparedness; (2) curriculum, education, and health policy; and (3) outreach and environmental capacity. We will explore their stories, the political actions that have contributed to the lack of resource management, and their role in shaping women's roles in addressing the impacts of climate change and environmental justice.

Keywords: womanist environmentalism, ecofeminism, diaspora, climate resilience, climate activism

1. Introduction

Sometimes, when we think about the environment, it becomes the movement of elite white women, who can be described as ecofeminist. In an article entitled, "6 Female Environmentalists Who Changed the World by Sabai Design" it highlights the accomplishments of six phenomenal women. These women are from a myriad of backgrounds; however, none are African American women [1]. Each woman featured in the article brought climate change and the efforts from grassroots beginnings to international prominence. This is not the only article that takes this approach to explaining environmentalism. Reading about their collective resilience, we started to think about the role that African American women play in this story of resilience and climate activism.

The father of environmental justice, Robert Bullard, is the pioneer and architect of the environmental justice moment [2]. He is the former dean of the Barbara Jordan-Mickey Leland School of Public Affairs at Texas Southern University and co-founder of the HBCU Climate Change Consortium, which he served as the founding director of the Environmental Justice Resource Center at Clark Atlanta University prior to his arrival at Texas Southern. Over his career, he has written 18 books that address sustainable development, environmental racism, urban land use, climate justice, housing, transportation, community resilience, regional equity and more. His book Dumping in Dixie: Race, Class and Environmental Quality is a standard text in the environmental justice field. Although, we could spend a tremendous amount of time looking at the role that black men play, we started to specifically address hidden figures in the HBCU community, specifically the role that African American women have played. We are not underplaying the role that other ethnicities and races contribute to this conversation, but in most social dynamics, black women are looked upon as counterproductive and the inequality are expressed enacted upon how their contribution is chronicled in the literature.

When we think of climate change in the United States, we should think of two leaders who have been instrumental in bringing attention to underserved communities that have experienced climate deficits: Drs. Robert Bullard and Beverly Wright. Both are sociologists, who have made environmental justice a focal point of their scholarship [3]. They have brought attention to environmental justice issues from Warren County, NC and to Cancer Alley, LA [4, 5]. It has been their seminal work that has brought attention to marginalized communities. They sounded the alarm when others ignored the problem. We refer to them as the father and mother of the environmental justice movement, particularly in the African American community. Although a recent article focused on celebrating the works of what has been termed "black justice heroes [6]," there are still others that have championed this cause by being ambassadors for social justice through their teaching, service, and research.

Table 1 illustrates a number of projects happening at HBCUs. In a report entitled "30 Colleges That Are Fighting Climate Change – This Is How American Colleges Are Working to Save the Planet" not one HBCU was mentioned. Giving the impression that there is no involvement in this quest to find climate solutions. We would expect Texas Southern University to get a mention since it is the home for the Father of the Environmental Justice movement and the Bullard Center for Environmental and Climate Justice. But not one of the HBCU Climate Change Consortium members is listed. We recognize that part of the issue is the methodology for making the listing included checking University websites. Although for our table we searched deeper and looked for grants, active Centers of Excellence, and curricular offerings. We also reached out to Beverly Wright and Mary Williams of the Deep Horizon Environmental Center, LA. In collaboration with TSU and Robert Bullard they received funding from the NIH to support the HBCU Climate Change Consortium (HCCC), which was established in 2011. The consortium was formed to raise awareness about the disproportionate impact of climate change on marginalized communities. The consortium is composed of 30 colleges and universities throughout the Southeast U.S. The organization has a mission to develop HBCU students, leaders, scientists, and advocates on issues related to climate change.

However, others have been toiling in this aspect of research and community activism for more than 20 years, and they have worked with historically black colleges and universities. Further, we have exposed our students to be the next leaders in

A. Risk reduction, sustainability and emergency preparedness	B. Education, health, curriculum and policy	C. Outreach and environmental capacity building
Karen Magid (Huston-Tillotson University) Center for Sustainability and Environmental Justice	Reginald Archer (Tennessee State University) GIS curriculum	Duncan M. Chembezi (Alabama A&M) - Director of the Small Farms Research Center, Public policy and applied economics
Paul B. Tchounwou, (Jackson State University), Director of Environmental Science program	David Padget (Tennessee State University) GLOBE curriculum	Dexter Wakefield, (Alcon State University) Interim Dean and Director of Land-Grant Program
Robert C. Wingfield, Jr., (Fisk University) Director of Environmental Toxics Awareness and Sustainability program	Ranjani W. Kulawardhana (Alabama A&M) GIS, Remote Sensing, and Environmental Science	Nicholas Panasik Jr., (South Carolina State University) Departments of Biology & Chemistry.
Michael A. Reiter (Bethune-Cookman College) Director and Chair Department of Integrated Environmental Science	Ruby Broadwell (Dillard University) Biology, GeoPath Urban Water Program	LaDon Swann (Mississippi Valley State University) director of the Mississippi-Alabama Sea Grant Consortium
Mintesinot Jiru (Coppin State University) Chair of the Department of Natural Sciences	Dagne Hill (Grambling State University), Associate Professor of Environmental Science	James Hunter (Morgan State University) Civil Engineering Storm water management
Fatemeh Shafiei, Ph.D., (Spelman College) Chair of the Political Science Department and Sustainability Committee	L. Faye Grimsley (Xavier University of Louisiana) Public Health.	Arnab Bhowmik, (North Carolina A&T State University) College of Agriculture and Environmental Sciences
	Olga Bolden-Tiller (Tuskegee University) Dean of Agriculture, Environment and Nutrition Sciences	Zhu H. Ning, Ph.D. (Southern University of Louisiana), James and Ruth Smith Endowed Professor, Urban Forestry and Natural Resources Educational
	Deidre M. Gibson (Hampton University) Associate Professor, and Chair, Marine and Environmental Science	
	Winston Thompson (Morehouse College of Medicine) Frontiers in Environmental Science and Health (FrESH)	

Table 1.
Climate champions and special projects.

developing leaders in the climate movement. So, the three women who were looked at in this chapter include Dr. Pamela Waldron Moore at Xavier University, Ms. Felicia Davis of the HBCU Green Fund and Dr. Mildred McClain, also known as Mother Bahati, from Savannah, Georgia. We researched these women, and when possible, we interviewed them. We categorized them as "womanist environmentalists" from their collective work. These women are at the nexus of environmental education and empowerment of minority communities.

In our conversation they did not identify as an ecofeminist. As was explained by Gross in "Buddhism and Ecofeminism: Untangling the Threads of Buddhist Ecology and Western Thought" [7], ecofeminism is a portmanteau that results by combining two words, "ecology" and "feminisms." The problem with simply giving every woman the title of ecofeminism, is that for women of color the concept of feminism does not exactly fit. We support Gross' argument "that feminism is in danger of ignoring the environmental crisis from the lens of anyone who is not white European ancestry. Leaving BIOPC woman viewpoints from the analysis of outcomes can make the definition skewed, incomplete, and often inadequate.

Dr. Waldron-Moore stated it best that

> *"Ecofeminist is an environmental movement. The men have messed up this world and this environment. They [men] have been the ones in place to make decisions. And they [men] have made decisions purely based on their own economic interest. Men have not found a way to profit from the environment or thinking about preserving the environment. And so, while their greed and their desire for power, has allowed them to sort of push, push, push, push, for those things. They have left women to clean up."*

Womanist environmentalism" is based in womanist theory, while diverse, holds at its core that mainstream feminism is a movement led by white women to serve white women's goals and can often be indifferent to, or even in opposition to, the needs of Black women. Thus, we describe the work of our participants as womanist environmentalist and not feminist. Feminism theory does not inherently render white women non-racist, while Womanism places anti-racism at its core. Both the empowerment of women and the upholding of Black cultural values are seen as essential to Black women's existence, which is an important distinction. In this view, the very definition of "feminine" and "femininity" must be re-examined [8].

Womanist environmentalist supports the idea that the culture of the woman, which in this case is the focal point of intersection as opposed to class or some other characteristic, is not an element of her identity, but rather is the lens through which her identity exists. As such, a woman's Blackness is not a component of her feminism. Instead, her Blackness is the lens through which she understands and experiences both feminist/womanist identity [9].

The current state of environmental discourse tends to view environmental issues as problems for science only, and not as issues of social justice. Such an approach ignores the fact that not all groups of humans are situated equally regarding ecological degradation and exposure to environmental toxins, as a direct result of histories of inequality and oppression:

> *"These histories are linked through processes of dualism, in which nature/humans, Anglo-European "whites"/people of color, and masculinity/femininity are placed into opposition. Such conceptual pairings are gendered, as well as raced, classed, and species. Ecofeminism directly interrogates the sources and effects of these pairings, exposing the ways in which sexist ideologies are connected to "naturism [10]."*

Therefore, this chapter argues, struggles for environmental justice that do not incorporate an explicitly gendered and ecofeminist analysis of ecological problems will not adequately understand the ways in which systems of oppression (such as racism, colonialism, gender discrimination, and environmental degradation) are interconnected and mutually reinforcing. For this reason, ecofeminism, a political movement, and theoretical stance which identifies and articulates these interconnections, is a necessary intervention into discussions and debates about how to alter the fact of environmental injustice.

Following the feminist movement, feminism is not described as the work African American women often do without notoriety. Hudson-Weems identify other differences between Womanism and feminism; Womanism is "family-oriented" and focuses on race, class, and gender, while feminism is "female-oriented" and strictly focuses on biological sex-related issues women and girls face globally [9]. They do the hard work and still disproportionately, they are acted upon by the things that happen in their environment and do not get credit for doing the work.

The work of African American women most particularly at HBCUS has not been celebrated. We highlight the work at Dr. Pamela Waldron-Moore, Political Science Professor at Xavier University; Ms. Felicia Davis affiliate with Clark Atlanta University and the HBCU Green Fund; and Dr. Mildred McClain (Savannah, Georgia) a.k.a. "Mama Bahati" of The Harambee House/Citizens for Environmental Justice—(HH/CFEJ), which was born out of a tremendous need for African Americans to develop collective strategies for the effective engagement of citizens in local decision-making. As such, their work serves as exemplars of African American women on the front lines of climate research and policy initiatives that is transformative in addressing environmental (in)justice and advocating for the needs of vulnerable communities. The result of these three women will be explored in this chapter. We will explore how they have established goals and objectives for their respective work. We will ask them to share how they have prepared for the next generation of climate activities. We will explore the perspective on how their work is valued in the academy. And we will close with each of their direct examination of the term "ecofeminism," what they have done, and their contributions to this genre.

1.1 Conceptual framework and method

Our conceptual framework is formed through the lens of intersectionality. The intersectionality of climate justice in our model attempts to address the fragmented scholarships of the role of HBCUs. The conceptual framework (**Figure 1**) that situates our work into three broad areas: (1) risk reduction and emergency preparedness; (2) health, education, curriculum, and policy; and (3) outreach and environmental capacity building.

The HBCU Climate Change Consortium (HCCC) is comprised of 30 member institutions. Who have been identified as members of the Deep South Center and the Bullard Center. Although there are 104 reported HBCUS we have focused on members of the HBCU-HCCC because they have been identified as having active membership.

We highlighted 20 of the Centers and Programs at these schools. In 2021, USBE (US Black Engineers) identified 15 ABET accredited HBCUs, however not all these schools identified in this report are a member of the HCCC. We have also included a map (**Figure 2**) showing the location of all the HCCC-HBCUs. We examine the websites of all HCCC members. We searched for the terms, HBCU Climate Consortium, Environmental Justice, Environmental Science degrees and Emergency Preparedness. We identified individuals who were listed as the person of contact and have identified them as Climate Champions (**Table 1**). In our evaluation of the websites, we found only two of the schools have HCCC affiliation identified on their website (i.e., Spelman and Huston-Tillotson University).

The HBCU thrust for climate justice was born from 1994 when President Clinton signed Executive Order 12898, which requires that the U.S. EPA and other federal

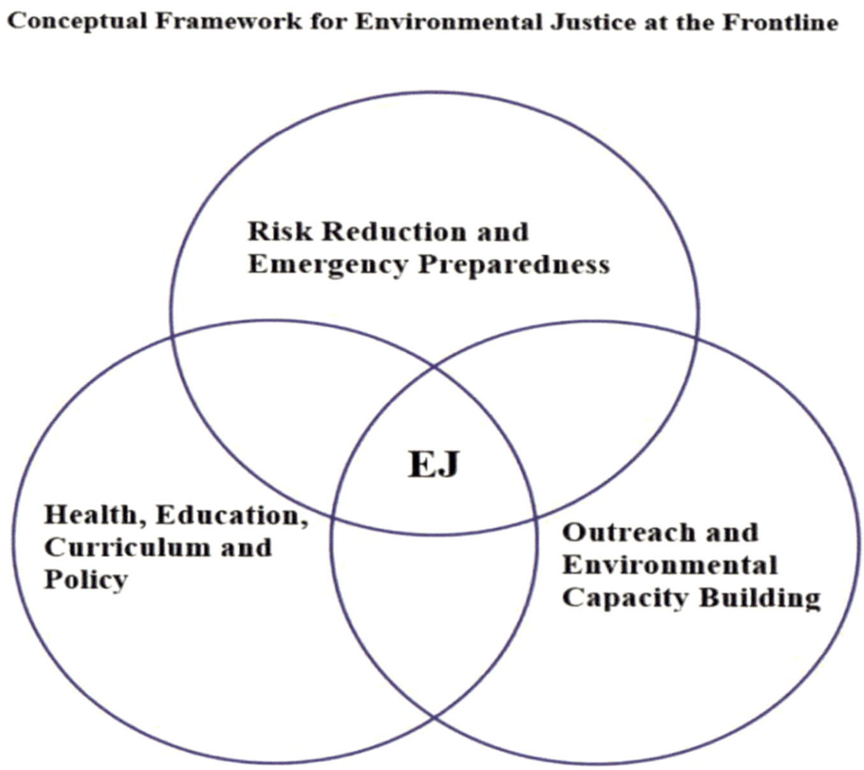

Figure 1.
Conceptual framework for environmental justice at the frontline.

agencies implement environmental justice policies [11]. These policies were designed to specifically address the disproportionate environmental effects of federal programs and policies on minority and low-income populations. The "environmental justice strategy listed programs, policies, planning and public participation processes, enforcement, and/or rulemakings related to human health or the environment that were revised to, at a minimum to: (1) promote enforcement of all health and environmental statutes in areas with minority populations and low-income populations; (2) ensure greater public participation; (3) improve research and data collection relating to the health of and environment of minority populations and low-income populations; and (4) identify differential patterns of consumption of natural resources among minority populations and low-income populations. In addition, the environmental justice strategy included, where appropriate, a timetable for undertaking identified revisions and consideration of economic and social implications of the revisions" [11].

In the almost 30 years since the creation of EO 12898, EPA and the Interagency Working Group (IWG) have not really focused on a concerted effort to coordinate the 11 other agencies so that the idea of a more equitable and just environment could be reached through the training of students in a deliberate and focused manner. There are 11 other agencies that comprised the IWG—Department of Agriculture, Department of Commerce, Department of Defense, Department of Energy, Department of Health, and Human Services, Department of Housing and Urban Development, Department of the Interior, Department of Justice, Department of

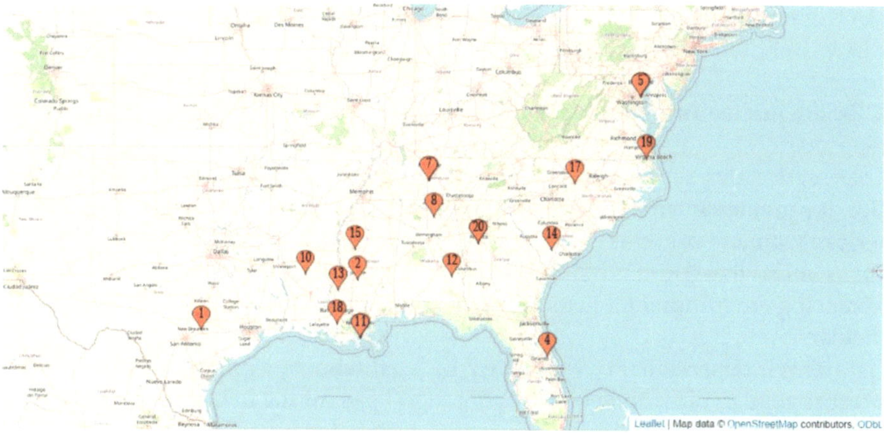

Figure 2.
Geographical location of the 20 of the HCCC-HBCUs discussed in **Table 1**. *This map was generated by using the Mapcusotmizer program created by Patrick Kaeding (https://www.mapcustomizer.com/). The numbers indicate the location of each consortium member: 1. Huston-Tillotson University, Austin, TX 78702; 2. Jackson State University, Jackson, MS 39217; 3. Fisk University, Nashville, TN 37208; 4. Bethune-Cookman College, Daytona Beach, FL 32114; 5. Coppin State University, Baltimore, MD 21216; 6. Spelman College, Atlanta, GA 30314; 7. Tennessee State University, Nashville, TN 37209; 8. Dillard University, New Orleans, LA 70122; 9. Grambling State University 403 Main St, Grambling, LA 71245; 10. Alabama A&M University Huntsville, AL 35811; 11. Xavier University of Louisiana 1 Drexel Dr., New Orleans, LA 70125; 12. Tuskegee University, Tuskegee, AL 36088; 13. Alcorn State University, MS 39096; 14. South Carolina State University, Orangeburg, SC 29117; 15. Mississippi Valley State University, Mississippi State, Mississippi 39,762; 16. Morgan State University, Baltimore, MD 21251; 17. North Carolina A &T University, Greensboro, NC 27411; 18. Southern A&M University Baton Rouge, Baton Rouge, LA 70807; 19. Hampton University, 100 E Queen St, Hampton, VA 23669; and, 20. Morehouse College, Atlanta, GA 30314.* **NOTE:** *When Colleges are in the same geographical area the markers are not seen. Examples of this are New Orleans (Dillard and Xavier), Atlanta (Morehouse and Spelman), Nashville (Tennessee State and Fisk).*

Labor, Department of Transportation, and Federal Emergency Management Agency. Historically HBCUs that are considered 1890 schools have found support from the Department of Agriculture. Other private HBCUs have found funding from NIH, NSF and if they have engineering programs Department of Energy and Department of Defense. The EPA small grants program has gone largely undersubscribed. In a review of the funding, only one HBCU (Shaw University in 2001) applied and received a small EPA grant to start an environmental justice education program. One of the problems with the small grants is that researchers who are generally the focus of the Centers and efforts at HBCUs to bring students into the fray of climate justice do not see these types of small grants as meaningful. The weight that they may be given for tenure and promotion is probably not worth the effort. Also, HBCUs leadership encourage its faculty to find opportunities that will maximize their indirect cost (IDC) which goes into the operation of the facilities so that faculty can support and train students. In reviewing the types of efforts that have been found at HBCUs in **Table 1**, we noticed that the emphasis was in policy, water quality, air quality, toxicity, resilience, renewable energy, and waste management. We noticed a trend that environmental justice has moved from an issue of social justice to one of science access. HBCUs play a vital role in this aspect of preparedness. The mission of HBCUs is to provide upward mobility and access to all races/ethnicities. The mission is rooted in the ability to educate all who desire and were denied education. HBCUs make up 3% of the nation's college and universities, but 22% of all African American college graduates who receive a bachelor's degree are educated via HBCUs [12]. According to a UNCF report, HBCUs are the institution of origin among almost 30% of Black graduates of science and engineering doctorate programs [13].

1.2 Climate justice and intersectionality

Intersectionality serves as the theoretical underpinning for our research on climate justice. It provides an analytic lens to explore and center the experiences of African American women's voices as central to the work on climate justice. Historically, women's voices have been missing in the discourse, yet they have played a critical role in advancing environmental justice in general, and for marginalized communities, in particular.

The term intersectionality was coined by legal scholar, Kimberle' Crenshaw, who posits that gender, race, and class are an integral part of the behavioral influences, role expectations, and the life experiences that structures the lived experiences of African American women and their responses to their environments [14]. As such, an intersectional lens, posits that our lives are shaped by aspects of our identification, including race, class, gender, and heteronormativity [15]. Moreover, Black women's lived experiences are situated in the context of economic, social, and political structures by interconnecting systems of *power* (e.g., patriarchy, race subordination, capitalism) and *oppression* (racism, sexism, classism) [15]. In essence, our social location as African American women is unique, and fosters a unique standpoint and analytical lens by which to examine the complexities of the social systems, the society and the world. It is in this context, that we bring an epistemological stance on how we make sense of what we learn (e.g. the environment) and how we know what we know (e.g. knowledge and truth). So, for climate justice, we center African American women's voices that have often been marginalized, despite the fact that African American women scholars, practitioners, and activists bring a holistic approach to our research, practice, and agency. This stance is not new for African American women, fighting for justice. For example, Anna Julia Cooper, known as the "Mother of Black Feminism" and the founder of Saint Augustine Normal School & Collegiate Institute (now Saint Augustine University) in her work *Voices from the South* (1892) notes, "Only the Black Woman can determine when and where I enter ... then and there the whole Negro race enters with me" [14]. As a sociologist, educator and activist, Cooper advocates from the activist/scholar tradition and advances a curriculum at Historically Black Colleges and Universities [16]. Environmental justice is part of the broader movement for civil rights and human rights. Historically and contemporarily, women have been the canaries in the coal mine, alerting everyone to the destruction of the environment. Currently, Spelman College, has the Anna Julia Cooper Endowed Professor in Woman's Studies, established in 2014 [17]. Thus, honoring the legacy of Black women fighting for justice for all people. Again, HBCUs have a long and rich history of developing leaders to be change agents. As Spelman's motto notes: A Choice to Change the World.

As documented, climate change has a differential impact on communities of color and other marginalized communities. According to the literature, there are two central themes of this differential impact [18]. That is: (1) climate change impacts everyone, but people of color suffer differential risks; and (2) people of color contribute less to climate change [18]. Communities of color are more vulnerable to extreme weather conditions of climate change, including sea-level rise, hurricanes, tornados, flooding, extreme heat and cold and other disasters that devastate the lives of all people, but particularly, the poor, low-income, and minority community Hurricane Katrina, in

2015, exemplifies the long lasting effects of displacement, loss of homes, jobs, health and well-being and the loss of life that impacts the African American community-lower ninth ward-in New Orleans, still to this day [18]. We, in the African American community, cannot forget the scenes from the convention center and our people standing on roof tops with signs begging for help, and children and the elders, languishing in horrible conditions, typifies the vulnerability of these communities. The displacement of students and faculty, and facilities damage to the two HBCUs-Xavier and Dillard had a differential and long-lasting consequences that Predominately White Institutions (PWIs) did not experience. HBCUs have been the safe space, physically and intellectually, for African American students and faculty. Hence, the economic, political, health, and social location of these vulnerable communities underscore the differential influence of climate change. In addition, marginalized community are least likely to contribute to climate change than major companies and corporations and the wealthy that put profits over people, particularly, black, brown and the poor [18] Environment crime, social harms, and justice has become a major concern across the global. The behaviors of the elite class such as industrial population, environmental hazards, global warming, and climate change is serious and pressing problem that is threatening the health and well-being of our planet [19]. Accordingly, there is an urgent need for research and policy that challenges the existing structures of power. The hegemonic power in theory and practice necessitates critical questions and viable solutions to affect change. As such, how do you evacuate the city when you have no money, no transportation? What do you do if you have to take care of the young and the old-a task which typically falls on women in the community? How do you recover when you have no insurance and no economic foundation to rebuild? Simply put, you do not. The challenge of transformative research and policy to protect the most vulnerable communities in a key factor of your climate champions and the HBCU legacy.

Systemic inequality—institutional racism, gender inequality, and economic inequality, health disparities, are at its core, the policies and planning or the lack thereof that disadvantage all people and communities, particularly marginalized communities. The intersectional framework, articulated here, speaks to and deconstructs systems of power, oppression, and privilege, which are taken-for-granted, in environmental discourse, policy, and practice. To this end, African American women are at the forefront of environmental justice movement—that all people have the right to live in safe communities. The climate champions, which we center in our discussion, are at the vanguard of the struggle for environmental justice.

1.3 Risk reduction and emergency preparedness

More than 80% (85 of the 104 HBCUs) are in FEMA regions 3, 4, or 6. The institutions in these regions face higher risks from the negative impacts of a changing climate. These impacts include increases in the occurrence and severity of extreme weather events, accelerating sea-level rise, more extreme heat days, poorer air quality, and higher energy demand. According to the Global Projects summary report produced by the Intergovernmental Panel on Climate Change (IPCC), it is projected by 2050, pervasive impacts on ecosystems, people, and infrastructure will result from increased frequency and intensity of climate and weather extremes, including hot extremes on land and in the ocean, heavy precipitation events, drought, and wildfires in these regions [20].

Johnson and Thompson make the salient argument that leaders at HBCUs must tailor their existing practices with tenets of a model that will help them effectively respond to new challenges [21]. One of the earlier definitions of what constitutes a crisis was provided by Pearson and Clair (1998): "An organizational crisis is a low probability, high impact event that threatens the viability of the organization and is characterized by ambiguity of cause, effects, and means of resolution, as well as by a belief that decisions must be made swiftly" Emergency planning is part of the broader disaster management cycle that all Universities, regardless of their USDA designation, must prepare.

These emergency plans identify prevention, mitigation, preparedness response, and recovery [11]. Risk assessment is one of the central tenants of the environmental justice movement. Adequate coverage and contingency planning are now part of the strategic, operational plans. The dictates of both state and local requirements affect these disaster-prone areas and the plans that each University ultimately plan. A prime example is Jackson State University and their on-going water crisis. No matter how the University leadership might strategically plan it is beholding to the dictates of the local government and state municipalities.

Universities must be proactive to ensure the safety of their stakeholders and to reduce the reduction in damage to infrastructure. HBCUs are even more vulnerable to these changes in the environment. They are no different from the population of people that they serve. In the book, "Dumping in Dixie," Robert Bullard makes the argument that brown and black people are the ones who are disproportionally affected by adverse climate conditions [4]. In addition to the climate, they are confronting socioeconomic inequalities. Providing the necessary knowledge and skills enables the whole community to contribute to national security.

According to the 2022 FEMA Higher Education College List, three HBCUs offer majors or programs related to risk and emergency preparedness: Howard University in Washington, D.C., offers a Bachelor of Science in Emergency Management and a Master of Science in Emergency Management; Jackson State University, Jackson, Mississippi, offers a Bachelor of Science in Emergency Management Technology and a Master of Science in Hazardous Materials Management; and, North Carolina Central University, Durham, North Carolina, which offers a Bachelor of Science in Homeland Security and Emergency Management. Other schools are associated with FEMA; however, two are considered part of the HBCU Climate Change Consortium.

FEMA has recognized a misalignment, mainly as most of these Universities serving large populations of socioeconomically disadvantaged communities have yet to be engaged in emergency preparation. In 2018, working with professors and associates of Howard University, FEMA supported the establishment of an HBCU-focused consortium, the HBCU Emergency Management Consortium (HBCU-EMC). HBCU-EMC recognizes the need for better preparation, and the ability to open career opportunities for HBCU graduates was also evident with the lack of significant presence in emergency management training. HBCU-EMC has started to address emergency response training needs among students at HBCUs. HBCU-EMC has made great strides in proposing actions to address the emergency response training void in the HBCU communities. In the beginning, the HBCU-EMC had a pilot project with several universities in North Carolina. After the lessons learned from their pilot projects in 2018, two HBCUs (Elizabeth City State University, NC, and Tennessee State University, TN) fully implemented Community Emergency Response Team (CERT) Training on their campus. One unintended outcome was that this CERT program was beneficial when these two Universities had to pivot and respond to COVID [12]. They had already gone through some of the best practices for emergency preparedness.

1.4 Health, education, curriculum and policy

Rooted in a civil rights tradition, environmental education and policy are the pillars and undergirding to the environmental justice movement started by Robert Bullard and Beverly Wright. As sociologist they took the initial approach to making the environment a human rights issue. Although it probably was not an orchestrated effort approximate 75% of undergraduate students at HBCUs do not major in STEM. Therefore, if we do not approach this topic of climate justice from multiple entry points, we will miss a cadre of students. The curriculum development in higher education has also not been targeted for job creation but for knowledge creation. Therefore, the approach to curriculum development is left to the individual instructor. Academic freedom has been a hallmark of higher education and having a certain autonomy is not foreign to the HBCU faculty community. HBCUs still have faculty lead efforts and curriculum that reflect their individual interest. Exploring gaps in environmental knowledge and behavior experts suggest integrating environmental issues into courses with human behavior and the social environment to assist students in analyzing and understanding the complex connections between environmental problems and human health and wellness. Other authors more broadly call for the incorporation of environmental sustainability strategies into humanities and social/behavioral sciences [22].

From its inception, EJ (environmental justice) is based on the understanding that all people deserve a healthy environment—clean air, clean water, healthy food, and a toxic free ecosystem. The goal of EJ is to ensure that environmental benefits and burdens are equitably distributed. The lack of a unifying theme to create a theoretical framework to study and use a systems approach to solving complex environmental problems that are both socio-politically acceptable to environmental justice issues has not been implemented.

Law and medicine have been two areas of higher education that HBCUs have formed a niche. Despite the fact that there are only being six HBCU law schools and four medical schools, HBCUs have continued to lead the landscape in exploring topics and preparing its students to compete for a more just and equitable future. Each of these institutions has a motto that conveys a commitment to social justice, community service and the public good served especially for underserved communities. The faculty at these institutions have taken the lead in making sure that their students know what it means to work for social justice of which climate change is an important factor. Since 2000, Public Health programs have exploded at HBCUs. However, historically the Council on Education for Public Health (CEPH) did not accredit a Master of Public Health degree program at a Historically Black College or University until 1999; therefore, no accredited public health courses of study were available to Black students until institutions serving White students began to desegregate under legal challenges [23]. Howard University and Morehouse College were among the first two programs to be accredited by CEPH. Health disparities were a consequence of a legacy of Jim Crow politics. Some sociologists have termed this a form of racism. Environmental racism is a form of systemic racism characterized by the disproportionate exposure of people of color to environmental hazards, such as when communities of color are forced to live near harmful waste sources like sewage works, mines, landfills, power plants, major roadways, and sources of airborne particulate matter. In the US, significant health issues—including cancer, lung disease, and heart disease-are far more common in communities of color. When using the EJ Screening tools provided the EPA you can overlay demographic information with race, income, and environmental conditions [24]. As a result, the user can screen where there are areas of poverty and race and if it collaborates with increase incidences of exposure to pollutants and toxins.

Ethical review of prior behavior that led to dire health consequences for African Americans were later brought to the public attention. Henrietta Lacks and the Tuskegee Syphilis experiment are two world famous cases that speak to how African Americans were treated and used for experimentation. These cases still strike mistrust in the minority community. Therefore, it is imperative that HBCUs work to create the knowledge capital and help to dismantle the distrust.

1.5 Environmental capacity and outreach

By their very existence, HBCUs are in the business of providing opportunity. Besides the Black church and Panhellenic organizations, they have been the place for social and upward mobility. One of the Civils right traditions they have taught its students is how to protest a system that was neither just nor humane. Institutions like Clark Atlanta, Fisk, Hampton University, Xavier of Louisiana, North Carolina A&T, Howard University and Morehouse where the places that Black leadership developed. It was through the instruction and guidance of great intellectual thinkers that young men and women flourished. The people we have come to admire John Lewis, Martin Luther King, Rosa Parks, Jessie Jackson, WEB Dubois were able to think about how to dismantle the Jim Crow era ideology through the capacity building that were instilled in them by their educators. George Kelsey was Martin Luther King's religion teacher, and, in his paper, he believed King was "conducting activities in the finest Mosaic and prophetic tradition" [25].

We propose that these programs are founded in advocacy and activism. HBCUs by their very inception have played roles in shaping their graduates to go into areas to serve the underserved. The strategic mission is to create environmentalism, environmental activism, environmental activists, and better position its graduates to participate in the green economy. The ability to create sustainable change in a resource limited conditions which most HBCUs find themselves it is vital that they find a way to persist. This transfer of knowledge and helping their students transfer their knowledge into real action is a hallmark. HBCUs have realized that their business schools are a way to teach entrepreneurship and incubate ideas that can lead to wealth and additional resources for the communities they serve.

Internationalization became another opportunity for HBCUs to be more involved in environmental justice. They also realize that their students will compete in a global economy. They have also moved to internationalizing their curriculum and co-curriculum opportunities like study abroad, which will ensure that all students are exposed to international perspectives and build global competence. The need to internationalize allows the opportunity to participate in ecoterrorism.

2. Climate champions

As **Table 1** illustrates there are many people at HBCUs doing the work and they stand at the frontline. We highlight the work of three women who for the last 20 years have provide training and education to generations of new climate champions. The concept of Climate Champions was coined from the 2021 Virtual Climate Conversation conference [26]. We asked them the following questions: Who are your collaborators? How or what sparked their interest in Climate Justice? What are the goals and objectives of their climate initiatives? Do they have publications of their work? How strongly do they agree that climate work is seen as white person's work? What key lessons have they learned from their work? What recommendations do they have for others, specifically women who are interested in climate resilience work? What is their feeling on ecofeminism?

We asked these questions to provide readers with insight that can help them to champion environmental justice issue. Also having this type of historical knowledge allows for the sustainability of what has already been established. While this discussion, of their incredible body of work, knowledge, and advocacy is not exhaustive, we highlight the salient themes that emerge from more in-depth interviews and review of their work via their websites. **Figure 3** shows two of the three Climate Champions that we shared their unique perspectives. Dr. Mildred McClain is not pictured here. Although we include her important contribution, we were never able to interview her and get permission to use her image. If you would like to see images of Dr. McClain, please visit http://www.theharambeehouse.net.

2.1 Voices from the frontlines-unique perspectives

2.1.1 Policy, law and creating knowledge economies using the environment (Pamela Waldron-Moore)

Dr. Waldron-Moore published a book entitled "Knowledge Economy and Sustainable Development in Post-Disaster Societies of the Black Diaspora." A native of Guyana an island nation in the Caribbean, she grew-up witnessing the effect of unfair profit margins. An essential component of aluminum manufacturing is bauxite which is mined from the soil in Guyana and other regions that are a majority Black and brown. The companies turn the mineral into more "in-demand" products, earning disparate and much higher profit margins than what is paid for the raw minerals. She has worked to expose her students in asking questions that can help inform policy. She was recently a part of a RAND proposal that is designed to generate knowledge that elucidates pathways to vulnerability and resilience in other disaster contexts. The data will permit empirical tests of interdependencies between different built and social environment factors at multiple levels. The project's findings will help

A B

Figure 3.
Two climate champions interviewed for this chapter—(A) Dr. Pamela Waldron-Moore (Xavier University of Louisiana, New Orleans, LA) and (B) Ms. Felicia Davis (HBCU Green Fund, Atlanta, GA). All pictures were used by permission.

policymakers to target assistance toward individuals and neighborhoods most in need of support during complex crises. The project leverages partnership between political science students and faculty at Xavier University of Louisiana (XULA).

2.1.1.1 Collaborations—working together

"I'm working with a group called disaster research in place (DRIP). So, we call it a drip group. It was started by the social science Research Committee and Research Council, which are located in New York. We have a workshop for the group. We have folks from the Caribbean. Certainly, especially Puerto Rico, and we have, other groups within the gulf and so forth that assist, most of these are professors at colleges in Texas and Colorado and, across the country, but they all focus on disaster as it relates to putting Puerto Rico. The activities that have taken place since then are the efforts to get Puerto Rico to benefit from funds that are out there that they may need to learn about. And certainly, get them the kind of assistance to take care of the disaster."

2.1.1.2 Caribbean nation-interest in climate justice

"I am interested in climate justice partially being from a Caribbean nation; however, Katrina sparked my interest. I identify so closely with where I was from, six feet below sea level, as is Louisiana, especially New Orleans. And so, after Katrina and all that we saw taking place in the city and the capital, we decided to go and do as much as possible. But my focus is a lot on the education that is missing when it comes to following up on what can be done, what has been done elsewhere, and how we can share this information so we can learn from each other because, as you must recognize that the Diaspora is just left behind in all these things. That's a lot to do with what Maslow called higher-order needs. When people are struggling to find the basics of shelter, food, and health care, they don't have time to think about, the existential questions, like climate change and what Top 10 or 15 years the pipe because they're looking to service their needs now. That's what comes from government or welfare programs or ambulatory promises. When those specific things, I mean the idea that the public needs to be taken care of by their local government officials and so forth that the fear of human-beings rests with government officials. And as a political scientist, I know that if you sit back waiting for the government to help you, you might be waiting for a while and missing out on many opportunities to initiate yourself. African Americans and people of the Diaspora and the resilience that come forth because that is the only way they can recapture their past and the resilience that we know what ancestors had still embedded within them. We have to find it, identify it, and work towards making that knowledge capital that is resident within our community that we work."

2.1.1.3 Service or scholarship—the value of our work

"Well, I would say it was service, mostly, and I never had to necessarily use it. In terms of scholarship, I have presented papers at a lot of conferences, especially at NYU's pedagogical conferences. I've talked about how to teach Climate. Just assaulted teach, you know, students who don't know anything about Climate understand where the deficiencies lie. I've had courses in my quantitative analysis where I focused on them going into the community and our culture centers, talking to individuals advising individuals on steps to take to resolve their environmental issues. First, they must recognize what the environmental issues are. So, we must identify them and have individuals recognize the part they need to play to get action taken within a community. I taught students and community leaders to look at policy papers. We submitted things to the City of New Orleans in the early days of 2003; this was even before Katrina. We also have conferences where we invite students from different places to present. These efforts were under the race gender and class projects in Louisiana."

Dr. Pamela Waldron-Moore went on to express her views on the overlapping themes in her work. She notes,

> *"I write on the political economy of development and most of my work is about development issues. So disastrous, just a branch off that topic. I'm looking at the question of where you find low Economic Development, by the very places you're likely to find disaster and inability to address the questions of disaster. So that correlation allows all my research scholarship to be seen holistically rather than just a focus on political science."*

2.1.1.4 Womanist environmentalist vs. eco-feminist

"Ecofeminist is an environmental movement. That men that have missed up this world and this environment. They have made because they have been the ones in place to make decisions. And they have made decisions purely based on their own economic interest. They have not, you know, the profit does not exist in thinking about preserving the environment. And so, while they their greed and their desire for power, has allowed them to sort of push, push, push, push, push for those things. They've left women to clean up. I think of Wangari Maathai. She was from Kenya, and she awarded a Nobel Peace Prize. I think it might have been 2006 or 2007 and one of the first thing she did with her team was to reach out to HBCUs. They reached out to me and as a result I invited her to Xavier. She was just one woman who planted a phenomenal number of trees—a million trees or something.

"She and her organization ensured that the climate was supported in a positive way because she saw the re-greening of areas. It's always been women who have to go back and clean up the mess that men have made men of cutting down the trees and so forth and so on because they're looking for building and lumber sources and so forth. They're not thinking about replanting it, and its women who have done that. So, from the perspective of women doing the cleanup, I see that as a womanist activity, but from this sense of The Men Who made the mess and therefore, the ecosystem requires a different approach ecofeminism works to push the idea that the political activism of women is what will create the kind of climate change of ecological change that one needs to survive in a healthy world. So, you can mention that ecofeminism is what it has been traditionally seen to be really on. The lying point is that without the ecosystem, addressing males and females and having a totally holistic approach to repairing the damage that has been done. It doesn't exist, but in the womanist feel the logical tradition, where are women who assume these roles, where justice and health care matter? You might also want to point out that is how the New Movement sees itself in the African diaspora."

2.1.1.5 Climate work is seen as a white person's work

"Well, you know, I say we because they are the ones who have the luxury of time the luxury of resources to go and get something done. When you think of the Sierra

Club, when you think of the green parties in most countries of the world, whether it be Europe, or the US what does this mean real for me, either? And all of that, it's always white people who have the luxury of pushing things for the environment because they are post materialist and by post-materialism mean there was a point in time in the world where materialism was most important and that people got to the stage where they have accomplished, most of the material things in life that they're looking for. And these are only white people...It's not us who helped. We've got all these things because we're still trying to catch up and so when you think about post-materialism which is a concept proposed by ... Ronald Englehart and he's a political scientist. He argues that until people get to the stage where they can think in higher order levels-meaning not just supporting themselves in the basic ways but when they can sit back and appreciate the spiritual, the spiritual elements of their lives. We're they're grateful for nature so forth. And so many start seeing nature as being more important than the running and seeking after material things. Then and only then will have a system where it becomes everybody's issue to take care of the environment and Black people can adjust. They would recognize that this is about existence, a very preservation of humanity, and once that humanity element is in place, I think that is the element that will lead us away from thinking that it is white people's work, not ours. If we don't get involved now, white people are going to continue, extracting from us, taking our natural resources because they have the technological and expert capacities to do so. Because they will, then pay us in pennies what they will hope to expand into millions...I talked about our aluminum at home, our bauxite which is what creates aluminum. We have about 17 countries in the world that recognize bauxite. We don't have the capacity to do that and the people who work in the bauxite industry at the lowest level, they see them as just basic laborers. But is this the same labor that converts this bauxite in the west? They drink their colas and stuff out of into this aluminum product that makes so much money. So, until we get to the stage where we see the difference between raw materials and the production of manufactured goods, we are going to stay poor."

2.1.1.6 The struggle is real: Having our say

"You know this is a struggle. The struggle for recognition, the struggle for respect this term for justice, this trouble for clarity is something that's got to go on. And so, I prefer to think of everything as once we share knowledge, everybody will know. Therefore, I wrote this book, really, I could have done without another thing on my plate. But you know I wanted to make sure that once I leave people still have something to refer to and you know with a little bit of luck and good health. I can continue writing because there are all kinds of things were not in my head.

2.1.2 HBCU green funds (Felecia Davis)

Sustainability leader and cofounder of the HBCU Green Fund, Felicia Davis, was one of three metro area women named "Atlanta Power Women" by award-winning actor Mark Ruffalo's (Incredible Hulk) ATL100 campaign. She co-authored a paper to discuss ways to increase the participation of HBCUs in the geosciences [27]. Ms. Davis started out collaborating with Drs. Beverly Wright and Robert Bullard. She and Dr. Bullard worked in Atlanta together. She was at the beginning of the discussion of environmental justice. The HBCU Green Fund partnered with the Harambee House/ Citizens for Environmental Justice (HH/CFEJ) to lead a delegation of 27 persons from the United States and Africa to participate in COP27 in Egypt last November and are planning to take a delegation to COP28 in Dubai in November of this year. Below are some highlights from the conversation we had with Ms. Davis.

2.1.2.1 Voices from the frontline—continuous collaboration

I started out with Bob and Beverly. Together we've been on the climate Road since 2000. The others go back to doing toxics and environmental justice. Of course, before that, I entered the climate, global warming space coming out of Civil Rights, voting rights. And that's because I had some visibility for that kind of organizing. I was recruited to work on global warming. I didn't even know what it was. I didn't trust what it was because they said, okay, this is true. I said, okay, the first thing it was to me was to see if it was happening in Africa. I'm going to see if it's real. The people there will tell me, and I did make that. I learned more than I could have imagined from people who are experiencing climate change already. And that was now 23 years ago. So, we collectively adjusted the theme in the US, mainly looking at air quality, like asthma and power plant. But work was driven by the extremely large environmental groups not what in my opinion of using because generally when an extra push was needed or their strategy was failing, tons of black people in the international spaces."

2.1.2.2 The politics of it All: HBCUs

"How best to make a quantum leap? And I have always been attached to a supportive belief in Black colleges as they anchor to the black community. Not to dismiss churches and other organizations within our community. But thinking black colleges attract, bright young Black people and it is, it could be a quantum leap come from there. So I set my sights or working to engage young Black people in the climate movement, seeing it as not only kind of the quintessential problem or talent of our lifetime. But knowing that the solution would necessarily be the next transformation and I've always regarded energy as one of the best determinants of development."

2.1.2.3 Perspective matters: Follow the money

"It was a lot of money to bring people together and to promote sustainability. But once again, it was from when I would say very credible good but still a very Anglo prospective, and that is always something a Black colleges modeling what others had done some of which is good for example Oberlin and Yale both provide certain exposure to what state of the art is and what ideas are important. In 2010, there was a project with the United Negro College Fund (UNCF). I targeted people at minority-serving institutions. And I brought HBCU people together with Hispanic-serving institutions, as well as Tribal Colleges, who do a lot of innovation, together. I expected to see the federal investment and the potential to upgrade HBCU infrastructure to add renewable energy to both research and development, as well as innovation."

2.1.2.4 Creating my own space: maintaining relationships

"I attended the sixth Conference of the Parties (COP-6). I said I must bring young people from Africa to this conference. And I brought 10 students from Senegal to the Hague (Netherlands) for the climate conference, which at that point, it was a kind of Greenpeace and Ozone activism. And they brought the young people, and I've maintained the relationship over all these years. They have maintained a commitment to working on climate change. So that piece to connect all the dots that have always been there in the backdrop, but even for Black America, our odd and interesting relationship with the continent-it's weird. You have the people doing the USAID development work, and then you have the [US Afro-centric] approach, which is not necessarily beneficial, especially to Africans. There is a need to connect young people of African descent wherever they might be."

2.1.2.5 Back to the beginning: the international and global space

"Beverly, Bob, and I were first to take HBCU students to COP. But Beverly and Bob elevated it to a new level by the time they arrived in Paris. I started somewhere along the line started working on centering justice and equity. Gender popped up for me in the international space. I was like a focal point with Gender Climate Conference Women for Climate Justice, which I found a very potent and powerful network within the UN.

Still concerned about Black people making the connection, fast forward to COP-26, which was in Scotland. I worked with Action for Climate Empowerment, and you know some of the networks I was a part of I felt the need to go to Scotland; however, we were coming out of COVID. I was explicitly struck by the absence of Black people. It is what I expected. The global South would be represented differently. I was on record. So, the little sessions that were on the first thing I say in the past, I have brought black African American women to COP. I never went alone. My crew tends to be Black women that are politically engaged. I was just determined that Egypt is in Africa, and there would not be a COP in Africa without African American people. We did a Pre-COP virtual summit in September on sustainability opportunities to feature the work of young people in Africa.

Now I have a network of young people in 20 sub-Saharan countries doing climate work and little micro-projects in 15 countries; it was not something by design. It was not a response to some proposal. It was organic and probably some of the most impactful work I've done, and I still had this commitment to HBCUs. What about these HBCU students? I have a little fellowship program; there's an Atlanta University Center with a couple of fellows at Howard. And so I decided to have a relationship with something called the Eco Villages. This is 100 villages along the Sahel and Northern Senegal, and Mauritania".

2.1.2.6 The role of HBCUs

"At Clark Atlanta they are having an environmental justice summit and I'm going to power up at the end of the day. It's just how I am open and very progressive. A Jewish scholar was at the summit and described his work at a solar farm in Burundi. He asked if anyone would be in Burundi and made the offer that they would be welcomed to join his project. I decided I was going to Burundi because I had already make an initial investment... Again, the most important thing that I know about the quintessential here, right now, in this moment, in the geosciences".

2.1.2.7 Eco-feminism vs. Womanism

"There was a point I would have considered myself an ecofeminism. I am concerned by new framing and power dynamics. So theoretically, I came with a feminist, almost high court, feminist. I would say that my base as a young professional, I own a constant reawakening to gender as the thing right. From a woman's perspective and how we don't get there without that. So, you know, I say [ecofeminism] is a challenge because I'm thinking who's driving the narrative. Who are the shero's. Look at what women do. Therefore, I have problems with how we set the narrative. There is the hardship at people at the bottom, which is generally women. However, we can acknowledge what the breakthrough comes there. The discipline comes there and so, and I look at how African women approach things. No, we don't have the luxury of leisure. And yet, we plant millions of trees, and we demand treaties to decarbonize you need to do what plants have. So, I come through early on as a feminist and then emerging with deep black woman's thought while hope for the whole. If there is going to be ecofeminism. We need to define it. I just take it all in and place us in a proper perspective because and I would say indigenous women everywhere who have existed throughout time, and they have done this work without the title of ecofeminism."

2.1.2.8 Climate work is everybody's work

The only way we are going to really make an in road is that we have more professional people who study the science. Basic science approaches to the world's problems. We have to have more students going into the sciences to understand the use of the technology. I've always regarded the energy sector as one of the best determinants of development. If you are burning dung, you probably doing worse than if you got solar panels. Knowing the role energy has played in the development of whole societies. We have this moment and let us work to push HBCUs forward.

2.1.3 Engaging local communities for emergency preparedness and capacity building (Mildred McClain)

2.1.3.1 The body of work-bring it all together

The academic and professional pedigree of Dr. Mildred McClain is impressive. More importantly, her commitment to Environmental Justice has a long and diverse history of planning and social policy, ensuring a seat at the table for those who have been excluded. Dr. McClain (aka) Mama Bahati's body of work is grounded in a cultural lens and framework of African traditions- the healing power of plants and herbs, and the protection of the children- that reflects nature. Thus, advocating for healthy lifestyles. As an educator for 40 years, Mildred McClain, Ed.D. is an activist/teacher, who has devoted her life and career to fighting environmental injustice, particularly, focusing initially on emergency preparedness.

In this regard, Dr. McClain is the Executive Director of the Harambee House in Savannah, GA., founded in 1990. Harambee House is a nationally recognized Environmental Justice community-based organization that is committed to capacity building of communities to problem solve and promote development. This organization serves, the local, state, regional and national and international communities. Working closely with diverse agencies, she has partnered with the Department of Energy, the Environmental Protection Agency, and for the Toxic Substances and Disease Registry, always keeping her eyes on the prize by advancing public health and environmental justice.

Dr. McClain has served as a delegate with the Conference against Racism and the World Summit on Sustainable Development in South Africa. Bringing together an impressive array of researchers, scholars, activists, governmental officials, and policy makers, to address the global impact of climate change, sustainability, and global racism and exploitation to the forefront of environmental justice. Always the teacher, Dr. McClain, shares her knowledge with the youth. Her commitment to youth is evidenced by her training programs for youth leaders and others, in capacity building that promotes health and wellness for marginalized communities. Indeed, she is a climate champion who has worked tiredly on behalf of the interest of such communities.

As Dr. McClain notes, regarding her interest and start in advocacy in environmental justice, "I came to the environmental justice movement through nuclear weapons. The federal government's Savannah River site, which produced radioactive materials used in nuclear weapons, contaminated the river, and environment near my hometown of Savannah, Georgia. Witnessing its impact on my community growing up ultimately cause me to expand my work beyond voting and civil rights." (Quote from The Progressive Magazine, October 18, 2021). Beyond the founding of the Harambee House, Dr. McClain, is the founder and executive director of the Citizens for Environmental Justice and Senior Fellow with Partnerships for Southern Equity. She has traveled extensively in Africa, Europe, North America, South American, the Caribbean and Japan, reflecting the international scope of her work.

Dr. McClain's body of work, activism, and policy advocacy, reflects African American women's commitment to empowerment and justice that often goes without notice, but is changing the discourse in environmental justice.

3. Conclusions

This chapter highlights the contribution that three African American women are playing in the global conversation on climate change. We outline their contributions in environmental justice as it relates to risk reduction and emergency preparedness, health, education, curriculum, and policy and in building environmental capacity. Many who have taken on the challenge of climate; however, their contribution has largely been overlooked. They have stepped up to prepare the next generation of climate champions to address questions associated with a rising sea, changing weather, and health inequities. These women give voice to the experiences of marginalized communities in particular. Their positionality as African American women provides a unique angle of vision which reflects their social location in the race, gender, and class hierarchy that enables a more expansive and inclusive discussion of environmental inequality and inequity. An intersectional lens allows for understanding the complexities and nuances of the lives of oppressed and marginalized that is intentional and unapologetic in challenging systems of power and privilege.

As womanist scholar, Alice Walker notes, "womanism is to feminism, as purple is to lavender." Simply put, a womanist perspective is grounded in the cultural understanding and lived experiences of African American women. It is this uniqueness that articulates an intersectional lens that makes womanist environmental approaches inclusive of multiple system of oppression (race, gender, class, heterosexism, etc.) that challenges power and inequality. It is in this intellectual tradition that we approach our work and advocacy in environmental justice, as the women we have highlighted here have demonstrated. Healing, collective action and empowerment are central tenants of HBCUs have been critical in providing a "safe space" to challenge hegemonic ideas that dominate the discourse on environmental concerns of vulnerable and marginalized communities.

From this research we make the following observations: (1) although the EPA has implemented a framework to address environmental justice concerns seems disconnected from the realities of how vulnerable and marginalized communities live. The Whitehouse Initiative on HBCUs could be a place to better orchestrate the efforts of EJ. The Whitehouse Initiative was created by the Carter Administration, to increase the capacity of HBCUs to provide the highest-quality education to its students and continue serving as engines of opportunity. (2) We noticed that 30 schools have been identified as members of the HBCU Climate Change Consortium. We would like to see that each school would add a badge to their respective homepages. Identifying them as members of the consortium. Besides participating in the Annual HBCU Climate Change Conference, which brings together noted scholars, climate experts, community leaders and HBCU students and faculty for a three-day program of intensive workshops, lectures, exhibits, and demonstration projects on climate change impacts and solutions, we would like to see this be more visible.

Acknowledgements

The authors wish to acknowledge Ms. Danika Salmans and Dr. Samelia Okpodu-Pyuzza for their help with the creation of the Table. We also like to thank Ms. Bailey Holmes Spencer, MPH, Drs. Arlene Maclin, Myron Williams, and Sasha Coleman-Johnson for reading this manuscript. We also wish to acknowledge the NSF GeoPaths program. This concept of Climate Champions resulted from NSF award number 2211914 from the Directorate for Geosciences.

Conflict of interest

The authors declare no conflict of interest.

Notes/thanks/other declarations

We would like to thank the climate champions-Dr. Pamela Waldron-Moore, Ms. Felicia Davis, Dr. Mildred McClain (aka) Mama Bahati for their tireless work in research and advocacy. We acknowledge their unweaving dedication to the cause of Environmental Justice. We also thank them for sharing their knowledge and wisdom through their interviews and body of works that inform and inspire the next generation of environmental leaders-our students.

Author details

Camellia Moses Okpodu[1*] and Bernadette J. Holmes[2]

1 Department of Botany, University of Wyoming, Laramie, Wyoming, USA

2 Department of Sociology and Criminal Justice, Norfolk State University, Norfolk, VA, USA

*Address all correspondence to: cokpodu@uwyo.edu

IntechOpen

© 2023 The Author(s). Licensee IntechOpen. This chapter is distributed under the terms of the Creative Commons Attribution License (http://creativecommons.org/licenses/by/4.0), which permits unrestricted use, distribution, and reproduction in any medium, provided the original work is properly cited.

References

[1] 6 Female Environmentalists Who Changed the World (3/31/2022). Available from: https://sabai.design/blogs/the-green-house/6-female-environmentalists-that-have-changed-the-world [Accessed: May 13, 2023]

[2] Robert Bullard, Father of Environmental Justice,' to give keynote lecture Oct. 1. States News Service, 18 Sept. 2019. Gale In Context: Biography. Available from: link.gale.com/apps/doc/A600025830/BIC?u=wylrc_uwyoming&sid=summon&xid=3268539a. [Accessed: May 13, 2023]

[3] Bullard RD. Race and environmental justice in the United States. Yale Journal of International Law. 1993;**18**(1):319-335

[4] Bullard RD. Dumping in Dixie: Race, Class, and Environmental Quality. 3rd ed. Boulder: Westview Press; 2000

[5] Wright B. Living and dying in Louisiana's cancer alley. In: Bullard RD, editor. The Quest for Environmental Justice: Human Rights and the Politics of Pollution. San Francisco: Sierra Club Books; 2005. pp. 87-105

[6] "Celebrating Black Environmental Justice Heroes". Available from: https://www.ecorise.org/environmental-justice/celebrating-black-ejheros/[Accessed: May 12, 2023]

[7] Gross Rita M. Buddhism and ecofeminism: Untangling the threads of buddhist ecology and western thought. Journal for the Study of Religion. 2011;**24**:17

[8] Phillips L. The Womanist Reader. New York and Abingdon: Routledge; 2006

[9] Hudson-Weems C. Africana womanism: The flip side of a coin. Western Journal of Black Studies. 2001;**25**(3):137-145. Available from: https://www.libproxy.uwyo.edu/login?url=https://www.proquest.com/scholarly-journals/africana-womanism-flip-side-coin/docview/200355102/se-2

[10] Mallory C. Environmental justice, ecofeminism, and power. In: Rozzi R, Pickett S, Palmer C, Armesto J, Callicott J, editors. Linking Ecology and Ethics for a Changing World. Ecology and Ethics. Vol. 1. Dordrecht.: Springer; 2013. DOI: 10.1007/978-94-007-7470-4_21

[11] Federal Register (1994) Federal Actions to Address Environmental Justice in Minority Populations and Low-income Populations (vol. 59, No. 32). Available from: http://www.archives.gov/federal-register/executive-orders/pdf/12898.pdf

[12] Downer G, Randolph Cunningham SM. HBCUs' preparedness planning and responses to reopening campuses safely during the COVID-19 pandemic: Needs, assets, and insights for future emergencies and disasters. Journal of Emergency Management (Weston, Mass.). 2022;**20**(4):287-299. DOI: 10.5055/jem.0700

[13] The Impact of HBCUs on Diversity in STEM Fields. Available from: https://uncf.org/the-latest/the-impact-of-hbcus-on-diversity-in-stem-fields#:~:text=HBCUs%20are%20the%20institution%20of,long%20before%20enrollment%20in%20college

[14] Crenshaw K. Demarginalizing the intersection of race and sex: A black feminist critique of antidiscrimination doctrine, feminist theory and antiracist politics. University of Chicago Legal

Forum. 1989;**1989**(1):8. Available from: http://chicagounbound.uchicago.edu/uclf/vol1989/iss1/8

[15] Krafft Erin K, Giustina JD, Krumholz ST. Gender, Crime, and Justice: Learning Through Cases. New York: Rowman & Littlefield; 2021

[16] Glass KL. Tending to the roots: Anna Julia Cooper's sociopolitical thought and activism. Meridians. 2005;**6**(1):23-55. Available from: http://www.jstor.org/stable/40338682

[17] Guy-Sheftall Beverly (2014). Bio. Anna Julia Cooper Professor of Women Studies. Available from: https://faculty.spelman.edu/beverlyguysheftall/bio-and-cv/bio/

[18] Andersen ML. Race in Society: The Enduring American Dilemma. New York: Rowman & Littlefield; 2017

[19] Lynch Michael J. & Raymond Michalowski (2010). Primer in Radical Criminology: Critical Perspectives on Crime, Power & Identity. Boulder, CO: Lynne Rienner Publishers, Inc.

[20] Global Climate Projections — IPCC. Available from: https://www.ipcc.ch/report/ar4/wg1/global-climate-projections

[21] Johnson MS, Thompson S. Covid-19 crisis management at historically Black colleges and universities (HBCUs): A contemporary approach to governance and leadership. Journal of Underrepresented & Minority Progress. 2021;**5**(SI):27-46

[22] Beltrán R, Hacker A, Begun S. Environmental justice is a social justice issue: Incorporating environmental justice into social work practice curricula. Journal of Social Work Education. 2016;**52**(4):493-502. DOI: 10.1080/10437797.2016.1215277

[23] Stefkovich JA, Leas T. A legal history of desegregation in higher education. The Journal of Negro Education. 1994;**63**(3):406-420. DOI: 10. 2307/2967191

[24] U.S. Environmental Protection Agency (EPA). EJScreen Technical Documentation; 2017

[25] Kelsey to King, 28 February 1956, in Papers 3:146. Available from: https://kinginstitute.stanford.edu/encyclopedia/kelsey-george-dennis-sale

[26] Okpodu CM, Holmes BJ, Williams MNV, Waldron-Moore P, Tyson P, Twesigye CK. Climate conversations: A one day virtual symposium on the impact that climate change has on the African diaspora. Environmental Sciences Proceedings. 2022;**20**:2. DOI: 10.3390/environsciproc2022020002

[27] Reginald A, Felicia D, Sue E, Richard G. HBCUs broadening participation in geosciences. A Journey Through Integration. 2019:361-378. DOI: 10.1007/978-3-030-03273-9_17

Section 2

Climate Change: Soil Health and Anthropogenic Role

Chapter 6

Role of Soil Health in Mitigating Climate Change

Isidora Radulov and Adina Berbecea

Abstract

Soil health plays an important role in mitigating climate change, soils being the main reservoir for sequestering carbon and reducing greenhouse gas emissions in the atmosphere. In poorly managed soils or cultivated with unsustainable practices, carbon can be released in the form of CO_2 into the atmosphere, contributing to climate change. The conversion of forests and pastures into agricultural land has led to large losses of carbon from the soil. The restoration of degraded soils and the use of conservation practices will determine the reduction of greenhouse gas emissions, increase of carbon storage capacity and ensure resilience to climate change. This chapter will present the principles of sustainable management of soil fertility with the aim of reducing greenhouse gas emissions and sequestering carbon in the soil, as well as the effective use of fertilizers to ensure soil health and reduce the impact of climate change.

Keywords: soil health, carbon sequestration, nutrient cycle, soil reaction, climate change

1. Introduction

Soil health has been defined as "the continued capacity of soils to support ecosystem services, in line with the Sustainable Development Goals and the Green Deal" (EU Mission Soil). Despite this undeniable ecological, economic and social role, holistic understanding and practice-oriented recommendations to maintain soil health are still limited. Climate change, land exploitation and controversial discussions about sustainable land management present great challenges in all industrial sectors. The vital role of soil is not only about the production of safe food, but also about healthy ecosystems and human well-being.

Soil health, as its continued ability to function as a living and vital ecosystem, supporting all living organisms, is maintained through the interconnection of agricultural science with policy, stakeholder needs and sustainable supply chain management. Although soil health was focused on crop production, it is also found in water quality, human health and the intensity of climate change. However, the quantification of soil health is still dominated by chemical indicators, despite the growing appreciation of the importance of soil biodiversity (**Figure 1**).

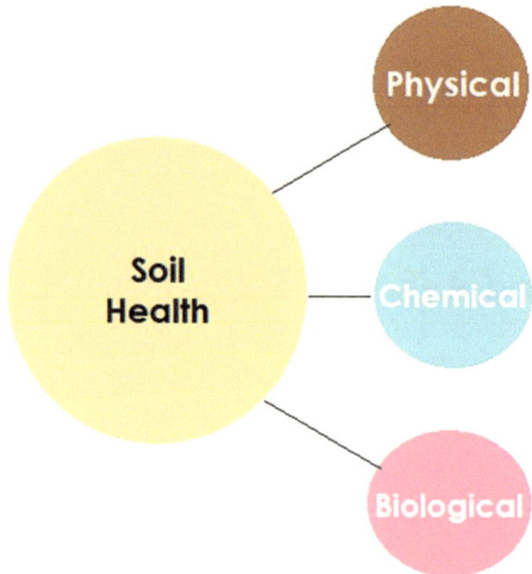

Figure 1.
Soil health diagram.

Soil degradation in Europe, but also worldwide, as a result of inadequate and unsustainable management practices, industrial contamination, soil compaction, air pollution and climate change, is linked to environmental impacts [1]. Around 60–70% of soils in the EU are unhealthy, with various forms of degradation. In agriculture, the inappropriate use of chemical fertilizers and pesticides along with unbalanced crop rotations and intensive mechanization cause soil degradation [2]. These factors affect the various properties of the soil and cannot replace its functions in organic matter cycling, nutrient cycles, aggregate formation and combating pathogens and pests. The irrational use of chemicals in agriculture contributes majorly to the pollution of environmental factors (soil, water, air) and to the reduction of biodiversity by harming non-target plants, insects, birds, mammals and amphibians [3]. Soil erosion, compaction, pollution, salinity and climate change are exacerbating soil health problems in Europe and around the world [4]. It is estimated that climate change will have an overwhelming impact on agriculture through direct and indirect effects on the evolution of agricultural production, soil quality, animal breeding and diseases.

Because of its direct and significant influence on agricultural productivity, the increasing need for fertilizer and amendment applications, and the profound influence of soil chemical properties on changing its biological and physical health, soil chemical and physical health is receiving more attention from both researchers, as well as the interested parties and will be discussed in this chapter.

2. Soil chemical health management for climate change mitigation

The increasing degradation of soils as a result of climate change can be reduced by maintaining the chemical health of the soil, especially by reducing nutrient deficiencies and the accumulation of pesticide residues in the soil, but also by sequestering carbon, because the organic carbon of the soil is one of the most important criteria for evaluating his health.

2.1 Soil organic carbon

Although natural climate changes are a slow process involving relatively small changes in temperature and precipitation patterns, amplified by anthropogenic factors, they have an overwhelming impact on soil fertility. The effects of climate change are expected to influence soil moisture conditions with effects on soil temperature and CO_2 levels. Agriculture, through plant cultivation and animal husbandry, besides carbon dioxide (CO_2) and methane CH_4), is an important source of nitrous oxide (N_2O). It is estimated that agricultural systems produce about one quarter of global N_2O emissions [5].

The organic carbon (OC) content of the soil is influenced by climatic factors, the biotic activity of plants and microbial communities, as well as the physic-chemical properties of the soil, all of which control its dynamics. Soil organic carbon (SOC) is an important component of the terrestrial carbon pool and plays a crucial role both in maintaining the functioning of ecosystems and in the global carbon cycle [6]. The amount of organic carbon in the soil is the result of a long-term balance process, between losses and accumulations [7]. The physical and chemical stability of terrestrial carbon deposits under the action of climate change is unclear. The uncertainty regarding the size and stability of the soil organic carbon reserve is due to the lack of knowledge of both the distribution and the cycle times of organic carbon from the depth, because the amount and age of C stored below 30 cm depth and especially below 1 meter is not very well known [8]. To estimate the impact of climate change on the carbon cycle, studies have been carried out on organic carbon in particles in rivers, which show that increased precipitation and increased temperature destabilize organic carbon stores. On the other hand, the increase in CO_2 content in the atmosphere has the effect of intensifying the photosynthesis process, increasing the efficiency of water use, obtaining high harvests, increasing the amount of plant residues and, as a result, increasing the content of organic matter in the soil. It also increases the availability of C for microorganisms and accelerates the nutrient cycle.

The increase in CO_2 content in the soil causes a change in the C/N ratio, resulting in an intensification of microbial activity manifested by an increase in the partial pressure of CO_2 in the gaseous phase and its concentration in the liquid phase of the soil, with a negative effect on the accumulation of organic C in the soil. Soil C deposits, which act both as a source of C and as a reserve for atmospheric CO_2, are massively influenced by climate change and soil water content. Global warming can directly or indirectly influence the organic carbon content of the soil, by acting on microbial and enzymatic activity [9, 10].

The response of soil organic matter decomposition to increasing temperature is an important aspect of ecosystem response to climate change. The impact of climate warming on the dynamics of soil organic matter decomposition has not been determined, due to the existence of the second fraction of C: a labile one, with a short cycle and high sensitivity to temperature increase, and a fraction with reduced sensitivity to temperature change, with a long cycle for decades or even centuries, which makes up the majority of soil C deposits [11, 12]. The existence of these two types of fractions limits the ability to forecast the influence of temperature changes on the organic C content of the soil [13, 14]. Even a small change in soil organic carbon content can substantially affect the stability of ecosystems,

Soil organic carbon has two important functions for drought resistance: it can store up to 10 times its weight in water, and it is used as a food source for soil microorganisms (bacteria, fungi, and other soil life) that build soil structure. This also creates a habitat for macrofauna, such as earthworms, which make wider pores in the soil for water to drain into the soil profile rather than the surface causing soil erosion.

The primary sources of organic carbon, of a biogenic nature (emissions of volatile organic compounds from vegetation, biological particles - pollen, plant debris, soil, dust, bacteria and viruses, forest fires, volcanic emanations, plankton activity), anthropogenic (burning and fossil fuel and ethanol production, biomass burning, household heating and cooking, solvent use), emissions from agriculture (such as pesticides) and natural gas exploration, are released into the atmosphere as gases and particles. Gases from the soil are in a permanent exchange with the atmosphere, the lack of this exchange leading to the creation of unfavorable conditions for microbiological activity and mobilization of nutrients in assimilable forms. Following the exchange with the atmosphere, the soil receives oxygen as it is consumed and eliminates CO_2 due to its higher concentration in the soil. CO_2 release from soil can be considered as an indicator of microbial activity and soil fertility. Thus, fertile soils release more than 60 mg CO_2/kg/day, and poorly fertile ones less than 30 mg CO_2/kg/day [15]. The intensity of gas exchange depends to a high extent on the air permeability of the soil; it is maximum in sandy soils and minimum in clayey, more compact soils. On average, annually, in the temperate zone, precipitation contains 0.82–2 mg organic C/L [16], inorganic C being generally lower. Dissolved organic carbon from precipitation infiltrates into the soil, being leached onto the soil profile.

Soil clay percentage is often used to determine variations in SOC stability and to model SOC turnover [17]. In principle, clay content determines the stabilization of soil organic matter both by favoring the adsorption process of organic molecules on the surface of minerals with dimensions 2 μm, and by their occlusion in aggregates [18, 19]. Fine-textured soils contain higher amounts of protected SOC compared to coarser-textured soils [20]. In areas with similar pedoclimatic conditions, soil organic carbon content tends to increase in direct proportion to clay content. Due to the interactions between SOC and mineral particles, montmorillonite clays reduce the accessibility of OC to degradation, thus controlling the susceptibility to mineralization; and implicitly the release of CO_2 into the atmosphere [21]. Fine soil mineral particles and soil aggregate structures control organic carbon content by adsorption of organic matter to mineral surfaces and occlusion in aggregates, protecting organic matter from further degradation [22]. The interaction between the soil organic carbon content and the percentage of colloidal particles, such as clay and loam, are used to approximate the capacity of soils to store organic matter and to model the organic carbon cycle in the soil. Climatic conditions were found to affect the relationship between OC stocks and clay content [23]. Correlations between clay content and environmental factors and between environmental factors, as well as the various interrelations between them, can confound the effects of fine particles in OC storage. By controlling these site factors, the analysis of the assessment of the influence of fine mineral particles on OM storage can be more precise [24].

One of the most important and distinct properties of the soil is the total cation exchange capacity, as it can represent a direct measure of the adsorption of the organic carbon content on the mineral surfaces of the soil under the existing pH conditions. A series of researchers have shown that polyvalent metal cations (Ca, Al) have an important role in SOC stabilization by forming metal bonds and exchangeable bridges between the surface of minerals and organic compounds [19, 20]. It has been suggested that the strength of the bond between the metal cation largely depends on the size of the hydrating shell and the valence of the cation. The formation of complexes through ionic bonds between organic molecules and metal cations present on the surface of minerals depends mainly on the displacement of the hydrating shell of the cations, as well as on the type of organic molecule from the soil solution [25]. In soils with acid pH, it was observed that Al^{3+} plays an important role in the formation of complexes between organic molecules and mineral surfaces. The divalent calcium ion Ca^{2+}, weekly polarizable, has a tendency to form ionic bonds with O-containing ligands, such as carboxylic acids, thus forming complexes with organic substrates [26]. The monovalent cation Na^+ does not form ionic bonds with organic ligands, and K^+ only participates in these complexes in the interlayers of certain phyllosilicates. The physical protection of SOM, produced by aggregation, is due to the ability of the metal ions on the surface of the minerals to form organo-metallic compounds by binding several organic compounds [27].

2.2 Modification of soil mineral nutrient status

Climate change may alter the annual and seasonal availability and cycling of nutrients. These climate changes, together with the modification of soil properties and the distribution of nutrients in the soil result in a complicated scenario that influences the microbial activity of the soil and therefore the availability of nutrients to the plant. Variations in temperature or precipitation caused by climate change alter nutrient cycles and, as a result, plant nutrient availability [28]. The research carried out by various researchers [29] show that increasing CO_2, increasing temperatures and water stress are the main factors that could change the demand and availability of nutrients.

The influence of climate change on plant nutrition is highly complicated because climate factors influence all phases of plant growth, including plant development, metabolism, physiology, and production [30]. Agroecosystems capture water, light, CO_2 and nutrients and use them in metabolic processes to form proteins, carbohydrates and starch (**Table 1**).

Nutrient imbalances in the soil occur due to the differential absorption of nutrients and the addition of fertilizers, without taking into account the plant's specific consumption of nutrients corresponding to a certain period of their vegetation. Multi-nutrient deficiency of the soil was increased due to the application of only primary nutrients (especially N and P), with complete dependence on soil nutrient reserves for other nutrients.

Soil is an important part of the natural carbon, nitrogen and sulfur cycle. In the nitrogen cycle, nitrate and ammonium ions from rainwater are retained in the soil, absorbed by plant roots and microorganisms and transformed into amino acids or molecular nitrogen N_2 and nitrous oxide N_2O which are then released into the atmosphere. Under natural conditions, losses of gaseous nitrogen from the soil are

Reservoir	**Carbon (×10^{12} kg)**	**Nitrogen (×10^{12} kg)**	**Sulfur (×10^{12} kg)**
Atmosphere	700 (CO_2)	3,800,000 (N_2)	4
Surface ocean (to the depth of 50 m)	800	1 (organic)	0,1 (organic)
Living biomass	500	14 (organic)	2 (organic)
Soils, to 1 m depth	1800 (organic)	180 (organic)	20 (organic)

Table 1.
Distribution of carbon, nitrogen and sulfur at the earth's surface [25].

approximately equal to the N_2 retained in the soil and converted to amino acids by symbiotic microorganisms.

Nitrogen supply is essential for maintaining soil fertility and increasing crop yield. However, nitrogen that is not absorbed by plant roots can be converted into nitrous oxide (N_2O), a greenhouse gas that has 298 times the warming potential of CO_2 [31]. Various studies have shown that the application of mineral nitrogen fertilizers tends to produce a higher intensity of N_2O emissions even when applied at half the rate of nitrogen application as organic fertilizer [32]. The amount of nitrous oxide released into the atmosphere depends on pedoclimatic conditions, prolonged wet weather can increase N_2O emissions due to anaerobic soil conditions [33].

N_2O dynamics is affected by labile carbon content, as it can create anaerobic conditions in the soil and thus increase N_2O emissions. N_2O emissions are also influenced by soil mineralization-immobilization and nitrification processes [34] as well as clay content [35]. The ideal C/N ratio in soil is 12. When this ratio is changed by applying nitrogen fertilizers, nitrogen uptake by plants, addition of low-carbon organic matter such as straw or wood, soil microorganisms will restore the C/N balance through carbon oxidation, nitrogen fixation or denitrification.

Soil nitrogen content can fluctuate with fertilizer application, including the cultivation of leguminous or non-leguminous plants. Only about half of the applied dose of nitrogen fertilizer is used by plants, the rest being subjected to denitrification or leaching. Fertilization tends to increase the N_2O/N_2 gas emission ratio during denitrification. Water, organic matter in the soil, such as sewage sludge, manure can load the soil with larger amounts of nitrogen than the plants can consume. When these organic matters are oxidized, they will release CO_2, N_2 and N_2O into the atmosphere. However, the amount of gases released by the oxidation of organic matter in normal beds is relatively low, increasing as nitrogen fertilizers are applied. Soil microorganisms can reduce the nitrate ion NO_3^- resulting from the application of mineral fertilizers, increasing the N_2O/N_2 ratio as the amount of nitrate ions increases. Nitrate ion reduction is an important pathway of N_2O and N_2 emissions to the atmosphere. To counteract these effects, researchers are developing a variety of strategies, for example, urease inhibition and capping technologies have been tried to reduce N_2O emissions from soil, with varying levels of success [36].

Although mineral nitrogen fertilization increases biomass production and input of plant residues to soil, as well as organic carbon (OC) accumulation, it can enhance the microbial oxidation of soil organic carbon, leading to loss of organic matter and total soil nitrogen. Acceleration of soil organic carbon mineralization induced by N fertilizer application occurs only when N fertilizer is applied at excessive rates [37].

Soil nitrogen is the most difficult element to characterize: it appears in inorganic and organic form (both anion and cation), in the soil solution and in gaseous form. Plants absorb it only in its inorganic form. The forms of nitrogen contained in organic and mineral fertilizers are: ammonia, urea and ammonium and nitrate ions. Ammonia NH_3 reacts quickly with soil water forming the ammonium cation NH_4^+. Urea $(NH_2)_2CO$ is quickly converted, by urease, to ammonia. When urea or fresh manure is applied to the soil surface, nitrogen is lost as ammonia, especially in dry soils with high temperature and pH. After incorporation and decomposition in the soil, urea quickly passes into the form of ammonia and then ammonium. The ammonium cation, as a result of the positive charge, is retained by the negative charges of the colloidal complex. This prevents the loss of nitrogen through leaching, except for soils with low cation exchange capacity.

Phosphorus is a major nutrient macronutrient required for many physiological and biochemical functions of plants. Climate change creates challenges in phosphorus management, affecting overall crop production The availability and absorption of phosphorus in plants, as well as its mobility in the soil, depend on temperature variations, soil reaction, soil humidity and the increase in CO_2 concentration. P absorption and translocation are reduced at very low or high temperature values. At alkaline pH, phosphorus can be retained in the soil in the form of tertiary calcium phosphates, insoluble compounds, from which phosphorus is not accessible to plants. Acid values of the soil reaction will reduce the intensity of soil microbial activity, but also the rate of transpiration, absorption and use of P. High concentrations of CO_2 in the soil will reduce the absorption of phosphorus by plants [38].

Potassium, calcium, magnesium, manganese, zinc, iron and copper appear in the soil in cationic form. Apart from potassium and ammonium, which have a positive charge (K^+ and NH_4^+), the rest of the elements possess two or more positive charges. The greater the charge of the cation, the stronger it will be retained by the negative charge of the colloidal complex. When the sum of cation charges exceeds the soil's ability to retain nutrients, potassium and ammonium will be released before cations with two or more positive charges, being made available to plants. K^+ leaching generally occurs on sandy soils with low cation exchange capacity. As a result of the high positive charges, micronutrients form in the soil compounds with very strong bonds (coordinative bonds). They cannot be leached under normal conditions.

Numerous studies show that increasing CO_2 levels reduced the levels of some nutrients (such as Ca, K, Mg and P) in the edible parts of plants when they are found in lower amounts than N concentrations. It has also been shown, that high CO_2 levels reduce the concentrations of Co, Fe, Mg, Mn, Ni, S and Zn in plants [39].

In areas with short growing seasons, increasing temperatures can improve crop yields. However, sudden increases in temperature will have a negative impact on growth, photosynthesis, respiration, reproduction and the water regime of plants. In addition, increased temperature can shorten the growth and grain filling periods of the crop. In addition, different crop varieties may show diverse responses to future climate change due to the length of different seasons [40]. Arndal et al. [41] states that the increase in temperature determines the extension of the vegetation period and the increase in the intensity of the mineralization rate of organic matter, increasing the availability of nutrients and their absorption. The increase in soil temperature, which occurs faster in high areas as a result of climate change, leads to the intensification of microbial activity in the soil, influencing its physiology. As a result, there is an acceleration of the organic matter decomposition process, increasing the amount of mineral nitrogen in the soil. This could result in soil nutrient release and availability and decreased nutrient limitation. This happens more in warmer temperatures than in colder ones [42].

2.3 Soil reaction

The reaction of the soil directly influences its health, neutral pH values being suitable for most crops and different soil properties; while sudden changes in soil reaction affect its chemical health. Modification of soil pH due to anthropogenic activities are observed at a very slow rate due to the buffering capacity of the soil. Environmental factors that cause changes in soil reaction include:

- meteorological factors, mainly rainfall and temperature, which cause alkalization and erosion phenomena in the soil;

- climatic factors that intensify the alteration of parent soil materials;
- topographical factors, such as: the topography of the land and the presence or absence of vegetation on the surface of the soil.

Most cultivated plants cannot support a $pH < 4.5$ or $pH > 8.3$, nor large variations thereof. Most plants grow well on neutral (pH = 6.8–7.2) or slightly acidic (pH = 6.3–6.8) soils. As a rule, plants tolerate an acidic environment better than an alkaline one. The maximum sensitivity to acidity, respectively to alkalinity, manifests itself at the beginning of the vegetation period, in the first phases, and especially at the seedling stage. The sensitivity of plants to extreme pH conditions increases, in principle, under conditions of water and nutrient stress.

Low values of the soil reaction, $pH < 5.8$, cause its acidification. If on strongly acid soils, numerous plant species develop weakly or die, this is not only due to the presence of a large amount of $H+$ ions, but to the entire soil complex, especially Al^{3+} and Mn^{2+} mobile in the soil solution above certain limits, to the deficiency of some nutritional elements, as well as the imbalances produced by acidity in their accessibility for plants. The causes of soil acidification are application of fertilizers with acidic physiological reaction, organic matter mineralization, root absorption of nutrients, root secretions, nitrogen fixation by leguminous plants and acidic rain [43]. Acidification negatively affects soil health by reducing plant nutrient availability and soil microbial activity. Toxicity phenomena for plants may also occur due to increased concentrations of one or more mineral elements.

The high values of the soil reaction, $pH > 8.2$, determine its alkalinization. This phenomenon is the opposite of soil acidification, as soil pH is increased by the predominance of sodium carbonates and bicarbonates. For plants that are extremely sensitive to the presence of sodium, a content of $\geq$5% exchangeable sodium causes sodium to accumulate in toxic amounts in plant tissues. A high sodium content in soils leads to Mg and Ca deficiency. The high pH of saline soils can accentuate the deficiency of several microelements (Fe, Cu, Zn or Mn). Also, high pH values lead to an increase in the amount of soluble aluminum. The accumulation of soluble salts in soils inhibits plant growth. They induce the phenomenon of plasmolysis, by which water is removed from the plant, being passed into the soil solution.

The source of nitrogen for plants is organic matter, its decomposition being reduced at pH values lower than 7. At pH values below 5.5, the nitrification process predominates. Ammonium ions will accumulate in acidic forest soils because microorganisms participating in the decomposition of organic matter are less dependent on pH than those participating in the nitrification process. Ammonium ion fixation between layers of clay minerals with an expandable structure decreases with increasing soil pH.

Phosphorus has greater mobility in soils with pH = 5.5–7.0. At $pH < 5.5$, the formation of insoluble iron and aluminum phosphates occurs, from which phosphorus is inaccessible to plants. At pH values >7.0 and up to 8.5, secondary and tertiary calcium phosphates are formed, hardly soluble, from which plants cannot use phosphorus. At $pH > 8.5$, sodium phosphates are formed, easily soluble and accessible to plants.

On acid soils, potassium is easily leached, in contrast to neutral and alkaline soils, and following calcium amendment, its mobility decreases. Ca2+ ions cause an accentuated desorption of K+ from the colloidal complex, and have an antagonistic effect on its absorption by plants.

Calcium and magnesium are more soluble at pH values higher than 6.0. Iron, copper and zinc have reduced mobility at $pH > 7.5$. Boron is very mobile at low pH, and molybdenum at neutral pH (6.7–7). Molybdenum is less accessible to plants at low pH values.

In acidic soils, aluminum partially passes into soluble forms, the high concentrations of Al^{3+} in the nutrient solution having harmful effects on the roots, which undergo morphological changes, turn black, and changes in the capacity to absorb and retain cations occur. Al^{3+} occupies the exchange positions of the root hairs, preventing the absorption of other nutrients: K^{+}, Ca^{2+}, Mg^{2+}, Fe^{2+}, etc. In the presence of Al3+, the phosphorus introduced into the soil through fertilizers is inaccessible to plants due to the formation of $AlPO_4$, insoluble and inaccessible to plants. As a result, on strong soils acidic, with a lot of aluminum, it is necessary to apply high amounts of phosphate fertilizers, in order to prevent on the one hand the lack of phosphorus, and on the other hand the harmful effect of aluminum through its precipitation in the form of tertiary phosphate. The negative effect of aluminum starts from 15 to 50 ppm exchangeable Al, and from 0.3 me/100 g soil (27 ppm Al) calcium amendment is required.

Along with aluminum, manganese also produces toxic effects at very varied concentrations, depending on the plants' sensitivity to this element. Thus, while some plants show important disturbances at concentrations of 1–4 ppm Mn, corn can tolerate concentrations above 15 ppm Mn, without growth retarding effects. Mn accumulates more in the aerial part of the plant, which causes disturbances in protein metabolism. Common symptoms of toxicity are the appearance of brown spots on the leaves. The excess of Mn in the soil solution sometimes causes symptoms that indicate a false lack of iron.

3. Soil physical health

The physical health of soils, manifested by the phenomenon of erosion or compaction, is determined either by intensification of meteorological factors (precipitation and wind), the aerobic-anaerobic cycles of the state of soil moisture, or by human activities, such as: increased intensity of soil work, intensive grain-based cropping systems (rice-wheat cropping system; [44]), reduction of soil organic matter content as a result of non-application of organic fertilizers and lack of vegetation.

3.1 Soil erosion

The intensity of water erosion is determined by the following factors: rainfall characteristics (intensity, distribution and frequency), soil erodibility, slope and slope length, cultivation practices and erosion control measures. Erosion due to wind manifests itself especially on loose, dry soils not covered by vegetation and is subject to the action of strong winds. These conditions are mainly found in arid areas, erosion being accelerated by poor land management. Wind erosion of agricultural lands results in the removal of topsoil, biologically active soil particles, rich in organic matter and nutrients, affecting the water retention capacity, chemical fertility and biological activity of the soil. Wind erosion of agricultural lands results in the removal of topsoil, biologically active soil particles, rich in organic matter and nutrients, affecting the water retention capacity, chemical fertility and biological activity of the soil [45, 46].

Soil erosion due to precipitation and wind contributes to the contamination of the environment with dust particles, microplastics, heavy metals and pesticides, which reach the soil after fertilization with sewage sludge, compost and plastic film mulching, chemical fertilizers, pesticides [47, 48].

The high temperatures associated with climate change cause the intensification of soil degradation processes. Increased evaporation rates lead to excessive drying of the soil, reducing its fertility and making it more prone to erosion [49, 50].

Climate change has a direct impact on crops, requiring the choice of new species of plants to be cultivated, changes in crop rotation, the vegetation period and increasing susceptibility to pests and diseases. Soil cover with vegetation plays a crucial role in preventing erosion. When the soil surface is adequately covered, it acts as a natural barrier against erosion by reducing the impact of raindrops, slowing water runoff and decreasing wind speed near the ground. Ground cover helps maintain soil structure, stability and fertility, as well as preventing the loss of valuable topsoil. Vegetation, such as grasses, shrubs or trees, is one of the most effective forms of ground cover. Plant roots bind soil particles together, making them less susceptible to erosion. The aerial parts of plants, such as leaves and stems, intercept raindrops and reduce their erosive force. Plant cover also helps to improve water infiltration into the soil, allowing it to be absorbed and reducing surface runoff [51].

Modern agricultural practices aim at changing the cultivation calendar, sowing period and techniques, soil preparation and conservation practices, changing the periods of agricultural activities, adopting crops resistant to climate change. Modern practices may also include choosing crops and species better adapted to the growing season and available water, adapting crops with the help of existing genetic diversity and new opportunities offered by biotechnology. The productivity of vulnerable soils, especially those on slopes, must also be protected against water erosion (the effects of rainfall runoff) through the use of good agricultural and environmental practices, including contour plowing.

3.2 Soil compaction

When the soil surface is subjected to pressure, soil compaction occurs, leading to changes in soil permeability and porosity. Gases and water circulation through the soil it is prevented by the interruption of the pores, determining the existence of a reduced amount of water and oxygen. Root growth is hampered.

Soil is removed by erosion much faster than it can be replaced by the process of soil formation. The loss of the surface layer results in reduced fertility, causing lower productivity.

Soil compaction can cause or accelerate other soil degradation processes, such as erosion or landslides. Compaction reduces the degree of infiltration, which leads to the intensification of spreading on inclined surfaces. Due to the presence of a layer with low permeability, the upper layer of the soil will be more prone to water saturation and therefore heavier, presenting the risk of landslides. In the sixth zone, compaction can cause water supersaturation, determining the destruction of aggregates and causing crust formation.

The structure of the soil is improved with organic matter, reducing its predisposition to compaction, erosion and landslides.

4. Conclusions

Healthy soils are essential for increasing resilience to climate change by maintaining or increasing their carbon content and reducing greenhouse gas emissions. The release of carbon from the soil into the atmosphere results in land degradation, also contributing to climate change. Sustainable management and maintenance of soil health are key factors in increasing resilience to extreme weather events such as hurricanes, floods and droughts.

Soil degradation in Europe and around the world is the result of inadequate and unsustainable management practices in agriculture and forestry, industrial contamination, soil sealing, air pollution, having a major impact on climate change. The pressure to meet the growing demand for food and unsustainable agricultural practices, such as: the use of intensive tillage techniques, the abandonment and neglect of agricultural land, the cultivation of monocultures, the application of chemicals in the form of fertilizers, amendments and pesticides, have an impact on soils health, causing their compaction, degradation or pollution, reduction of microbial activity, reduction of bioavailability of nutrients or their loss through leaching or volatilization. One of the main factors of sustainable agriculture is the soil quality, which is determined by the influence of climatic and anthropogenic factors on its physical, chemical and biological properties. Degradation of soil quality is the result of current agricultural land management practices, but sustainable soil management can significantly contribute to achieving climate neutrality by eliminating anthropogenic emissions from organic soils and by increasing the amount of carbon stored in mineral soils. The use of fertilizers in excess can create nutritional imbalances, resulting in water and air pollution with negative effects on the environment and human health. However, the application of insufficient amounts of nutrients in the soil can have a negative impact on the microbial activity, reducing the organic matter content of the soil and its fertility. In order to maintain a healthy soil ecosystem, a sustainable soil management is required, namely: increasing the organic matter content of the soil, maintaining vegetation on the soil surface, rational use of fertilizers, applying crop rotation, eliminating plowing after harvesting. Concerned with these challenges, the Green Deal, especially the Circular Economy Action Plan, the Farm to Fork and the Bioeconomy strategies of the European Union set ambitious goals for the ecological transformation and sustainable adaptation of agriculture to the challenges of climate change. By 2030, the use of fertilizers is to be reduced by 20% and the use of chemical pesticides by 50%. At the same time, nutrient losses are to be reduced by 50%, while ensuring that soil fertility is not impaired.

Acknowledgements

This paper is published from the project 6PFE of the University of Life Sciences “King Mihai I” from Timisoara and Research Institute for Biosecurity and Bioengineering from Timisoara.

Conflict of interest

The authors declare no conflict of interest.

Acronyms and abbreviations

SOC	Soil organic carbon.
OC	Organic carbon.
OM	Organic matter.

Author details

Isidora Radulov* and Adina Berbecea
University of Life Sciences "King Mihai I" from Timisoara, Timisoara, Romania

*Address all correspondence to: isidora_radulov@usvt.ro

IntechOpen

© 2023 The Author(s). Licensee IntechOpen. This chapter is distributed under the terms of the Creative Commons Attribution License (http://creativecommons.org/licenses/by/4.0), which permits unrestricted use, distribution, and reproduction in any medium, provided the original work is properly cited.

References

[1] Borrelli P, Van Oost K, Meusburger K, Alewell C, Lugato E, Panagos P. A step towards a holistic assessment of soil degradation in Europe: Coupling on-site erosion with sediment transfer and carbon fluxes. Environmental Research. 2018;**161**:291-298

[2] Ferreira CS, Veiga A, Caetano A, et al. Assessment of the impact of distinct vineyard management practices on soil physico-chemical properties. Air, Soil and Water Research. 2020;**13**:26. DOI: 10.1177/1178622120944847

[3] Brühl CA, Zaller JG. Biodiversity decline as a consequence of an inappropriate environmental risk assessment of pesticides. Frontiers in Environmental Science. 2019;**7**:177. DOI: 10.3389/fenvs.00177

[4] Lehmann J, Bossio DA, Kögel-Knabner I, et al. The concept and future prospects of soil health. Nature Reviews Earth & Environment. 2020;**1**:544-553. DOI: 10.1038/s43017-020-0080-8 5

[5] Choudhary DK, Mishra A, Varma A, editors. Climate Change and the Microbiome. Soil Biology. Vol. 63. Cham: Springer; 2021. DOI: 10.1007/978-3-030-76863-8_25

[6] Shibabaw T, Rappe George MO, Gärdenäs AI. The combined impacts of land use change and climate change on soil organic carbon stocks in the Ethiopian highlands. Geoderma Regional. 2023;**32**:e00613

[7] Wan Y, Lin E, Xiong W, Li Y, Guo L. Modeling the impact of climate change on soil organic carbon stock in upland soils in the 21st century in China. Agriculture, Ecosystems & Environment. 2021;**141**(1-2):23-31

[8] Liu Y, Zhou X, Du C, Liu Y, Xu X, Ejaz I, et al. Trade-off between soil carbon emission and sequestration for winter wheat under reduced irrigation: The role of soil amendments. Agriculture, Ecosystems & Environment. 2023;**352**:108535

[9] Zhao F, Wu Y, Hui J, et al. Projected soil organic carbon loss in response to climate warming and soil water content in a loess watershed. Carbon Balance and Management. 2021;**16**:24

[10] Yigini Y, Panagos P. Assessment of soil organic carbon stocks under future climate and land cover changes in Europe. Science of The Total Environment. 2016;**557-558**:838-850

[11] Conant RT, Drijber RA, Haddix ML, Parton WJ, Paul EA, Plante AF, et al. Sensitivity of organic matter decomposition to warming varies with its quality. Global Change Biology. 2008;**14**:868-877

[12] Davidson E, Janssens I. Temperature sensitivity of soil carbon decomposition and feedbacks to climate change. Nature. 2006;**440**:165-173

[13] Knorr W, Prentice I, House J, et al. Long-term sensitivity of soil carbon turnover to warming. Nature. 2005;**433**:298-301

[14] Fang C, Smith P, Moncrieff J, et al. Similar response of labile and resistant soil organic matter pools to changes in temperature. Nature. 2005;**433**:57-59

[15] Crista F, Berbecea A, Radulov I, Lato A. Compendiu agrochimic.

Eurostampa Timisoara. 2020;**87**:93. ISBN 978-606-32-0888-1

[16] Jackson RB, Jobbágy EG, Avissar R, Barrett DJ, Cook CW, Farley KA, et al. Trading water for carbon with biological carbon sequestration. Science. 2005;**1944**:1947. DOI: 10.1126/science.1119282

[17] Coleman K, Jenkinson D. RothC-26.3-A model for the turnover of carbon in soil. In: Smith JU, Smith P, Powlson DS, editors. Evaluation of Soil Organic Matter Models. Cham: Springer; 1996. pp. 237-246. DOI: 10.1007/978-3-642-61094-3_17

[18] Lützow MV, Kögel-Knabner I, Ekschmitt K, Matzner E, Guggenberger G, Marschner B, et al. Stabilization of organic matter in temperate soils: Mechanisms and their relevance under different soil conditions – A review. European Journal of Soil Science. 2006;**57**:426-445. DOI: 10.1111/j.1365-2389.2006.00809.x

[19] Solly EF, Weber V, Zimmermann S, Walthert L, Hagedorn F, MWI S. A critical evaluation of the relationship between the effective cation exchange capacity and soil organic carbon content in Swiss Forest soils. Frontiers in Forests and Global Change. 2020;**3**:98. DOI: 10.3389/ffgc.2020.00098

[20] Rasmussen C, Heckman K, Wieder WR, Keiluweit M, Lawrence CR, Berhe AA, et al. Beyond clay: Towards an improved set of variables for predicting soil organic matter content. Biogeochemistry. 2018;**137**:297-306. DOI: 10.1007/s10533-018-0424-3

[21] Doetterl S, Stevens A, Six J, Merckx R, Van Oost K, Pinto MC, et al. Soil carbon storage controlled by interactions between geochemistry and climate. Nature Geoscience. 2015;**8**: 780-783. DOI: 10.1038/ngeo2516

[22] Inagaki TM, Possinger AR, Grant KE, Schweizer SA, Mueller CW, Derry LA, et al. Subsoil organo-mineral associations under contrasting climate conditions. Geochimica et Cosmochimica Acta. 2020;**270**:244-263. DOI: 10.1016/j.gca.2019.11.030

[23] Bailey VL, Bond-Lamberty B, DeAngelis K, Grandy AS, Hawkes CV, Heckman K, et al. Soil carbon cycling proxies: Understanding their critical role in predicting climate change feedbacks. Global Change Biology. 2018;**24**:895-905. DOI: 10.1111/gcb.13926

[24] Schweizer SA, Mueller CW, Höschen C, et al. The role of clay content and mineral surface area for soil organic carbon storage in an arable toposequence. Biogeochemistry. 2021;**156**:401-420. DOI: 10.1007/s10533-021-00850-3

[25] Kunhi Mouvenchery Y, Kučerík J, Diehl D, Schaumann GE. Cation-mediated cross-linking in natural organic matter: A review. Reviews in Environmental Science and Bio/Technology. 2012;**11**:41-54. DOI: 10.1007/s11157-011-9258-3

[26] Bohn HL, McNeal BL, O'Connor GA. Soil Chemistry. 2nd ed. New York: John Wiley & Sons; 2001. p. 307

[27] Newcomb CJ, Qafoku NP, Grate JW, Bailey VL, De Yoreo JJ. Developing a molecular picture of soil organic matter–mineral interactions by quantifying organo–mineral binding. Nature Communications. 2017;**8**:396. DOI: 10.1038/s41467-017-00407-9

[28] Shahane AA, Shivay YS. Soil health and its improvement through novel agronomic and innovative approaches.

Frontiers in Agronomy. 2021;**3**:680456. DOI: 10.3389/fagro.2021.680456

[29] Macdonald CA, Anderson IC, Khachane A, Singh BP, Barton CV, Duursma RA, et al. Plant productivity is a key driver of soil respiration response to climate change in a nutrient-limited soil. Basic and Applied Ecology. 2020;**50**:155-168

[30] Hou D. Sustainable soil management and climate change mitigation. Soil Use and Management. 2021;**37**:220-223. DOI: 10.1111/sum.12718

[31] Jansson JK, Hofmockel KS. Soil microbiomes and climate change. Nature Reviews. Microbiology. 2020;**18**(1): 35-46. DOI: 10.1038/s41579-019-0265-7

[32] Giordano M, Petropoulos SA, Rouphael Y. The fate of nitrogen from soil to plants: Influence of agricultural practices in modern agriculture. Agriculture. 2021;**11**:944. DOI: 10.3390/agriculture11100944

[33] Chirinda N, Carter MS, Rost AK, Ambus P, Olesen JE, Porter JR, et al. Emissions of nitrous oxide from arable organic and conventional cropping systems on two soil types. Agriculture Ecosystems & Environment. 2009;**136**(3-4):199-208. DOI: 10.1016/j.agee

[34] Song X, Zhang J, Peng C, Li D. Replacing nitrogen fertilizer with nitrogen-fixing cyanobacteria reduced nitrogen leaching in red soil paddy fields. Agriculture, Ecosystems & Environment. 2021;**312**:323. DOI: 10.1016/j.agee.2021.107320

[35] Miller CMF, Waterhouse H, Harter T, JGl F, Meyer D. Quantifying the uncertainty in nitrogen application and groundwater nitrate leaching in manure based cropping systems. Agricultural Systems. 2020;**184**:198. DOI: 10.1016/j.agsy.2020.102877

[36] Ribeiro PF, Apraku BB, Gracen V, Danquah EY, Afriyie-Debrah C, Obeng-Dankwa K, et al. Combining ability and testcross performance of low N tolerant intermediate maize inbred lines under low soil nitrogen and optimal environments. The Journal of Agricultural Science. 2020;**158**(5):351-370. DOI: 10.1017/S0021859620000702

[37] Soinne H, Keskinen R, Räty M, et al. Soil organic carbon and clay content as deciding factors for net nitrogen mineralization and cereal yields in boreal mineral soils. European Journal of Soil Science. 2021;**72**:1497-1512. DOI: 10.1111/ejss.13003

[38] Maharajan T, Antony Ceasar S, Krishna TPA, Ignacimuthu S. Management of phosphorus nutrient amid climate change for sustainable agriculture. Journal of Environmental Quality. 2021;**50**:1303-1324. DOI: 10.1002/jeq2.20292

[39] Pilbeam DJ. Breeding crops for improved mineral nutrition under climate change conditions. Journal of Experimental Botany. 2015;**66**:3511-3521

[40] Elbasiouny H, El-Ramady H, Elbehiry F, Rajput VD, Minkina T, Mandzhieva S. Plant nutrition under climate change and soil carbon sequestration. Sustainability. 2022;**14**:914. DOI: 10.3390/su14020914

[41] Arndal MF, Merrild MP, Michelsen A, Schmidt IK, Mikkelsen TN. Beier C,: Net root growth and nutrient acquisition in response to predicted climate change in two contrasting heathland species. Plant and Soil. 2013;**369**:615-629

[42] Soares JC, Santos CS, Carvalho SMP, Pintado MM, Vasconcelos MW. Preserving the nutritional quality of crop plants under a changing climate: Importance and strategies. Plant and Soil. 2019;**443**:1-26

[43] Rittwika M, Supatra S. Role of biological nitrogen fixation (BNF) in sustainable agriculture: A Review. International Journal of Advancement in Life Sciences Research. 2021;**4**(3): 1-7. DOI: 10.31632/ijalsr..v04i03.001. Available from: https://ssrn.com/abstract=3891277

[44] Chauhan BS, Mahajan G, Sardan V, Timsina J, Jat ML. In the indo-gangetic plains of the Indian subcontinent: Problems, opportunities, and strategies. In: Sparks DL, editor. Chapter Six - Productivity and Sustainability of the Rice–Wheat Cropping System. Delaware, USA: Academic Press; 2012. pp. 315-369. DOI: 10.1016/B978-0-12-394278-4.00006-4

[45] Zhang H, Song H, Wang X, Wang Y, Min R, Qi M, et al. Effect of agricultural soil wind erosion on urban PM2.5 concentrations simulated by WRF-Chem and WEPS: A case study in Kaifeng, China. Chemosphere. 2023;**323**:138250

[46] Yang M, Tian X, Guo Z, Chang C, Li J, Guo Z, et al. Wind erosion induced low-density microplastics migration at landscape scale in a semi-arid region of northern China. Science of The Total Environment. 2023;**871**:162068

[47] Qin M, Jin Y, Peng T, Zhao B, Hou D. Heavy metal pollution in Mongolian-Manchurian grassland soil and effect of long-range dust transport by wind. Environment International. 2021;**177**(023):108019

[48] Tian M, Gao J, Zhang L, Zhang H, Feng C, Jia X. Effects of dust emissions from wind erosion of soil on ambient air quality. Atmospheric Pollution Research. 2021;**12**(7):101108

[49] Ma X, Zhao C, Zhu J. Aggravated risk of soil erosion with global warming – A global meta-analysis. Catena. 2021;**200**:105129

[50] Cho R. How Climate Change Will Affect Plants, January 27, Climate, Earth and Society. Columbia Climate School. 2022. Available from: https://news.climate.columbia.edu/2022/01/27/how-climate-change-will-affect-plants/

[51] Zhao J, Feng X, Deng L, Yang Y, Zhao Z, Zhao P, et al. Quantifying the effects of vegetation restorations on the soil erosion export and nutrient loss on the loess plateau. Frontiers in Plant Science. 2020;**11**:573126. DOI: 10.3389/fpls.2020.573126

Chapter 7

Crops and Ecosystem Services, a Close Interlinkage at the Interface of Adaptation and Mitigation

Julian Schlubach

Abstract

Beyond the occurrence of extreme events, heat waves and increasing climate, interseasonal instability is expected to affect more frequently field crops and more broadly ecosystems. Ecosystem services will be at the core of adaptation to a steadily evolving situation. The role of biodiversity is crucial in this regard building the resilience of crops and ecosystems. Understanding how the changing climate, in different parts of the world, will affect plants according to their eco-physiological limits is challenging. Ecosystem services planned at a territorial level are part of the answer, mitigating local climate, regulating hydrological cycles, allowing soft pest control, and contributing to carbon sequestration. Technical solutions are part of the equation, but the potential of genetic optimization should not be overestimated, against the limits of the existing genetic diversity.

Keywords: heat waves, climate instability, crops, ecosystem services, adaptation

1. Introduction

Extreme climatic events, such as storms, hail, droughts, or megafires, mark the spirits and are the most visible events related to the effect of climate change. They mark the spirits and are as such highlighted in media, as well as in the Intergovernmental Panel on Climate Change (IPCC) report [1] and are subject to political communication. However, the longer-term progressive impact on ecosystems, which will in turn as well affect agricultural production systems, is more complicated to apprehend. Beyond the occurrence of extreme events, increasing climate interseasonal instability is expected to affect more frequently field crops and more broadly ecosystems.

It is indeed more complicated and more uncertain to communicate upon impacts that are more pernicious not only due to slow and continuous deterioration but also based on predictive models whose temporality cannot be precisely predicted. Risk analysis remains mainly based on past meteorological data analysis and often on metadata like average or mean temperature. Yet, as an increase in greenhouse gas emissions on a global scale continues year after year, the world is entering a period of chronic climate instability unparalleled on the scale of human history.

IntechOpen

Several structural phenomena will thus combine and affect agricultural production systems both locally and on a global scale. This will result in a structural progressive deterioration of food production and ecosystems over time.

Climate change is often reduced to a change from an initial situation to a new one, to which we need to prepare ourselves to adapt. However, the idea that there will be no new stable situation as such, but instead a permanent instability and change over time, is destabilizing and not well understood in most cases. The idea that climate change would induce benefits to colder regions of the world and thus also mainly affect warmer ones is as well widespread and can be reassuring [1]. This is reflected in the IPCC report [1], as well as in the Notre Dame Global Adaptation Index [2], suggesting between lines that developed countries have the means to face the consequences of climate change while the main issue is about supporting vulnerable ecosystems and poor populations. The concept of vulnerability is correct, but the idea that some parts of the world would not be affected or would have sufficient means to cope with any generated situation is misleading. Looking more closely at the impact on ecosystems, agricultural production and food security may provide useful insight to get a better understanding of upcoming challenges.

2. Past and present global change: What can be learned from earth history?

The accumulation of Green House Gas (GHG) in the atmosphere is the main factor triggering the global warming phenomenon. However, other factors are involved in reinforcing or mitigating the warming or cooling effects over time.

The Green House effect has been crucial in the earth's history allowing the atmosphere to be warm enough, while the Sun activity was much weaker than it is nowadays. Over time, huge quantities of carbon have been stored in the soil during different geological periods. Earth's global temperature has been varying over geological times according to the atmosphere's composition, the land cover evolution, the earth's circumvolution around the sun, and the long-term sun radiative intensity, among other factors [3].

2.1 The Permian mass extinction

At the end of the Permian era about 252 million years ago, massive eruptions to the North East of the Pangea, in an area now located between the arctic circle and Central Asia, released massive quantities of GHG resulting in a steep increase in air temperature. Research on fossils in geological layers shows that approximately 95% of living species disappeared by that time.

Besides rising temperatures affecting terrestrial life, the main factors triggering the mass extinction of marine life have been analyzed as declining oxygen levels in the water, rising water temperatures, and most likely also ocean acidification [4, 5]. Those factors are similar to trends observed nowadays.

Geological probes show that algae bloom occurred [6]. Those events are favored by carbon dioxide (CO2) concentration in the atmosphere, high temperatures, and abundant nutrients. The decomposition of the algae consumes the oxygen, leaving other living organisms to die. Such events can be observed nowadays in rivers, lakes, and tropical sea hot spots like the Gulf of Aden. Another parallel with the ongoing global warming is the increase in the frequency of wildfires [6, 7].

Very important sedimentary layers witness the erosion that prevailed during this period [7]. The assumption is that the loss in land cover, which resulted from the disruption of the ecosystem jointly with heavy rain episodes, resulted in an erosion process rarely observed in Earth's history.

2.2 The quaternary glaciation cycles and the Holocene period

Over the quaternary, the Pleistocene period beginning 2.6 million years ago included more than 50 large-scale climatic oscillations between cold periods persisting for as much as 100,000 years and interglacial periods of about 10,000 years [8]. The Holocene, beginning about 12,000 years ago, is the most recent period of the Quaternary era characterized by the development of human settlements. At the end of the Pleistocene, 20,000 years ago, most of the Northern European and American continents were covered by glaciers. With the ensuing warming period, the sea level has risen by about 120 m [9]. The temperature increase over the period is estimated at 1.5°C by four different models and about 5°C by one model. According to all five models, most of the warming occurred between 12,000 and 10,000 years ago and reached a maximum around 6500 years ago. A cooling period with a global average loss of 0.5°C ensued until the industrial era. The trend reversed again at the end of the nineteenth century, with a steep temperature increase of 1°C on average.

2.3 Global warming and ongoing challenges

Even though, the warming pace over 2000 years has been quick, the ongoing global warming is much faster and is evolving toward a higher temperature increase than the warming that happened since the last glaciation period. From this perspective, the consequences of the ongoing global warming might rather be compared to the event at the end of the Permian era.

The human action impoverishing and reducing natural ecosystems and the scope for the services they can provide is a cofactor that creates negative synergies with global warming.

The faced changes are expected to be progressive, at the scale of human life, even though at an increased speed over time and in any case at a too-fast pace to allow ecosystems and living beings to adapt. The consequence will be the extinction of species intervening in different regulation cycles and as a consequence, the impoverishment and loss in the resilience of ecosystems.

Besides, other factors influence the local climate adding to global warming, reducing ecosystems and human settlements' resilience, in turn reducing the carbon storage capacity and contributing to the release of additional GHG.

The relationship between ecosystems, land cover, climatic extreme events, and erosion is at stake with detrimental consequences for life on earth and for humanity.

The fact that in a temperate environment, as is the case in part of Europe, there is a greater margin for adaptation in the short or medium term than in other warmer regions of the world, does not mean that the ambient conditions will not deteriorate in a similar manner, in the longer term. Thus, in addition to the risk of frost in winter, heat waves are likely to become an increasing matter of concern. Climate instability is also likely to increasingly affect ecosystems and food production.

Meteorological prospective models take into account the occurrence of high temperatures [10]. However, downscaling models, at the local level, taking into account the interrelations of global and local factors remain insufficient. This is especially true for open-field agriculture.

3. Factors affecting the climate at the local level

3.1 Geophysical and environmental factors

Local climate conditions are influenced by other factors that differentiate them from each other. The latitude, the proximity of a coast or continental position, oceanic currents, and winds influence meteorology; the same applies to land cover and urban extension.

Climate change is likely to affect drastically already warmer areas in the short term, while temperate regions are likely to face more climatic instability in interseason periods and face more progressively high-temperature limits.

From an agricultural perspective, the limiting factors, at a given point in time, are not the same in different parts of the world.

Between Ecuador and the South and North Tropics, the high soil exposure to Ultraviolet (UV), the high temperatures, and strong rains are important factors distinguishing the local conditions from those that prevail in temperate regions. Under tropical latitudes, soil's organic layer is quickly degraded as soon as it is in direct exposure to the sun, limiting the scope for carbon sequestration and fertility conservation. High temperatures limit plant growth in the absence of permanent land cover.

In temperate regions, cold remains a limiting factor especially under high latitudes, while snow cover increases the soil albedo in winter and releases water over a longer period in spring into summer.

Besides, the local climate is influenced by ocean currents warming the climate in Northern Europe and cooling it in North America and South Africa's Atlantic coasts, while continental areas are more prone to cold winters and hot summers. Land cover also plays a crucial role in mitigating climate and regulating rainfall [11]. In continental areas, rain forests play a similar role to oceans regulating climate. Those influences will remain, but large-scale modifications in land cover result in climate changes at the local and regional level, unrelated to the global warming phenomenon, while in many cases worsening its effect. Urban sprawl and deforestation are both affecting local climate conditions.

3.2 Urban sprawl

Urbanization increases the warming effect locally by increasing infrared irradiation toward the atmosphere, thus increasing the warming effect at the local level. Urban areas expand often replacing agricultural earth and forests reducing at the same time the carbon sequestration potential.

Besides the loss of food provisioning potential, the soil absorption is modified so that flood risk is increased and groundwater inflow is decreased. The local climate is also modified influencing temperature and rain parameters including over the surrounding countryside.

Where urbanization happens on rangelands, like arid or semiarid areas, the triggered change can be marginal. The impact on the local climate and hydrological cycle can also considerably vary according to the latitude and environment.

However, the overall urbanization of increasing surfaces at a quick pace is a worsening factor in the background of global warming, which should not be underestimated from a climate and food security perspective.

By 2100, the urban land could range from about 1.1 million to about 3.6 million km^2 (roughly 1.8–5.9 times the global total urban area of about 0.6 million km^2 in 2000). Under a middle-of-the-road scenario, new urban land development amounts to more than 1.6 million km^2 globally, an area 4.5 times the size of Germany. Global per capita urban land is expected to more than double from 100 m^2 in 2000 to 246 m^2 in 2100 [12].

Urban growth is triggered by the increase of population worldwide as well as by the trend of populations searching for a greener environment. Since the mid-1950s, European cities have expanded on average by 78% for a population growth of 33% [13].

Beyond the impact on local climate and hydrological cycles of soil artificialization, permanent land cover over large areas, especially in the vicinity of urban areas, is of utmost importance in mitigating local climate.

3.3 Land cover and role of forests

At the local level, temperature is influenced by the land cover albedo and evapotranspiration. The albedo varies according to soil wetness, color, structure, and type. Sand reemits less infrared than some stones or rocks, while a dense tree cover offers a stronger cooling effect through evapotranspiration, varying according to the latitude and related local meteorological conditions.

The results of a study carried out in the North-East of China [11] show that forest cover regulates the local climate allowing slightly warmer temperatures in winter and lower in summer. An almost negligible average temperature increase of 0.04 +/− 0.02 Celsius degrees over the year is reported. Besides, there is a yearly increase in rainfall of 17.49 +/− 3.88 mm, which is predominantly increasing (93.8% of the yearly increase) in spring and summer when it is the most useful for agriculture.

The forest cover has a stronger cooling effect under higher temperatures. The role played by forests across Europe and the United States is thus expected to increase with global warming. The plants' evapotranspiration capacity is measured according to the Leaf Area Index (LAI). The higher LAI and evapotranspiration, the higher the potential cooling effect. Forests have a warming effect at night and in the winter period. In Europe, the forest cover climate mitigation is stronger from North to South and from West to East where continental conditions prevail [14]. Under high latitudes, forests play an important role in mitigating cold events in spring.

In tropical forests, the highest trees forming the canopy regulate the climate for lower layers of vegetation. Trees have different capacities to cope with air temperatures (Ta). This is used in models measuring the difference between mean canopy leaf (Tc) to air temperature (Tc–Ta) [15]. The higher the difference, the more resilient is the tree. More resilient trees present smaller leaves and larger stomatal conductance. *Pometia tomentosa* presents a high leaf temperature tolerance, which can reach 31.8°C, while *Mezzettipsis creaghii* records a maximal measured leaf temperature that can reach 21.6°C [15]. The difference of 10°C between the canopy and lower forest layers is characteristic of tropical forest microclimate. At the canopy level, photosynthesis is active until 2 p.m. when temperatures reach the day's height and then reduce drastically. The soil moisture has a strong influence on the plant's surface temperature and the capacity to sustain high air temperatures. However, beyond a threshold that varies according to plants, biomass productivity drops drastically and can endanger the plants' survival capacity. The forest density and intermediary layers are also of utmost importance to protect the soil from desiccation, preserving the soil moisture. Tree species diversity also contributes to increasing the forest's resilience to high temperatures.

The extent of deforestation, including the size of forest patches and spatial distribution, changes the impact of orest regression. In Latin America and in the Congo Basin, deforestation is increasing. In South America, 919,000 km^2 has been lost between 2000 and 2013 [16]. Carbon sequestration is affected by forest regression with a global effect expected in the medium- and long term, but the local impact on regulating services is affecting more directly local populations.

Where the permanent vegetation cover regresses, the climate is modified at the local level [17]. Soil heating increases air temperature and contributes to reducing rainy periods [18, 19], which are concentrated in more violent events, while the soil loses its ability to retain water.

Whether under tropical or temperate climatic conditions, spatial forest fragmentation threatens their resilience and provided ecosystem services. In Amazonia, a shift from large- to small-scale deforestation is taking place, while new hot spots open in Peru and Bolivia and deforestation has accelerated in Colombia [20]. Fragmentation threatens forest resilience and provided ecosystem services. The main drivers of forest fragmentation are wildfires, agriculture and herding pressure, timber logging, and urban sprawl. At the forest edges, conditions related to wind, sunlight exposure, temperature, and humidity are modified [20].

Besides, fragmentation endangers biodiversity, as different species communities are not being able to remain in contact and lose internal biodiversity generation after generation when becoming too small. In this regard, the focus on sole species on the red list of protected species is misleading as ecosystem services broadly depend on smaller barely visible fungi, insects, and bacteria. However, large enough preserved core areas and aggregation of patches in small distances help reduce the negative impact of forest cover regression [20].

Thus, land cover and land cover modification are key cofactors considering the effect of global warming on the local climate. Agriculture is especially sensitive to extreme climatic conditions and climatic instability.

4. Impact of changing climate conditions on ecosystems and crops

4.1 Field crops' physiological limits

Beyond extreme climatic events, more subtle changes concern, in particular, the disruption of ecosystems and of agricultural production systems due to increasing climatic instability, prolonged episodes of high temperatures beyond the tolerance threshold of plants [21, 22], and the occurrence of events affecting the plants' physiology at critical times in crop development [23].

In particular, the so-called tropical nights during which the temperatures no longer drop below 20°C disrupting the nocturnal respiration mechanisms necessary for plants [24] are likely to become more frequent. While the frequency of day heat waves and tropical nights will increase the probability that those changes will happen at a critical stage of crop development. Therefore, crop losses are likely to happen more frequently.

The predominant idea is that plants grow best with water, sufficient nutrients, and warmth. This assessment is a little simplistic but nevertheless takes on a form of reality in the relatively stable climatic framework that has prevailed over the last millennia while agriculture has developed. The cold has been one of the main constraints faced by field crops in temperate latitudes in northern Europe and the United States. However, the rapid change over a comparatively short timescale calls into question what seemed acquired.

At a different pace, in different areas of the world, raising temperatures for prolonged periods, beyond the plants' maximum tolerance, are likely to gradually erode yields [25]. Beyond a maximum temperature, the yield of the plant begins to drop, due to the so-called "zero" production threshold.

The Day Degrees (DD) methodology has been developed to attribute a plant's growth potential according to the difference between the minimum temperature required by the plant development and the meteorological maximum temperature at day [26]. The Day Degrees concept has been adapted to take as well into account limits in plant growth due to high temperatures [25].

In the West of Uzbekistan (Central Asia), the Day Degree (DD) presents a progressive yield erosion of the main field crops (**Figure 1**). The Day Degrees concept is applied, taking into account the crops' high temperatures' maximal tolerance [25].

Besides, the yield erosion in regions where high temperatures affect increasingly crops' development and extreme events at the time of critical stages of plant development can additionally lead to a drop in production or even a total loss of harvest. The more the climate becomes erratic in interseasonal periods, the higher the probability of the occurrence of such events.

High temperatures have an impact at different stages of a plant's development. For instance, high temperatures of 38°C during the daytime and 28°C at night cause the following changes in sorghum [27]:

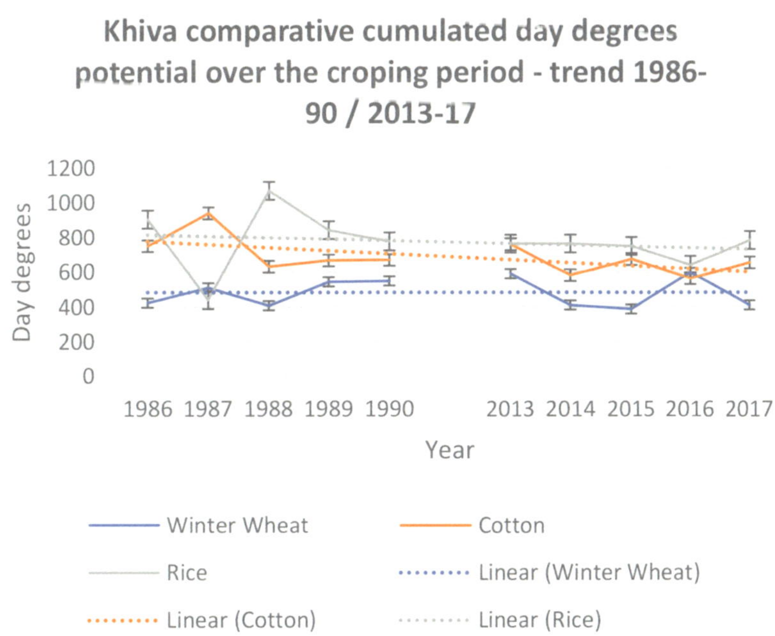

Figure 1.
Trend in the day degrees (DD) production potential for the cropping season of winter wheat, cotton, and rice in the vicinity of Khiva (West Uzbekistan), over 1986–1990 and 2013–2017 periods [25].

- reduced germination;
- reduced panicle development (flower);
- decreased pollen starch content and potential fertility;
- reduced crop yield (seeds per panicle and seed weight);
- reduced seed fertility potential; and
- reduced leaf development.

Sorghum leaf extension development reaches a maximum level from 32 to 35°C and decreases drastically at temperatures of 37–40°C and above [27]. Some plant development stages are more sensitive to temperature than others. In sorghum, the 10 days prior to flowering (panicle development) are especially critical; high temperatures at this stage reduce pollen viability and thereafter the potential yield.

The second sensitive stage takes place during flowering and the 10 subsequent days. At this stage, high temperatures also result in lower pollination and decreased production of viable grains.

Night temperatures above 20°C, called tropical nights, also affect the plants' respiration disrupting physiological mechanisms complementary to the day photosynthesis.

Each drop in production over the growing period is an episode where temperatures raise, reaching the zero-production threshold over a week period. In the present case (**Figure 2**), those episodes did not happen at crop-critical development stages, allowing a good harvest for the region [25].

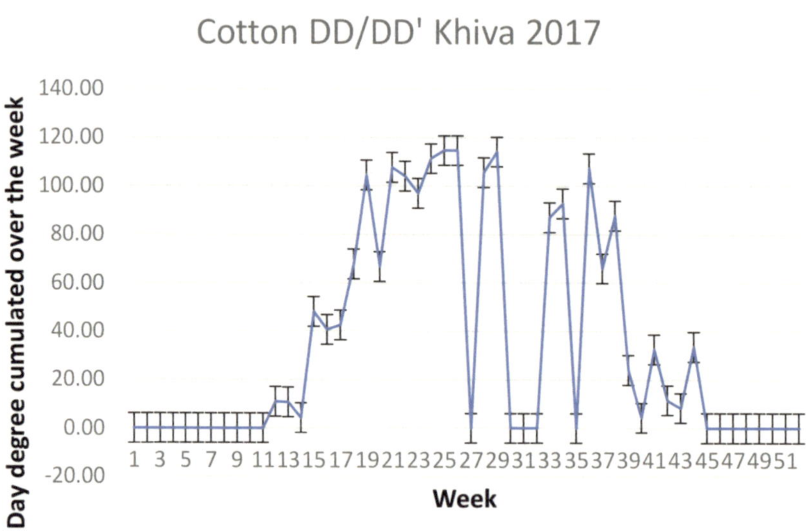

Figure 2.
Day degrees (DD/DD') cotton production potential per week over the year 2017 following Khiva (west of Uzbekistan) meteorological data [25].

However, high temperatures at critical stages of crop development are likely to become an increasing risk jeopardizing the expected yields. This risk is expected to increase over time while increasing temperatures will also limit the plant's vegetative development, as well as reducing yields.

The analysis of data from the World Bank open portal on climate change, for the period until the end of the twenty-first century for Uzbekistan, shows a similar trend in increasing temperatures (**Figure 3**) observed as the tendency observed in the West of Uzbekistan for the period 1986–2017 [25].

The risk of high temperatures happening at plants' development critical stages is expected to increase over the next decades, even though at a different pace according to regions of the world while increasing temperatures will also progressively limit the plants' vegetative development.

Regardless of the frequency of occurrence of temperatures that exceed the physiological limits of plants at critical development stages [22], increasing instability of air masses [28–30] could further endanger field crops. Intermediary seasons whether spring or autumn are expected to become more unstable with varying temperatures destabilizing crops as well as ecosystems.

4.2 Overall impact of climate change on food production systems

Ecuadorian Africa with particularly fragile tropical ecosystems and agriculture runs a major risk related to high-temperature episodes' occurrence.

However, it would also be misleading to believe that Northern countries will be spared from food security issues [1]. Food sovereignty in Europe is eroded by urban sprawl [13] and gradually by climatic events that affect production more frequently than in the past; even though, the physiological limits of field crops are yet still rarely reached. The vision that solutions are only a matter of investment, while climate

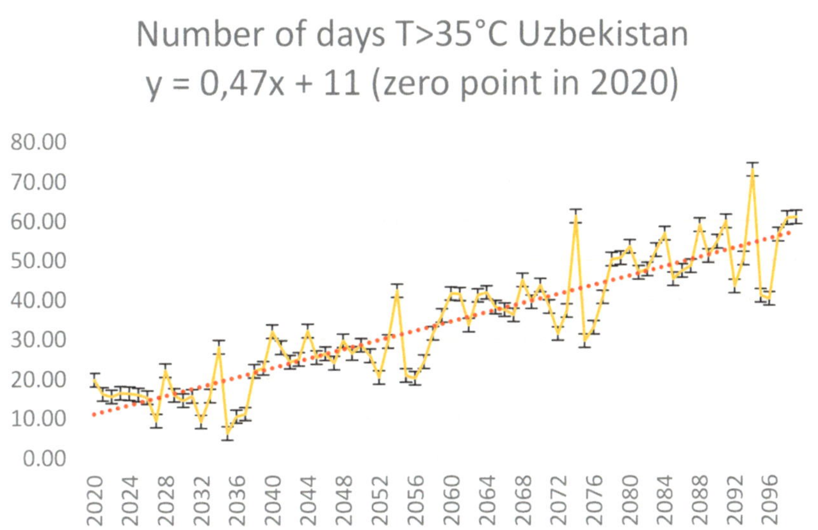

Figure 3.
Trend in the number of hot day events over the twenty-first century on average in Uzbekistan, according to the representative concentration pathway (RCP) 8.5 scenario (World Bank open portal data) [25].

change would not be a direct threat to the richer countries and to the privileged population among those, is misleading.

Besides, while temperatures are increasing, the water demand for irrigation will also increase. The increase in water demand by 2050 in Uzbekistan is expected to increase between 9 and 14% for cotton and between −1 and 5% for winter wheat [31]. At the same time, the inflow into the downstream area is expected to decrease by 26–35% by 2050. The situation will further deteriorate over the second half of the twenty-first century, while the water demand will grow and the Amu Darya River discharge will get radically reduced.

However, drawing a global picture regarding the world food production perspective over time would require a dedicated in-depth analysis.

For the time being, assessments of the impact of climate change on field crops developed among temperate climate countries still rely on the paradigm that droughts are the main risk while high temperatures are rather beneficial as long as enough water is available.

More complex impact assessment models involving high- and low-temperature availability, as well as soil and air humidity parameters are barely available. The interrelation between different parameters including land cover and ecosystems' resilience might as well require more attention.

The European meteorological data basis, Copernicus [32], provides only metadata, with weekly average temperatures being the smallest presented time unit. The data thus provided do not allow to develop models assessing climate impact on ecosystems and crops. More fine data can be retrieved from meteorological stations, but often require to be purchased.

Land cover will be of increasing importance mitigating the climate for the purpose of food production across the world. Under changing climatic conditions triggered by global warming, this implies resilient ecosystems.

4.3 Role of ecosystems and resilience

4.3.1 Ecosystems and ecosystem services

An ecosystem should be understood as a complex system, which is more than the sum of the species that compose it. The multiple interactions allow positive and negative feedback loops that provide some degree of flexibility to adapt to changing conditions. Parallel mechanisms allow the system to continue to autoregulate itself when a species makes default. It should be understood as a self-sufficient complex system achieving a dynamic equilibrium.

Following the same logic, ecosystem services or nature-based solutions are defined in broad terms, including benefits which, in practice, can be contradictory. This applies, in particular, to provisioning, whether it is about wood logging, crops, or herding, as the ecosystem equilibrium is deeply destabilized as soon as resource production and collection is intensified. Agricultural systems are based on a strongly reduced biodiversity which limits autoregulation mechanisms and thus resilience, even in the case of bioproduction which is more respectful of nature.

Ecosystems as defined herein play a crucial role in providing various services to humankind. They contribute to mitigating local climate, increasing soil water absorption, allowing more and longer water availability for plants, while contributing to refilling groundwater, reducing flood risk, and allowing regulation between species populations, including pests presenting risks for crops.

Land cover plays a crucial role in mitigating the climate at the local level. Dense vegetation offers a higher resilience potential by creating its own microclimatic conditions. The role it plays in mitigating the local climate, including by maintaining the atmosphere humidity, plays a crucial role also allowing the provision of other ecosystem services.

However, global warming and local human actions threaten forest ecosystems, while in tropical areas, the forest canopy is increasingly endangered by rising temperatures. The risk of reaching a critical point beyond which the trees' canopy survival could be endangered is not excluded in the medium- or long term. In this background, the reduction of forest density and biodiversity affecting the forest ecosystems' resilience represents a high risk.

It is expected that global warming will increasingly challenge plants' survival capacity and impact the water cycle. Biodiversity is a central element contributing to the resilience of ecosystems and of the services they provide.

4.3.2 Ecosystems' resilience and biodiversity

Biodiversity plays a crucial role in ecosystems' resilience, as well as in populations' capacity to adapt to changing conditions. In the background of global warming, biodiversity is an important factor to be considered to understand the challenges triggered by global warming and local changes.

Interspecific and intraspecific biodiversity, playing interrelated complementary roles, should be distinguished.

Interspecific biodiversity is related to a number of different living species dwelling within an ecosystem. It allows to regulate mechanisms between species, including pest control. Synergies are also important like mycelium prolonging a plant's roots while also providing food for insect larvae that burrow tunnels aerating soil layers, the same way earthworms do. Heterogeneity in trees will also limit the speed at which pests or diseases may spread, thus also participating in regulation mechanisms.

It is also necessary to make the link between the loss of biodiversity of ecosystems on the one hand and the loss of biodiversity within cultivated species on the other. Seasonal temperature changes and climate instability also affect ecosystems which, as they become impoverished, lose resilience and potential for climate mitigation and population regulation among different species.

Intraspecific biodiversity reflects the diversity of physiological expression among different individuals within the same species. The expression of genetic diversity is complex because it is not the result of the simple sum of genes and of the proteins, they are coding for, but as well of multiple interactions among the molecules thus produced. This diversity allows individuum to resist disease while others do not, but have other comparative advantages; the genetic pool among the species is preserved as long as the population remains important enough.

The expression of the biodiversity in plants is, for example, illustrated by the fact that not all seeds will germinate at the same time according to external conditions and not all in the same year, thus increasing the plant's chance to survive and spread even after a few drought years in the Sahel.

On the contrary, productivity implies that all seeds germinate at the same time, make flowers, and are ready for harvest in a synchronized manner. Modern agriculture, whether for crops or for herding, is therefore efficient at the cost of a higher vulnerability to external factors, like climate, pests, or disease. It is especially vulnerable to the expected unstable climatic environment.

The very low intraspecific biodiversity of cultivated plants reduces the resilience of crops and their ability to resist diseases or extreme events. This trend is especially challenging in the background of climate change.

Thus, the reduction at the same time of interspecific and intraspecific biodiversity strongly reduces the resilience of ecosystems. At the same time, preserving natural ecosystems is of utmost importance to face the growing unstable situation generated by global warming.

Answering the challenge faced by global warming, in the short and medium term, requires solutions developed in synergy, avoiding to the possible extent contradictory measures.

5. Adaptation

5.1 Land cover

Land cover is an important element in mitigating global climate change, as well as the climate at the local level. The IPCC refers to Land Use, Land Use Change and Forestry (LULUCF). Permanent vegetation and organic soil layers allow carbon sequestration, but they play as well an even more important role at the local level in regulating the hydrological cycle, mitigating the local climate, and regulating living populations. Natural ecosystems are more effective in serving those purposes than cultivated forests or fields. They are also more resilient, offering regulatory mechanisms that can replace one another when one fails, while also presenting a biodiversity which limits the risks faced from pests.

Along the Amu Darya in the West of Uzbekistan and along the border of Turkmenistan, the remains of the Badai Tugai Poplar Forest still play some role, mitigating floods, reducing dust in the air, and moderating temperature extremes to a limited extent. The river plays a similar role, as well, influencing temperatures in areas in the vicinity of water retention points.

However, the degraded forest strips reduced in size over time reduced drastically the scope of those services.

Prairie strips, as well as forests and bushes, provide shelter to various insects, which can be beneficial in regulating other harmful insects for crops or contributing to flower pollination, reducing insecticide exposure, and increasing water quality and availability [33]. A bocage landscape will play the same role, while also contributing, to some extent, to climate regulation. However, edges along fields are not sufficient to regulate the local climate providing full benefits like from a natural forest. Forest fragmentation reduces the ecosystem \services' benefit affecting rain pattern, temperature, wind regime, and biodiversity [20].

5.2 Irrigation techniques as the mitigating factor

Regarding crops, a widely held dominant idea is that there is no high temperature that cannot be compensated by sufficient irrigation and selection of suitable plant breeds, or with the help of genetic engineering. This perspective seems to be predominantly commercially motivated. The bias thus created is not incompatible with the search for solutions, but the food security topic also requires a strategic perspective, establishing the limits of proposed solutions.

Indeed, the limits faced by plants' physiological limits can be pushed back using irrigation techniques, in particular, by sprinkling in fine droplets.

However, the availability of water is expected to decrease, in particular, due to the reduction in the volume of snowmelt [34] and rainfall events that are less spread out, more violent, and therefore less able to replenish groundwater. Water reservoirs will remain marginal solutions with regard to the loss of mobilized water volumes. In this regard, it should be noted that drawing from groundwater to supply surface water reservoirs, which will also be subject to increasing evaporation, might not be a rational solution. This could locally result in considerable changes in the level of the groundwater with the key to entire ecosystems which could collapse when the roots of the plants no longer reach the water. Wanting to irrigate more to compensate for temperature increases is a headlong rush that will only be possible for a time and in no case can be extended to all existing agricultural production systems.

While high temperatures will gradually and regularly exceed the physiological limits of cultivated plants, thinking of being able to compensate for high temperatures by increased irrigation is a headlong rush. Available water resources are likely to become an increasingly limiting factor while, at the same time, crops' physiological limits will be more frequently reached.

5.3 Genetic engineering

Genetic engineering can contribute to providing answers to specific faced issues, along the existing genetic pool. However, beyond the risk of reduced intra-specific biodiversity, the interactions between genes are not mastered and the long-term side effects of introduced genes are broadly unknown. In most cases, the transaction cost is not known and not subject to specific studies either. Resistance to some pests or pesticides, as well as crops offering better characteristics for transformation, can be achieved but it might be at the cost of the nutritional value, of a vulnerability to future pests, and of a higher dependency on pesticides.

Thus, genetic engineering implies a specialization with narrow intraspecific biodiversity, which results in a higher vulnerability to a changing environment. Besides, the pool of genes and their expression present a broad diversity which nevertheless does not mean that there are no physiological limits.

Besides, genetic engineering can accelerate the optimization of characteristics present within existing species and to some extent between living species, but it does not create anything new; pushing the limits does not mean the absence of limits. Thus, there are indeed physiological limits, even if these differ from one species to another. Plants fixing carbon on a four-carbon molecule, called C4, offer better heat and drought stress characteristics than plants fixing carbon on a three-carbon molecule, called C3. C4 plants, like sorghum or maize, have the advantage of absorbing more carbon with openings in the leaves, stomates kept smaller, thus also limiting water loss through evapotranspiration. However, those plants are also less resistant to cold temperatures. Besides, it would be rather illusory to think that C3 plants can be transformed into C4 ones, should such a change represent an overall advantage.

6. Conclusions

Humanity is facing an unprecedented challenge triggered at the same time by global warming and by concurrent actions endangering resilience factors. Ecosystems and the biodiversity they embody are crucial to contribute to mitigating the consequences of the faced changes. Ecosystem services contribute to mitigating local climate, regulating hydrological cycles, allowing soft pest control, and contributing to carbon sequestration.

However, ecosystems are themselves endangered at the same time by global warming and by human pressure. The biodiversity implied by ecosystems' services and resilience still requires a better understanding by decision-makers.

Under temperate climatic conditions, crop growth and production limitations due to high temperatures are new factors which have been marginal in most of Europe and North America until recent years. The tendency in temperate countries is to think that climate change can be coped with or even be beneficial. Those beliefs are misleading and may slow down political decisions crucial in international fora. It is thus of utmost importance to collect factual data and develop research to properly understand challenges and adaptation options.

In this regard, providing solutions to some climate change challenges might be seen as an opportunity to innovate and unlock new sources of profit, but they often have a transaction cost which can bear negative consequences.

Nevertheless, solutions likely to mitigate global warming exist, but those are not universal and are also likely to face limits. Living systems could increasingly be jeopardized if different thresholds should be bypassed. Therefore, a precautionary principle should preferably be applied.

From this perspective, the path of specialization of living production systems, whether for crops or for livestock, at the cost of a loss of intraspecies biodiversity, presents high risks. Having a homogeneous population of plants or animals considerably reduces the resilience of species. Impoverished ecosystems are as well vulnerable.

Genetic engineering is promoted as a possible solution. But the related communication mostly ignores the intrinsic limits of intra- and interspecific genetic potential, and thus of the industrial food production's inherent fragility.

In arid and semiarid areas, maize, sorghum, or millet presents a higher temperature resistance. However, maize requires more water availability, which may not be suitable for the area. Sorghum presents not only better characteristics and lower water demand but produces also lower yields. Millet presents not only a higher resilience but also even lower yields. In areas facing strong water deficit, a shift toward high-value crops, including fruit and vegetable resilient plants, like pomegranate, on smaller surfaces might also be assessed locally in replacement of less resilient crops.

Worldwide, water resources management will require a long-term perspective, implying difficult arbitrations not only between domestic use and food production, but also between agricultural productions and land areas to receive water.

Land cover, ecosystems' services, water use arbitration, and technical solutions, among other aspects, require a systemic approach. For achieving such an approach, public policies should play a key role in supporting measures for ecosystem conservation serving the common interest of the public including private actors. Related political choices would require to be supported by detailed impact and transaction costs analysis according to the issue and to the solutions looked at.

Tackling the issue is of utmost importance in order to avoid periods of instability resulting from poorly controlled food security which would lead the world into spirals of degradation.

Acknowledgements

The present publication is based on the work of the author, bibliographic references, and experience drawn from experience in different parts of the world.

Conflict of interest

The author declares no conflict of interest.

Appendices and nomenclature

CO2	carbon dioxide
DD	day degrees – method to estimate plant productivity according to favorable temperature for plant growth
GHG	green house gas
IPCC	intergovernmental panel on climate change
LAI	leaf area index—measuring leaf surface available for evapotranspiration
LULUCF	land use, land use change and forestry
RCP	representative concentration pathway
UV	ultraviolet

Author details

Julian Schlubach
Independent Expert, Aix en Provence, France

*Address all correspondence to: julian.schlubach@gmail.com

IntechOpen

© 2023 The Author(s). Licensee IntechOpen. This chapter is distributed under the terms of the Creative Commons Attribution License (http://creativecommons.org/licenses/by/4.0), which permits unrestricted use, distribution, and reproduction in any medium, provided the original work is properly cited. (cc) BY

References

[1] Schlubach J. IPCC 2022 report assessment from a food security perspective. International Journal of Environmental Science. Juniper Publishers. Opinion Volume 31 Issue 5 - April 2023. Available from: https://juniperpublishers.com/ijesnr/IJESNR.MS.ID.556327.php

[2] https://gain.nd.edu/our-work/country-index/

[3] https://www.bgs.ac.uk/discovering-geology/climate-change/what-causes-the-earths-climate-tochange/#:~:text=Geological%20records%20show%20that%20there,carbon%20dioxide%20(CO2)

[4] William JF, Georgy A, Jannes M, Tabea R, Niklas HK, Terry TI, et al. Machine learning identifies ecological selectivity patterns across the end-Permian mass extinction. Paleobiology. 2022;1. DOI: 10.1017/pab.2022.1 University of Hamburg. "Three critical factors in the end-Permian mass extinction." ScienceDaily. ScienceDaily, 1 March 2022. Available from: <www.sciencedaily.com/releases/2022/03/220301131200.htm>

[5] Andrew H Knoll, Richard K Bambach, Jonathan L Payne, Sara Pruss, Woodward W. Fischer. Paleophysiology and end-Permian mass extinction. Science Direct - Earth and Planetary Science Letters. 2007;**256**:295-313

[6] Combined Reports - Animals Died in 'Toxic Soup' during Earth's Worst Mass Extinction, a Warning for Today. University of Connecticut - UConn Today – Research & Discovery, Communications. Sep 2021

[7] Fox CP, Whiteside JH, Olsen PE, Grice K. Flame out! End-Triassic mass extinction polycyclic aromatic hydrocarbons reflect more than just fire. Earth and Planetary Science Letters. 2022;**2022**(584):117418. DOI: 10.1016/j.epsl.2022.117418

[8] Elias SA. The Quaternary. Royal Holloway, University of London, Egham, UK. Elsevier Inc; 2013

[9] Zong C. Late Pleistocene Sea Levels and Resulting Changes in Global Land Distributions. Submitted to the graduate degree program in Geography and the Graduate Faculty of the University of Kansas in partial fulfilment of the requirements for the degree of Master of Science. Chair: Jerome E. Dobson, Stephen Egbert, Xingong Li. University of Kansas; May 2015

[10] World Bank. Climate Change Knowledge Portal. Available from: https://climateknowledgeportal.worldbank.org/

[11] Jiao Y, Bu K, Yang J, Li G, Shen L, Liu T, et al. Biophysical effects of temperate forests in regulating regional temperature and precipitation pattern across Northeast China. Remote Sensing. 2021;**2021**(13):4767. DOI: 10.3390/rs13234767

[12] Gao J, O'Neil BC. Mapping Global Urban Land for the 21st Century with Data-Driven Simulations and Shared Socioeconomic Pathways. Nature Communications. 2020. DOI: 10.1038/s41467-020-15788-7

[13] European Environmental Agency. Urban Sprawl in Europe - the Ignored Challenge. EEA Report No 10/2006

[14] Tang B, Zhao X, Zhao W. Local Effects of Forests on Temperatures across

Europe. © 2018 by the Authors. Basel, Switzerland: Licensee MDPI; 2018 http://creativecommons.org/licenses/by/4.0/

[15] Song Q, Sun C, Deng Y, Bai H, Zhang Y, Hui Y, et al. Tree Surface Temperature in a Primary Tropical Rain Forest. © 2020 by the Authors. Basel, Switzerland: Licensee MDPI; 2020

[16] Potapov P, Hansen MC, Laestadius L, Turubanova S, Yaroshenko A, Thies C, et al. The last frontiers of wilderness: Tracking loss of intact forest landscapes from 2000 to 2013. Science Advances. 2017

[17] YanTao S, DaoWei Z, Xiang ZH, Guang L, YingHua J, Qiang L. Effects of vegetation height and density on soil temperature variations. Chinese Science Bulletin. 2013;**58**:907-912

[18] Li K, Zhang J, Wu L, Yang K, Li S. The role of soil temperature feedbacks for summer air temperature variability under climate change over East Asia. Earth's Future. 2022;**10**:e2021EF002377. DOI: 10.1029/2021EF002377 10.1029/2021EF002377RESEARCH ARTICLE 1 of 20 – pp 1-20

[19] Zhang H, Liu B, Zhou D, Wu Z, Wang T. Asymmetric Soil Warming under Global Climate Change. International Journal of Environmental Research and Public Health - MDPI. Mar 2019

[20] Slattery Z, Fenner R. Spatial analysis of the drivers, characteristics, and effects of Forest fragmentation. In: Ronald CC, editor. Sustainability. 2021;**13**:3246. DOI: 10.3390/su13063246

[21] Rahman SU, Arif M, Hussain K, Hussain S, Mukhtar T, Razaq A, et al. Evaluation of Maize Hybrids for Tolerance to High Temperature Stress in Central Punjab. American Journal of Bioengineering and Biotechnology. Columbia International Publishing. 2013;**1**(1):30-36. DOI 10.7726/ajbebt.2013.1003

[22] Hatfield JL, Prueger JH. Temperature extremes: Effect on plant growth and development. Elsevier Weather and Climate extremes. 2015;**10**:4-10

[23] Vara Prasad PV, Djanaguiraman M. Response of floret fertility and individual grain weight of wheat to high temperature stress: Sensitive stages and thresholds for temperature and duration. Functional Plant Biology. 2014;**41**:1261-1269

[24] Sadok W, Krishna Jagadish SV. The hidden costs of Nighttime warming on yields. Trends in Plant Science. 2020. pp. 644-651

[25] Schlubach J. Downscaling model in agriculture in Western Uzbekistan climatic trends and growth potential along field crops physiological tolerance to low and high temperatures. Heliyon. 2001 https://authors.elsevier.com/sd/article/S2405844021011312

[26] Miller P, Lanier W, Brandt S. Using Growing Degree Days to Predict Plant Stages. Montana State University, Agriculture and Natural Resources (Farm Management) – Extension service; 2018

[27] Prasad PV. High-Temperature Tolerance in Sorghum – What Do we Know and What Are the Possibilities? Conference: 1st Australian Summer Grains Conference at: Gold Coast, Australia. 2004;**1**:1-20

[28] Ryan A. Zamora, Robert L. Korty, Matthew Huber. Thermal Stratification in Simulations of Warm Climates: A Climatology Using Saturation Potential Vorticity. University of New Hampshire; 2016. pp. 5083-5102. DOI: 10.1175/JCLI-D-15-0785.1

[29] Lindsey R. Understanding the Arctic Polar vortex. Climate.gov – Science and Information for a Climate-Smart nation. 2021

[30] Lolis CJ. Climatic features of atmospheric stability in the Mediterranean region (1948-2006): Spatial modes, inter-monthly and interannual variability. Wiley InterScience. Meteorological Applications. 2007;**14**:361-379. DOI: 10.1002/met.36

[31] Boehlert BB, Droogers P, Neumann, JE, Srivastava JP, Sutton WR. Reducing the vulnerability of Uzbekistan's agricultural systems to climate change: Impact assessment and adaptation options (English). A World Bank study. Washington, DC: World Bank Group; Available from: http://documents.worldbank.org/curated/en/485571468318338846/Reducing-the-vulnerability-of-Uzbekistans-agricultural-systems-to-climate-change-impact-assessment-and-adaptation-options

[32] https://cds.climate.copernicus.eu/#!/home

[33] Lisa AS, Jarad N, Matthew JH, Chris W. In: David T, editor. St. Paul, MN: University of Minnesot; 2017;**114**(42):11247-11252. DOI: 10.1073/pnas.1620229114

[34] Mikko P, Peter D, Walter I, Natalia K, Arthur L, Ari V. ADB Central and West Asia working paper series: Climate Change and Sustainable Water Management in Central Asia – No 5, May 2014

Chapter 8

Fashion Footprint: How Clothes Are Destroying Our Planet and the Growing Impacts of Fast Fashion

Stelios Andreadakis and Prince Owusu-Wiredu

Abstract

From agriculture and petrochemical production to manufacturing, logistics, and retail, the textile and fashion industry is the second most polluting industry in the world, responsible for between 8 to 10% of total carbon emissions and 20% of global wastewater, with a predicted increase of 50% in greenhouse gas emissions by 2030. To gain a better understanding of the state of the academic literature on the environmental impact of fast fashion, we systematically identified and analysed 30 publications published between January 2000 and December 2022. In the end, we discovered that there is a growing research interest in fast fashion, especially in relation to its devastating environmental impacts, which range from the cultivation of raw fibres to the recycling of fashion waste. Subsequently, we provide a summary of the key findings, including the carbon and water footprints, as well as some sustainable practices believed to reduce the industry's negative environmental impacts.

Keywords: fast fashion, climate change, greenhouse gas emissions, global warming, environmental impact, sustainable fashion, fashion footprint, textile industry

1. Introduction

In recent years, discussions of global warming and the resulting issue of climate change have become increasingly popular in political and scientific discourses. It is not only the most talked-about topic of the day, but also the focal point of political and social campaigns in countries across the globe. Today, one of the most obvious ways in which the clothes we wear contribute to global warming is by littering developing countries, which lack the infrastructure to properly manage such waste, with the massive amount of fast fashion waste consumed in countries of the Global North, referred to in the following text as the West [1]. Individually, determining the most efficient approach to addressing climate issues can be overwhelming. That is, at the individual level, climate action involves changing habits and routines by making choices that have less harmful effects on the environment [1]. In addition, individuals may demonstrate an interest in how they can influence system-wide changes within organisations. Notwithstanding its intricacies, the clothing and fashion sector represents a highly consequential industry that affords us all the opportunities to exert a

 IntechOpen

positive influence on the environment [2]. According to a recent report, the industry is responsible for between 8 and 10 percent of global emissions, a figure that exceeds the combined emissions of aviation and shipping [1]. In addition to the predicted increase of 50% in the industry's carbon footprint by 2030, the industry emits 1.2 billion tons of CO_2 annually [3].

Consequently, this chapter provides a welcome opportunity to discuss the environmental cost of fast fashion, establishing how the clothes we wear significantly contribute to global warming and the issue of climate change as a whole, from the cultivation of raw fibres to the recycling of textile waste, including the vast quantities dumped in developing countries, particularly those in Africa, where these clothes end up in landfills and are burned on open frames, emitting twice as much greenhouse gas as other well-known sources [4, 5]. Although rising temperatures are associated with the vast majority, if not all, extreme weather events, such as droughts, floods, and heatwaves, the effect our clothing has on rising temperatures and thus global warming has not been adequately brought to the public's attention; therefore, the industry continues to flood the market with what has been labelled cheaply-produced-disposable clothing [3, 6].

2. What is fast fashion and why is it a problem?

Nine years ago, the issue of where our clothes come from went from being a matter of curiosity to something more urgent. It was on April 24, 2013, when the Rana Plaza disaster occurred on the outskirts of Dhaka in Bangladesh; more than 1130 individuals perished when the factory in which they worked collapsed [4]. Today, worse than this is the potential for the industry to contribute to a serious global catastrophe that could wipe out the entire human race if 'precautionary measures' are not taken to regulate its operations [3, 7]. According to a recent study, not only does the global apparel industry contribute more to climate change than international aviation and shipping combined, but the purchase of a single white cotton shirt produces the same amount of emissions as driving 35 miles in a car [3]. Given the veracity of these findings, we have no choice but to reevaluate the environmental impact of the industry in terms of the responsibility of corporations to protect the environment.

To begin, the term "fast fashion" refers to affordable yet fashionable garments that are cheaply produced and quickly abandoned in favour of new styles [8, 9]. Internationally, while China is the world's top textile exporter and leads the fashion industry, the U.S and the EU consume more than half of the industry's output, and Africa, particularly East and West African nations, is drowned under the weight of second-hand goods dumped at their shores [6, 9]. Despite the fact that clothes are an integral part of our daily lives, many people fail to consider their origins and how they ended up in their closets. In other words, from the cultivation of fibres that will become textiles to the dying processes and final consumption, the clothes we wear impact our environment at every step [3, 8]. Today, these effects are exacerbated by the rapid production of inexpensive clothing that mimics the ever-changing trends of high-fashion brands. According to one report, the industry is the second-most polluting, just behind the oil/fossil fuel industry [5].

While waste production can contribute to a variety of environmental issues, including greenhouse gas emissions, the production of waste from fast fashion is no exception. In other words, while the clothing and textile industry is notorious for contributing significantly to contemporary environmental problems such as climate change and/or global warming, the advent of fast fashion has exacerbated the problem [7]. In the age of cameras living in our pockets, it seems as though every moment demands us to be "picture-perfect" [8, 9]. This demand for perfection has greatly benefited the fashion industry, especially the niche industry of "fast fashion" [9]. From T-shirts and shoes to accessories, fast fashion brands have emerged as a more affordable alternative to high-end designers such as Louis Vuitton, Gucci, Chanel, Dior, Balenciaga, Armani, and Yves Saint Laurent.

While fast fashion currently dominates the industry in terms of speed, price, and marketing, these companies are not known for their innovative or ground-breaking designs, nor for their ethical business practices [9]. On the other hand, High-end designers such Louis Vuitton, Gucci, Chanel, and others are known for setting trends and fashion standards while remaining within the limits of their operations, especially in terms of the environment. Thus, given the fashion industry's limited avenues for innovative designs, as well as the possibility that different companies will mirror similarities in products, there should be a protective line that guards the industry to prevent unnecessary competition at the expense of the environment [6, 9]. The ability of fast fashion brands to produce clothes at a fraction of the cost of their more expensive counterparts has contributed not only to an increase in clothing consumption, but also to an increase in the environmental impacts of clothing.

As consumers actively seek trendy clothing at affordable prices, fast fashion brands such as Zara, H&M, UNIQLO, GAP, Forever 21, Topshop, Esprit, Primark, Fashion Nova, and New Look have imitated the distinctive looks of high-end designers and sold their products at a fraction of the price [8]. Consequently, the current fashion market is characterised by ever-changing trends. In other words, what is fashionable today can change in an instant based on a variety of factors, such as social media effects, celebrity couture looks, and popular culture, among others [8, 9]. The Spanish company and leader in fast fashion, Zara, rose to prominence in the 1990s by delivering twice-weekly shipments of new clothing [9, 10]. The company established the standard for fast fashion when it began producing clothing under its own brand, which imitated the newest fashion trends while being mass-produced. A product was designed and mass-produced rapidly, and it was only available for a limited time. The production and retail methods were based on limited sale periods, in which a product was designed and mass-produced quickly and only sold for a limited time [10].

Today, 30 years after the invention of Zara's fast fashion marketing strategy, companies such as Forever 21, H&M, Primark, and many others have adopted a similar strategy, now with daily rather than weekly introductions of new clothing [8, 11]. In contrast, higher-end brands such as Louis Vuitton, Gucci, Yves Saint Laurent, and others, release new designs four times a year; consequently, the term "52 season companies" was coined to describe brands that dominate the fast fashion industry. That is, despite its economic benefits, fast fashion is in many ways detrimental to the human environment, as not only does the industry pollute the environment through its manufacturing processes, but the market also generates so much waste that the law must intervene against such a barbaric business model. While some of the largest names in the industry have a physical retail presence, fast fashion is widely known to dominate the e-commerce realm, i.e., online shopping [10, 11]. The perception that wearing an item only a few times is acceptable has led to an increase in discarded items, which has fueled overconsumption and overproduction. Consequently, to reduce the environmental impact of the industry, this misconception must be eradicated. Within the first 15 years of the twenty first century, clothing production doubled, and since the year 2000, most brands have released as many as 24 collections per year, for both the summer and winter seasons [9]. In the past, there were two clothing-buying seasons (Spring/Summer and Autumn/Winter), but now there are 52 microseasons, one for each week. That is, to remain profitable, fast fashion brands produce vast quantities of clothing, which not only pollute the environment, but requires a substantial amount of resources [11].

Big fashion brands grow in popularity for one simple reason: people want the latest styles, which are frequently pioneered by celebrity actors, musicians, or models [9]. The main issue is that these big fashion brands tend to charge a lot for their clothes, with huge price differences between, say, a jacket from Primark and one from Gucci. Actually, the term 'fast fashion' was coined when some companies realised they could imitate the latest fashion trends from major fashion brands and sell them at a fraction of the price [9, 10]. Why is this an issue? The issue is that large fashion brands rely on a rapid turnover for their expensive items, with a relatively smaller group of wealthy individuals regularly purchasing expensive items. Because fast fashion is constantly seeking to imitate major fashion brands, a similar situation of relying on a quick turnover occurs, although the profit per item is significantly lower [8, 11]. Consequently, fast fashion requires a significantly larger audience to generate revenue. If a much larger audience constantly purchases copies of whatever the most recent fashion trends are, a large quantity of clothing, footwear, etc. must be produced [11]. This partially explains why fast fashion poses a significant environmental threat, as in addition to the pollution caused by the production of such large quantities, the improper management of fashion waste, such as dumping in developing countries, has a detrimental impact on our planet.

To gain a better understanding of the academic literature on the environmental impact of the industry, we systematically identified and analysed 30 publications published between January 2000 and December 2022, from an analysis of 115 papers in order to compile a comprehensive collection of articles pertinent to the topic of study. Drawing on text analysis, the systematic review process involved four essential steps. Originally, the literature to be reviewed was identified using a topic-specific Google Scholar search. The objective was to conduct a search that was large enough to be exhaustive but also completely focused on the topics at hand. The following is the search query that was used to retrieve the primary dataset:

Clothing OR Apparel OR Textile OR Fashion OR Fabric OR Cloth* AND (Impact OR Sustainab* OR Environment* OR Effect OR Climate OR Pollution OR Emission* OR Greenhouse, CO_2 OR Landfill*).

The search was conducted on April 5, 2023, yielding a total of 225,000 articles for analysis. Next, the titles and abstracts were screened for relevance to the scope of the review using inclusion and exclusion criteria. Articles were excluded if they were unrelated to fast fashion, centred solely on solid waste treatment methods, or made no connection between the textile industry and greenhouse gas emissions or environmental impacts. Out of the 115 papers initially deemed relevant, a total of 30 were selected for analysis: 10 from Heinonline, 15 from Web of Science, and 5 from highly cited papers on ResearchGate, i.e., resulting in a dataset of 30 relevant articles. Finally, the content of the selected studies was analysed for major themes related to fast fashion, such as waste pollution, Greenhouse gas emissions, environmental impact, the effect of landfills, sustainability, climate change, global warming, waste trafficking, i.e., the dumping of second-hand clothes in developing countries, and recommendations to slow the negative environmental impacts of fast fashion, such as the transition to Circular Economy and the promotion of sustainable fashion.

3. Discussion

Through this systematic review we found that the negative effects of fast fashion on the environment are manifested through carbon emissions, water consumption, and energy footprints, particularly in the use phase of clothing, the saturation of international clothing markets, and an increase in the end-of-life textile waste, the majority of which are dumped in Global South countries, particularly those in Africa, emitting twice as much greenhouse gas as other well-known sources through the Greenhouse gas effect of landfills and the open incineration of damaged clothes [7, 8]. We discovered that, in comparison to other stressors on climate change, such as the oil/fossil fuel industry and transportation (air, water, and land), research interest in the effects of fast fashion on global environmental sustainability has increased in the last 5 years (74% of articles published) [8]. While the term "textile industry" referred to the production of both raw materials such as yarn and finished products such as

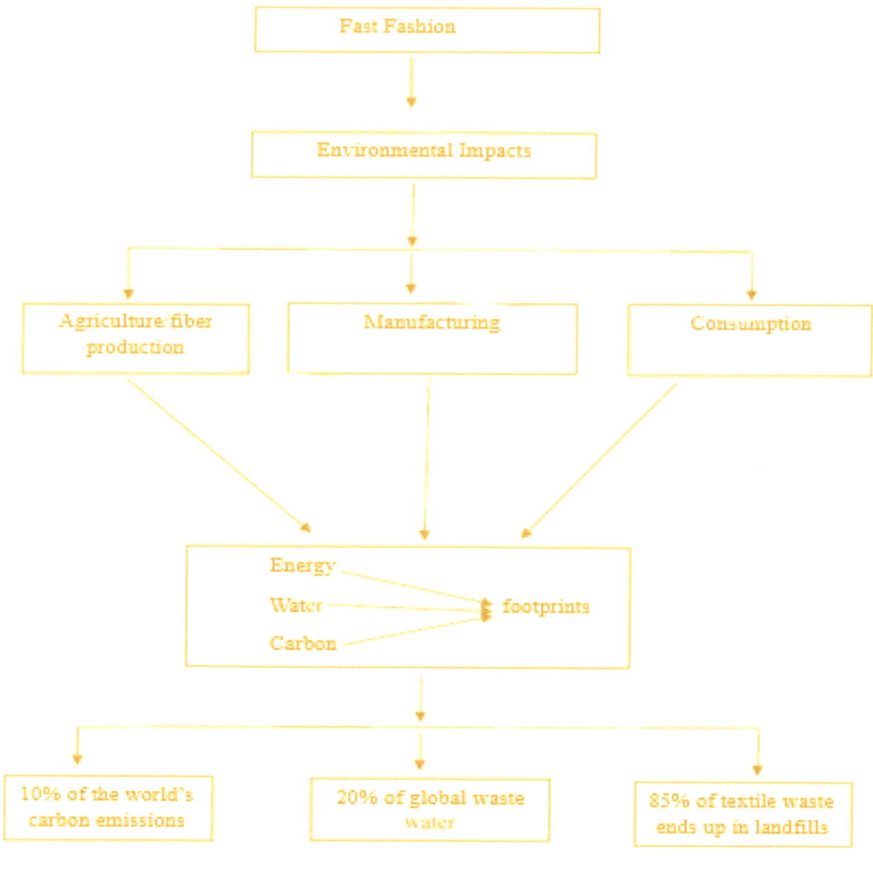

Figure 1.
The environmental impacts of fast fashion.

clothing, the terms "fast fashion industry" and "fashion industry" referred more specifically to issues involving finished garments and articles of clothing (**Figure 1**).

Today, when environmental issues are at the top of the news, major industry players, such as Zara, H&M, and Primark have played significant roles in increasing the industry's carbon footprint by accelerating fashion cycles [12]. Thus, this review revealed, among other things, that the textile industry not only accounts for 8 to 10% of total carbon emissions, but it also accounts for 20% of global wastewater production [8]. The shortening of fashion cycles creates a constant need for new concepts and designs, as well as the constant need to dispose of "old" items to make room for new ones. Among other things, "sustainability," "fast fashion," "textile industry," "greenhouse gas emissions," "carbon footprint," "climate change," "global warming," and "waste pollution," i.e., the dumping of used clothing in developing countries, particularly in Africa, were identified as major emerging topic areas in the literature. As they serve as the foundation for mitigating environmental impacts, concepts relating to sustainability were found to be prevalent in the texts. Elaborated below are some of the most predominant areas of study in this rapidly growing field.

3.1 Energy, water, and carbon footprints

With 1.7 million tons of CO_2 emitted annually, which is 10% of all greenhouse gas emissions, the textile industry is a major contributor to global warming [12, 13]. At the European level, clothing is identified as the fourth most ecologically sensitive consumption category, with food, housing, and transport ranking higher in terms of environmental impact [12]. Currently, there is a notable surge in the demand for clothing, with projections indicating a rise from 62 million tons in 2015 to 102 million tons in 2030 [12, 13]. There are many sources of pollution in textile manufacturing; for instance, wastewater can be produced in agricultural cultivation, textile pre-treatment, cleaning of machines after use, as well as wet or laundering processes [14, 15]. The fast fashion industry relies on mass production, which requires a lot of resources because of the sheer volume of clothing that must be produced. To produce just one cotton T-shirt, for instance, it takes between 10,000 and 20,000 litres of water, or about 3000 litres of water per kilo of raw cotton [3]. Consider this in the context of a global industry that sells around 2 billion cotton t-shirts annually. You can begin to comprehend the staggering quantities of water required to sustain this industry.

To say the least, calculating the carbon footprint of fast fashion is daunting. However, we will begin with the production of a single pair of jeans. From the harvesting of the cotton to the production of the jeans to their final delivery to a store, a substantial amount of carbon emissions must be accounted for. The production of a pair of jeans requires approximately 3700 litres of water [3, 5]. This results in approximately 33 kilogrammes of carbon dioxide equivalent (CO_2e) emissions [3]. And this is only for one pair of jeans; imagine how much water is required for your entire wardrobe. When considered on a global scale, the implications are mind-boggling. Each year, the fashion industry is responsible for: 93 billion cubic metres of water, which is enough to meet the needs of 5 million people; 10% of all carbon emissions, which is more than all international flights and shipping combined; 20% of the world's wastewater from fabric dyeing and treatment [3, 12]. Based on these reviews, there are no indications of a slowdown in the growth rate of fast fashion, meaning that these statistics will almost certainly worsen. Fast fashion emissions are expected to rise by 50% by 2030 [8, 12].

With constant new trends and ostensibly affordable prices, the temptation to purchase the newest products can be overwhelming. Albeit how does this affect our carbon footprint? Considering the entire lifecycle of a garment, from production to transportation to disposal in landfills, the fashion industry releases 1.2 billion tons of carbon emissions annually [12, 15]. Between 2003 and 2018, clothing utilisation decreased by 36%, with a third of young women believing that a garment is considered 'old' after being worn once or twice [14, 15]. This shift in values has been exacerbated by the use of inexpensive materials in clothing, which reduces their durability and makes them more difficult to repair [15]. The United Kingdom, for instance, spends an annual average of £59.3 million on clothing and exports £8.2 billion worth of clothing [14, 15]. Today, due to the combination of low prices and poor quality, many individuals believe that clothes are disposable. In spite of the fact that 85% of the garments sent to landfills could be recycled, the United Kingdom sends 300,000 tons of clothing to landfills each year due to poor quality, making it the fastest-growing waste economy [14, 16].

Not only is the fashion industry's carbon footprint affected by the amount of waste sent to landfills, but CO_2 emissions during the manufacturing and transportation processes, as well as water pollution from dyeing processes and microfiber pollution, also contribute to the industry's enormous carbon and environmental footprint, highlighting the industry's urgent need for change. For instance, some garments may travel the globe multiple times during the manufacturing process, contributing to an increase in air travel-related emissions [3]. Often, clothing is manufactured in developing nations where pollution regulations are less stringent. Thus, changes such as switching to renewable energy in factories and reducing the use of polyester will have some positive effect on reducing the carbon footprint of the fashion industry. The majority of fashion's environmental impact stems from the use of raw materials: cotton for the fashion industry, for example, uses approximately 2.5% of the world's farmland; synthetic materials like polyester require an estimated 342 million barrels of oil annually; and production processes such as dying require 43 million tons of chemicals annually [8, 17]. In summary, the production of clothing consumes a significant quantity of natural resources and generates greenhouse gas emissions that contribute to global warming. According to one study, the fashion industry could account for 26% of carbon emissions by 2050 if nothing changes [14]. Even so, to address the environmental impact of fast fashion, both individual and collective measures aimed at modifying the behaviour of these brands are required.

According to one study, the high carbon footprint of the fast fashion industry may be due to the industry's high energy consumption, which is often influenced by the energy source that is used [3]. In China, for instance, textile production is dependent on coal-based energy, resulting in a 40% greater carbon footprint than textiles produced in Turkey or Europe [3, 9]. High energy requirements and CO_2 emissions are not only associated with textile production processes, but also with textile consumption (mainly laundry) and distribution, especially when air freight is utilised. Nowadays, in addition to fibre type, the method of textile production may also impact energy consumption and environmental effects [3, 8]. For instance, conventional cotton cultivation can generate 3.5 times more CO_2 than organic cotton production, which creates twice as much CO_2 in India as organic cotton cultivation in the United States [3, 15]. As natural fibres have a smaller carbon footprint than synthetic fibres, it is thought that one effective way to reduce CO_2 emissions related to fibre production would be to replace polyester with natural fibres [3]. Albeit it is believed that the lower carbon footprint of natural fibres during production is later offset by the higher energy requirements of washing, drying, and ironing compared to synthetics during the usage phase [3, 17]. Thus, to lessen the fashion industry's effect on greenhouse gases, production volumes and non-renewable energy use must be lowered; polyester manufacture should be replaced with renewable plant-based textiles; and sustainable shipping and garment usage must be considered [3, 8].

3.2 The impact of low-quality second-hand clothes dumped in developing countries

As a modern business strategy, fast fashion encourages people to view clothing as disposable [18, 19]. According to Bick, the average American, for example, discards over 80 pounds of clothes annually, which accounts for approximately 5% of landfill space [19]. Items that are not immediately transported to landfills often find their way into the second-hand clothing market [6]. However, owing to their substandard quality, they ultimately end up in landfills, exacerbating the environmental crisis and contributing to the greenhouse gas emissions associated with these sites [5, 6]. These days, textile waste presents a considerable environmental challenge on a global scale [6, 16]. The current state of affairs, characterised by the extensive production and consumption of fast fashion products, necessitates a worldwide approach to addressing this challenge. This is particularly crucial given that waste management remains one of the most pressing environmental concerns worldwide [2, 5]. Throughout history, countries located in the Global North have predominantly addressed the issue of fashion waste by exporting surplus clothing to developing nations, with a particular focus on those situated in Africa [1, 6].

Contemporary research indicates, however, that these practices have long since ceased, as the improper disposal of fashion waste in these regions poses a significant ecological risk not only to these societies, but to the global community as a whole. This mismanagement includes the burning of unwanted textiles and the overflowing of landfills, resulting in the emission of greenhouse gases and the contamination of water, particularly from the washing of synthetic materials into our ocean [18, 20]. Consequently, it is imperative that we expand upon current industry efforts and fundamentally rethink the way clothes are manufactured, right down to the fibres that are used [20]. Garments should be designed with durability in mind, ensuring that they do not fall apart at the seams and can be repurposed through recycling even after prolonged use [20, 21]. While the marine environment is the ultimate destination for various anthropogenic pollutants, the problem is exacerbated by microfibers released from synthetic clothing and other textile materials [22]. According to Sunanda, approximately 150 million microfibers are introduced into the Atlantic Ocean on a daily basis due to poor management of textile waste, which includes fibres that are washed away from landfills [22].

According to one study titled "Come on EU! The massive dumping of discarded clothing in Ghana and Chile must stop", Western countries are responsible for exporting significant amounts of fashion waste to low-income countries, particularly those in West Africa, and the reason is simple: Western communities overconsume cheap, low-quality clothing, and they dislike the waste [23]. This practice makes the textile industry and, by extension, fast fashion substantially more polluting than they already are. Made by low-paid workers in China or Bangladesh, consumed in the Global North, rarely worn, and quickly discarded, fashion waste has a number of negative effects on the global environment [9, 24]. Nowadays, clothes are purchased in large quantities online, tried on, returned, and then not resold. The synthetic materials used to make these garments, most commonly polyester, make them difficult to recycle [3]. Globally, an estimated 92 million tons of textile waste are produced annually, with the equivalent of a dump truck's load of clothing ending up in landfills every second because so little gets recycled [23].

While people are purchasing more clothing than ever before, and the average consumer today buys 60% more clothing than 15 years ago, and only 12% of these items are recycled on a global scale [23, 24]. In the context of clothing consumption, the United Kingdom stands out as the European country with the highest rate of purchases per minute. While the global annual purchase of clothing is estimated to be around 56 million tons, it is projected to rise to 93 million tons by 2030 and 160 million tons by 2050 [24]. Where does all of this waste end up if only 12% of it is recycled? According to Spijkers, while it has become common practice for Western communities to dump their unwanted clothing in developing nations in an effort to reduce pollution, such practices are merely a relocation of the problem and never a solution, as environmental issues affect all nations equally regardless of their geographic origins [25]. According to two recent studies, for instance, the Kantamanto market in Accra, Ghana, is currently the largest second-hand clothing market in West Africa [1, 23]. Each week, 20 million items of second-hand clothing arrive here; yet 40% are rejected from the market due to their poor quality [6, 26]. Unsold clothing is frequently abandoned in sewers and landfills before being washed into the ocean, where it is washed up on beaches by the waves and buried in the sand, in addition to the large quantities that are taken away to destroy marine life and food sources [23].

Using Ghana as a case study, it is estimated that 10,000 articles of second-hand clothing arrive in Accra, Ghana's capital, every 5 minutes from the United Kingdom alone [1, 6]. However, due to defects and poor quality, a substantial amount is typically consigned to landfills. While this chapter addresses the grave environmental impact of these clothes, traders in the second-hand clothing business in Accra lament the recent decline in their business as more and more of their stock is dumped as waste due to low quality [1, 6]. In 2013, the government commissioned a massive landfill in Kpone with a daily capacity of 700 tons in an effort to address this issue. The Accra Metropolitan Assembly (AMA) was tasked with collecting up to 70 tons of waste (unsold second-hand clothing) every day from the Kantamanto market alone [6]. The procedure was initiated in 2016, and 4 years later, Kpone was overflowing with textile waste. Due to a lack of funds, the Accra Metropolitan Assembly was unable to continue with this project [6]. Today, in addition to the massive volumes dumped in landfills and burned in open flames, a significant portion of these garments are routinely dumped in sewers and carried into rivers, ultimately ending up in the ocean, posing a grave threat to the marine ecosystem and significantly contributing to rising sea levels [1, 23].

While the focus has always been on the health risks associated with these practices, this chapter focuses on the ecological repercussions of these practices, specifically how they contribute to the issue of climate change and global warming through the Greenhouse gas effects of these sites and the vast amount of CO_2 emitted through open burning. Today, despite the lack of significant progress, countries such as Ghana are seeking international cooperation to prohibit the export of second-hand clothes given its negative environmental effects. Today, the dumping of low-quality second-hand clothing is a major problem not just in West Africa, but also in East African countries such as Tanzania and Kenya [27]. When consumers donate their used clothes to a charity, a take-back box at a brand's store, or a recycling station, they often hope that the clothes will be resold or recycled into new garments. Nonetheless, due to damage and poor quality, only a tiny proportion (between 10 and 30 percent) are typically resold in the locations where they are first collected. Consequently, some are downcycled into lower grade materials such as rags, and more than half are exported for resale [27].

To learn more about what happens to these second-hand clothes, in their study, Greenpeace, an independent international campaigning network dedicated to safeguarding the environment, travelled to Kenya and Tanzania, two of the world's top five net importers, to speak with traders in the second-hand clothing business to find out more about the fate of these clothes after they are exported [27]. According to Greenpeace, the Global North has found a way around its textile waste problem by exporting used clothing to countries in the Global South, forcing them to deal with the consequences of fast fashion despite lacking the infrastructure to do so [27]. As they descended from the Gikomba market to the Nairobi River, the researchers were shocked to discover that they were literally walking on textile waste [27]. The waste was accumulating along the riverbanks, falling into the water, and flowing downstream to the ocean. According to their observations, the Gikomba market in Nairobi is covered in layers of textile waste. In attempt to address this issue, the locals burn these textiles on open fires, primarily in the evening, causing severe air pollution that affects not only the residents of these communities, but also the entire global community via the massive amounts of CO_2 emissions.

Similar to the situation in Ghana, where 30 to 40% of these second-hand textiles have no market value, 150 to 200 tons of waste (60 to 75 truckloads) are abandoned, burned, or transported daily to Dandora and other overflowing dump sites in Nairobi [27]. According to Wohlgemuth, the fast fashion industry has reduced clothing to the status of disposable cups [27]. Thus, slowing down fast fashion is the only option to reduce the flood of textile waste being dumped on the Global South. Global fashion brands must drastically alter their linear business models and begin creating fewer clothes that are higher quality, longer lasting, repairable, and reusable [14, 27]. In addition, there must be a shift away from the neo-colonial attitude of Global North countries that use Global South countries as dumping grounds for their unwanted waste, while doing little or nothing to support or develop the clean manufacturing of local textiles, employing the same high standards and best practices required in Europe.

The aforementioned findings demonstrate unequivocally that it is insufficient for global fashion brands to concentrate just on cleaning up their supply chains. The clothing and fashion business must intensify its efforts to reduce the enormous end-of-life consequences of its products. That said, according to Wohlgemuth, the global fashion industry will be unable to rectify its negative environmental impact in the absence of laws [27]. An evaluation of 29 leading brands, including H&M, Nike, Adidas, G-Star, and Primark, reveals that voluntary commitments are insufficient to slow down the rising volumes of textiles and alter the destructive trajectory of fast fashion, with the majority of impacts being felt in countries of the Global South, where clothes are made and dumped [9]. Although it is expected that trade policies and regulations will be the most effective solutions for bringing about widespread change in the clothing and fashion industry, consumers, particularly those in high income countries, have a responsibility to support companies and practices that reduce the negative environmental impacts of fast fashion.

3.3 Moving towards a more sustainable fashion industry: alternate business models

Our systematic review identified a number of tools that could be used to move towards a more sustainable fashion industry, such as a business model in which society moves from a linear economy, based on the concept of take–make–dispose, to a circular economy that aims to retain all resources or products in the system for as long as possible to reduce end-of-life textile waste [8]. The circular model optimises the process of reducing waste by maximising resource reuse or producing new resources from waste materials [7, 13]. Among other things, the product service system has been suggested as a prospective business model aimed at reducing the environmental impact of the clothing and fashion industry [27]. In the sharing economy, we see product-service systems as forms of co-consumption. Within this framework, consumers are able to pay for the services rendered by a product without the need to own the product. This proposed system has the potential to reduce resource consumption, specifically water, by improving the quality and durability of clothing. Additionally, it advocates for the adoption of practices such as lending, renting, redesigning, and upgrading to minimise the number of individually owned items. Nevertheless, the question of whether or not this is feasible in terms of clothing warrants further investigation [28, 29].

Ethical fashion, Ecofashion, and sustainable fashion all aim to reduce the negative environmental impact of fast fashion models [28, 29]. Ecofashion is based on designing garments that are better for society and the environment, ethical fashion is based on fair trade and environmental standards, and sustainable fashion focuses on tailoring the clothing life cycle to align with the ideas of sustainable development, taking into account design, material, production conditions, and the consumer [8, 30]. Jacometti provides a summary of the current measures taken by the European Union to create more sustainable practices in the fashion industry by transitioning from a linear to a circular economy [31]. Throughout the literature, numerous studies have emphasised the significance of changes in consumer behaviour that can significantly reduce the environmental impact of the clothing and fashion industry [8, 32]. As the negative environmental impacts of the fashion industry extend beyond the production stage, the utilisation phase of garments is a significant contributor to environmental degradation [8, 33]. This can be seen in the transportation of clothing to retail outlets as well as the subsequent usage phase [16, 34]. As the weekly laundry of a single household could potentially discharge thousands of microfibers, using machines with high efficiency ratings, lower washing temperatures, air drying, using front loading, and full load machines can reduce the energy and carbon footprints of the global apparel industry [30, 35].

4. Conclusion

As discussions of global warming and the resulting issue of climate change have grown in popularity in recent years, this chapter aims to identify fast fashion as a growing threat to the current state of affairs, focusing on how fast fashion contributes to greenhouse gas emissions from the cultivation of raw fibres to the recycling of second-hand clothing, especially the vast quantities dumped in developing countries. Ultimately, the chapter reveals that the rise in fast fashion consumption is influenced by a variety of socio-cultural and economic issues. Consequently, it is crucial to consider not only the environmental impact associated with the production of clothing, but also the reliance on fast fashion, a business model that produces affordable yet fashionable garments that are cheaply produced and quickly abandoned in favour of new trends.

According to the available sources, clothing accounts for between 8 to 10% of total carbon emissions and 20% of global wastewater, a figure that exceeds the combined emissions of aviation and shipping. In addition to the predicted 50 percent increase in the industry's carbon footprint by 2030, the industry emits 1.2 billion tons of CO_2 annually. Thus, the study highlights the emergence of a new field aimed at bridging the gap between fast fashion, the textile industry, and their environmental impact. That is, in addition to addressing the carbon and water footprints of the global apparel industry, this field seeks to investigate its contribution to climate change and global warming.

We discovered, among other things, that fast fashion's energy and water use, as well as its carbon footprint, contribute significantly to today's environmental issues, notably climate change; however, studies have frequently treated these topics in a broad manner, leaving room for additional study. Thus, we identified "fast fashion," "sustainability," "textile industry," "carbon footprint," and "water footprint" as the five major emerging themes that the literature could explore in greater depth. In addition, to assist the fashion industry in becoming more sustainable and in line with the principles of the circular economy, it is recommended that additional research be conducted on the evaluation of the textile industry's carbon and water footprints, recycling initiatives, and methodologies. That is, to influence the future of the global apparel industry, we recommend that consumers reevaluate their purchasing habits in favour of more sustainable options. This has the potential to incentivise manufacturers to implement environmentally friendly production methods and conform to more stringent regulations concerning the handling of post-consumer waste, particularly that of clothing or textiles.

Author details

Stelios Andreadakis* and Prince Owusu-Wiredu
Brunel University, London, UK

*Address all correspondence to: stelios.andreadakis@brunel.ac.uk

IntechOpen

© 2023 The Author(s). Licensee IntechOpen. This chapter is distributed under the terms of the Creative Commons Attribution License (http://creativecommons.org/licenses/by/4.0), which permits unrestricted use, distribution, and reproduction in any medium, provided the original work is properly cited. (cc) BY

References

[1] Manieson LA. Castoff from the west, pearls in Kantamanto?: A critique of second-hand clothes trade. Journal of Industrial Ecology. 2023;**27**(3):811-821. DOI: 10.1111/jiec.13238

[2] UNEP. ActNow for Zero-Waste Fashion. ActNow [Internet]. 15 August 2019. Available from: https://www.un.org/sustainabledevelopment/blog/2019/08/actnow-for-zero-waste-fashion/. [Accessed 06 April 2023]

[3] Niinimäki K. The environmental price of fast fashion. Nature Reviews Earth & Environment. 2020;**4**(1):189-190. DOI: 10.1038/s43017-020-0039-9

[4] Blanchard T. Fashion Revolution Week: seven ways to get involved. The guardian [Internet]. 24 April 2018. Available from: https://www.theguardian.com/fashion/2018/apr/24/fashion-revolution-week-seven-ways-to-get-involved. [Accessed 07 April 2023]

[5] Herbst H. The price of fashion: The environmental cost of the textile industry in China. Fordham International Law Journal. 2022;**45**(5):907-958

[6] Priya A. Impact of second-hand clothing waste in Ghana. International Journal of Law Management & Humanities. 2022;**5**:1679-1680. DOI: 10.10000/IJLMH.112981

[7] Rao P. Battling the damaging effects of fast fashion: A call for sustainability in the fashion world, as effects of climate change continue to bite. Africa Renewal [Internet]. 24 December 2019. Available from: https://www.un.org/africarenewal/magazine/december-2019-march-2020/battling-damaging-effects-%E2%80%98fast-fashion%E2%80%99. [Accessed 08 April 2023]

[8] Bailey K, Basu A, Sharma S. The environmental impacts of fast fashion on water quality: A systematic review. Water. 2022;**14**(7):1073-1081. DOI: 10.3390/w14071073

[9] Abdel-Jaber H. The devil wears Zara: Why the Lanham act must Be amended in the era of fast fashion. Ohio State Business Law Journal. 2021;**15**:234-240

[10] Annamma J, Sherry JF, Venkatesh A, Wang J, Chan R. Fast fashion, sustainability, and the ethical appeal of luxury brands. Fashion Theory. 2012;**16**(3):273-295. DOI: 10.2752/175174112X13340749707123

[11] Mukherjee S. How Zara became the undisputed king of fast fashion?. The strategy story [Internet]. 09 November 2020. Available from: https://thestrategystory.com/2020/11/09/zara-fast-fashion-case-study/. [Accessed 09 April 2023]

[12] Gill T. What is the carbon footprint of fast fashion?. The eco experts [Internet]. 18 August 2022. Available from: https://www.theecoexperts.co.uk/blog/carbon-footprint-of-fast-fashion. [Accessed 11 April 2023]

[13] Marques AD, Marques A, Ferreira F. Homo Sustentabilis: Circular economy and new business models in fashion industry. SN Applied Sciences. 2020;**2**(306):1-5. DOI: 10.1007/s42452-020-2094-8

[14] Climate Seed. Textile industry: environmental impact and regulations. Climate Seed [Internet]. 28 September 2022. Available from: https://climateseed.com/blog/secteur-du-textile-impact-environnemental-et-r%C3%A9glementation. [Accessed 10 April 2023]

[15] Clarke R. Fast Fashion's Carbon Footprint. Carbonliteracy [Internet]. August 2021. Available from: https://carbonliteracy.com/fast-fashions-carbon-footprint/. [Accessed 12 April 2023]

[16] Choudhury RK. Environmental impacts of the textile industry and its assessment through life cycle assessment. In: Muthu S, editor. Roadmap to Sustainable Textiles and Clothing. Textile Science and Clothing Technology. London: Springer; 2014. pp. 1-5. DOI: 10.1007/978-981-287-110-7_1

[17] Pil-Ju P, Kiyotaka T. Quantifying producer and consumer-based eco-efficiencies for the identification of key eco-design issues. Journal of Cleaner Production. 2008;**16**(1):95-97

[18] Sohail Y, Danmei S. Propelling textile waste to ascend the ladder of sustainability: EOL study on probing environmental parity in technical textiles. Journal of Cleaner Production. 2019;**233**:1451-1456. DOI: 10.1016/j.jclepro.2019.06.009

[19] Bick R. The global environmental injustice of fast fashion. Environmental Health Fashion. 2018;**17**(92):1-2. DOI: 10.1186/s12940-018-0433-7

[20] Jimenez A. The Effects of Fast Fashion in West Africa. Borgen project [Internet]. Available from: https://borgenproject.org/fast-fashion-in-west-africa/. [Accessed 10 April 2023]

[21] Mathews B. New report calls on fashion industry to end greenwashing. Apparelinsider [Internet]. September 2018. Available from: https://apparelinsider.com/new-report-calls-on-fashion-industry-to-end-greenwashing/. [Accessed 11 April 2023]

[22] Sunanda M. Marine microfiber pollution: A review on present status and future challenges. Marine Pollution Bulletin. 2019;**140**:188-192

[23] PS Foundation. Come on EU! The massive dumping of discarded clothing in Ghana and Chile must stop. Plastic Soup Foundation [Internet]. Available from: https://www.plasticsoupfoundation.org/en/2022/03/the-massive-dumping-of-discarded-clothing-in-ghana-and-chile-must-stop/. [Accessed 12 April 2023]

[24] Nayak R, Panwar T, Nguyen VT. Sustainability in Fashion and Textiles: A Survey from Developing Country. Cambridge, USA: Woodhead Publishing; 2020. pp. 1-10. DOI: 10.1016/B978-0-08-102867-4.00001-3

[25] Spijkers O. Friends of the Earth Netherlands (Milieudefensie) v Royal Dutch Shell. Chinese Journal of Environmental Law. 2021;**5**(2):237-256. DOI: 10.1163/24686042-12340073

[26] Gupta L, Saini HK. Achieving sustainability through zero waste fashion-a review. Current World Environment. 2020;**15**(2):154-156. DOI: 10.12944/CWE.15.2.02

[27] Wohlgemuth V. How Fast Fashion is using the Global South as a dumping ground for textile waste. Greenpeace [Internet]. 2022. Available from: https://www.greenpeace.org/international/story/53333/how-fast-fashion-is-using-global-south-as-dumping-ground-for-textile-waste/?. [Accessed 14 April 2023]

[28] Armstrong CM, Niinimäki K, Kujala S, Karell E, Lang C. Sustainable product-service systems for clothing: Exploring consumer perceptions of consumption alternatives in Finland. Journal of Cleaner Production. 2015;**97**:30-36. DOI: 10.1016/j.jclepro.2014.01.046

[29] Samira I, Geiger SM, Schrader U. Collaborative fashion consumption–a cross-cultural study between Tehran and Berlin. Journal of Cleaner Production. 2019;**212**:313-323. DOI: 10.1016/j.jclepro.2018.11.163

[30] Pensupa N, Leu SY, Hu Y, Du C, Liu H, Jing H, et al. Recent trends in sustainable textile waste recycling methods: Current situation and future prospects. Chemistry and Chemical Technologies in Waste Valorization. 2017;**375**(76):189-228. DOI: 10.1007/s41061-017-0165-0

[31] Jacometti V. Circular economy and waste in the fashion industry. Laws. 2019;**8**(4):27-28. DOI: 10.3390/laws8040027

[32] Austgulen MH. Environmentally sustainable textile consumption—What characterizes the political textile consumers? Journal of Consumer Policy. 2016;**39**:441-466. DOI: 10.1007/s10603-015-9305-5

[33] Chen F, Zhu J, Yang Y, Wang L. Assessing environmental impact of textile production with water alkalization footprint. Science of The Total Environment. 2020;**719**:137522-137525. DOI: 10.1016/j.scitotenv.2020.137522

[34] Bhardwaj V, Fairhurst A. Fast fashion: Response to changes in the fashion industry. The International Review of Retail, Distribution and Consumer Research. 2010;**20**(1):165-173. DOI: 10.1080/09593960903498300

[35] Peters G, Li M, Lenzen M. The need to decelerate fast fashion in a hot climate-a global sustainability perspective on the garment industry. Journal of Cleaner Production. 2021;**295**:126390-126395. DOI: 10.1016/j.jclepro.2021.126390

Section 3

Climate Change: Eco-Regions and Other Influences

Chapter 9

Young Students' Knowledge, Attitude and Practice towards Climate Change

Nahed Salman, Mariam Al-Mannai, Asma Abahussain and Mahmood Saeed Mustafa Alalawi

Abstract

Climate change is undisputedly one of the most significant problems in the world. CO_2 emissions are getting higher by the hour. Bahrain's kingdom has obligations to reach zero emissions by 2060. The Bahrain climate championship was conducted in Manama to raise the awareness of the community as a whole. The target is to bring the youth's attention to the alarming matter. This paper examined students' knowledge, attitudes and practices related to climate change using pre-and post-questionnaires. The majority of the surveyed students have poor knowledge about climate change. However, the post-questionnaires naturally show reasonable improvements. This means that students have benefited from the climate championship conducted in Bahrain and the enrichment unit provided. Similarly, few significant improvements were noticed in students' attitudes and practices towards climate change. This research utilised the climate change champions conducted in Bahrain to disseminate the KAP questionnaire of this research.

Keywords: climate champions, knowledge, attitude, practice, university students

1. Introduction

Climate change is considered one of the most challenging problems facing people and the planet. The Intergovernmental Panel on Climate Change (IPCC) defines climate change as how the atmosphere changes over sometime, decades or longer, whether natural processes or human activity are the cause or not. The United Nations Framework Convention on Climate Change (UNFCCC) defines climate change as a shift in climate and evolution in the atmosphere's composition due to human activity. Climate change has long-term effects caused by industrialisation and rapid global development. This can impact man, animals and plants adversely. The latest Human Development Report (HDR) calls on world leaders to make bold steps to reduce the tremendous pressure on the environment and the natural world. Otherwise, humanity will pay a high price. The importance of implementing sustainable, eco-friendly solutions is not overstated. Major nations and international leaders, such as the United States, have proposed tight regulations and encouraged individuals to engage in environmentally friendly actions to reduce pollution. Governments and giant technical corporations should not be the only ones involved in this endeavour. It is a team effort; everyone plays his part to make a massive difference, to reach a solution.

The Arabian Gulf countries are concerned about climate change, especially the adverse effects on agriculture, food and water security, infrastructure resilience, public health and the environment. The World Bank designated the Kingdom of Bahrain and the other Arabian Gulf countries as the highest per-capita emitters of CO_2. In the case of Bahrain, the CO_2 emissions reached 21.6 metric tons per capita in 2020. This high rank is essentially due to Bahrain's massive energy uses compared to its population. The population of the Kingdom of Bahrain is 1.5 million [1].

Governments were urged at COP 26 to provide comprehensive and global climate change education for their citizens, particularly those in their youth. Climate change poses the greatest threat to youth in small island states and other vulnerable countries. Young people can provide innovative solutions to climate change and should participate in decision-making [2].

A youth council for climate change was created in the Arab countries in 2021 to amplify the youth's voice and empower youth engagement in assessing climate change issues and generating innovative solutions to the problem [3]. In addition, the 27th COP is planned for Egypt, and the 28th COP is slated for the United Arab Emirates [2, 4].

His Highness Sheikh Salman bin Hamad Al Khalifa, Bahrain's crown prince and prime minister, represented Bahrain at the 26th Conference of the Parties (COP26). The Kingdom of Bahrain is responsible for only 0.07% of global emissions. His Highness emphasised that Bahrain will set an example. Among the plans Bahrain has for 2060 is a net-zero goal. By 2035, Bahrain intends to reduce its CO_2 emissions by 30 per cent, double the amount of renewable energy it uses, and expand the coverage of mangroves and trees [5].

Additionally, Bahrain is making significant efforts to engage youth in climate change awareness activities. Fifty (50) students from different universities across the kingdom of Bahrain were invited to participate in a "Climate Champions" event. This is a part of the active citizenship program to support climate change solutions with creative ideas [6]. This research intends to investigate the level of knowledge, attitudes and practices regarding climate change among university-level students.

2. Literature review

RFD et al. [7] conducted a cross-sectional study that adopted a knowledge and attitude questionnaire using a convenience sample of 1059 nursing students. The students' representatives were from four Arab countries (Saudi Arabia, Palestinian, Egypt and Iraq). The results indicated a moderate knowledge about climate change and an average attitude towards the environment. Saudi and Palestinian students scored significantly better in knowledge than Iraqi and Egyptian students [7].

Kumar et al. [8] distributed a survey to 120 students in the Allahabad district in India to evaluate the level of global warming awareness. The results indicated that the vast majority (61.67%) showed an average awareness of global warming. At the same time, there was also a low-slung level (21.67%) and a high level (16.66%) of awareness of global warming. A significant relation was observed between the level of awareness and education. Ultimately, the study highlighted the need to impart knowledge and encourage a positive attitude towards ecological issues among university students [8].

Ibrahim et al. [9] applied a descriptive cross-sectional research design to assess Assiut University students' global warming knowledge and attitude. A convenient sample of 1300 students from different faculties was included. Their study utilised a modified environmental issues questionnaire with a Likert scale. The study found that most of the respondents had poor knowledge. Also, most of the students had a positive attitude towards global warming. This gap in knowledge of global warming must be addressed! Ibrahim et al. [9] highly recommend integrating various environmental themes into the curriculums for all students in the education cycle.

On the other hand, a study was conducted in Japan and Caribbean countries in 14 communities. The study applied mixed methods. That study aimed to assess that region's Knowledge, Attitude and Practice. A KAP survey was conducted in 10 districts; in addition to group discussion in four communities, 88% of the respondents were either highly concerned or somewhat concerned about climate change phenomena. Most (93%) were familiar with climate change, but not all could define its meaning correctly. The respondents lacked comprehensive knowledge of climate change, its causes and its impact [10].

Additionally, several KAP studies were conducted in GCC, in the United Arab Emirates. Two studies were conducted at universities. The first study aimed to investigate university students' KAP level towards Education for Sustainable Development (ESD) and the environment. The data were analysed using descriptive analysis and a t-test. The study revealed that UAE university students expressed high environmental knowledge and attitudes. Also, they showed good environmental behaviour [11]. The second study was conducted on 280 students from three universities in the Emirate of Ajman. The KAP survey was analysed using Wilcoxon Rank Sum, Kruskal -Wallis and Spearman Correlation tests. The results showed that the majority (82%) had sufficient knowledge; (78%) had good environmental attitudes, and only (8.2%) showed eco friendly practices. The study proved a positive correlation between knowledge and attitudes and a negative correlation between knowledge and practice [12].

Furthermore, Freije et al. [13] conducted a study to assess global warming awareness among science students at the University of Bahrain. One hundred and forty-three students were surveyed. The participants studied four majors: Biology, chemistry, mathematics and physics. The number of students was 51, 28, 40 and 24, respectively. Most of the study samples (133) were females aged 18–24. Their findings indicate that biology students studying in related academic curricula had the highest knowledge of global warming about its three claims, causes, effect and solutions. The study emphasised the need to expose all students to a wide range of environmental theories in their core curricula, regardless of their specialised area, to make them more aware of ecological issues [13].

The third communication report assigned an education group to conduct a particular study about climate change knowledge. The study involved 391 students in their primary stage, fourth and fifth grades. The study elaborated on pre-and post-questionnaires to measure the knowledge of climate change. The questionnaires were conducted pre and post-an enrichment programs. The study was conducted following the implementation of six enrichment programs, tutor training and teaching of students. A total of 16 teachers (eight males and eight females) were trained to be ready to apply for the six units during the second semester of 2017 and 2018. Student awareness improved significantly after completing the enrichment units [14]. Although a limited number of initiatives engaged the youth, the number of climate change awareness events is still limited to date. This research calls for an increase in the number of activities and events to enhance the awareness of youth and individuals' climate change topics.

3. Methodology

This research paper uses the KAP questionnaire to measure the awareness of climate change among the participants. The questionnaire contains three main parts: knowledge, attitude, and practice. The paper also refers to the "Personal Information" part at the beginning. The personal information part requests information about gender, age, nationality, university name and specialisation at university and high school. This is followed by twenty-four (24) questions regarding knowledge of climate change employing yes/no questions and selecting from a list and occasionally free-response questions. Open questions are obtainable in four scenarios. The scenarios are the number of Sustainable Development Goals (SDGs), the number of Bahrain National Determined Contributions (BHR-NDCs), the National Communication Report (NCR) and the explanation of carbon footprint.

Thirteen (13) questions about attitudes towards climate change are included in the third part. A Likert scale that ranges from 1 to 5 was utilised, with one (1) representing strongly disagree and five (5) representing strongly agree. Disagree, unsure and agreed were the remaining two (2) to four (4) scales, respectively.

Again, Likert scale (always- sometimes- never) is used in the fourth part, which includes fifteen (15) questions—the questions contract with "practice" towards climate change. Additionally, a closed-ended question is added towards the end of this part to investigate students' desire to learn topics related to climate change. The question provides several topics to pick from. The list included eleven (11) climate change topics. In addition, A KAP questionnaire is administered at the beginning and end of the climate champion event (pre and post).

Google Forms are used to encode the KAP questionnaire in Arabic and English versions. KAP questionnaire and statistical analysis treatments are also performed. Data is collected and analysed using an Excel spreadsheet and Statistical Package for the Social Sciences (SPSS) Version 26.0 (Armonk, NY: IBM Corp). SPSS offers frequencies and percentages for the items in each KAP aspect. T-tests, Chi-squares, cross-tabulations and the latest version of Microsoft Excel (MS) Sheet are used to gather and analyse the data.

4. Results and discussion

4.1 The study sample

38 out of 50 students from 10 universities participated in the pre-and post-KAP questionnaires. Around half (48%) of the participants of Climate Champions were from the University of Bahrain (UOB), 13% from Polytechnic Bahrain (PB) and the rest from nine other universities. The percentage of female participants was higher than that of males; 66% of the sample were females, and 76% of participants were Bahrainis. All participants graduated from secondary school; most of them studied was unified track–science (84%) (**Figure 1**). All participants were between 19 and 38 years old, and 86% were less than 26 years old. Therefore, the sample of the study is from the youth segment.

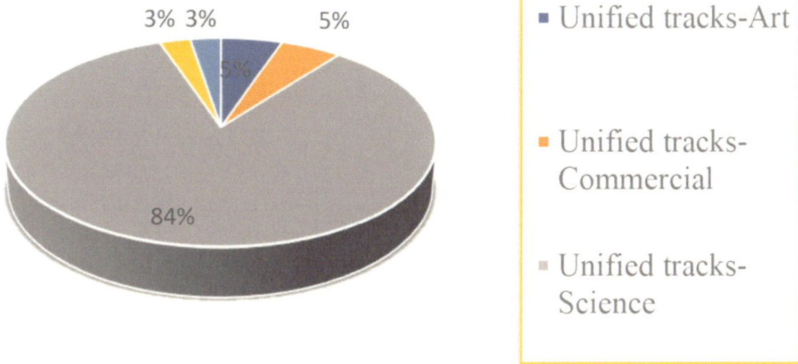

Figure 1.
Characteristics climate champions participants based on their major in secondary schools.

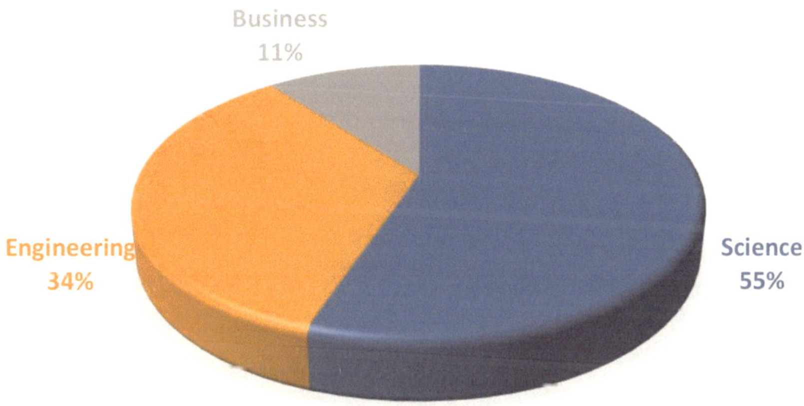

Figure 2.
Characteristics of the climate champions participants based on their major in university.

Also, the sample of this study targeted multiple specialisation disciplines in different academic years. 55% were students in the science field (pharmacy, physics, biology and information technology), and 34% were engineering students (**Figure 2**). Third-year university students were the highest percentage (53%). In contrast, the fourth-year- and second-year-level university students were 26% and 21%, respectively. 100% of participants thought they had a good or excellent knowledge level of climate change (**Figure 3**).

4.2 The level of *Knowledge* regarding climate change

The results of the pre-event KAP questionnaire showed that 84% of the respondents used the Internet as the primary source for obtaining information related to climate change. Social media comes second with 81.5%. School or university ranked

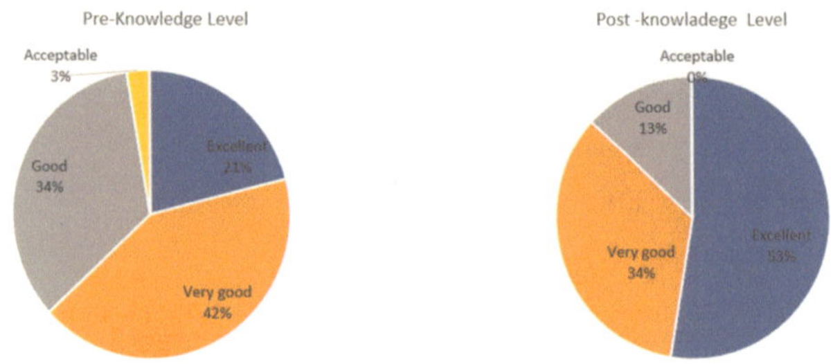

Figure 3.
Climate champions' responses regarding their climate knowledge level from their point of view.

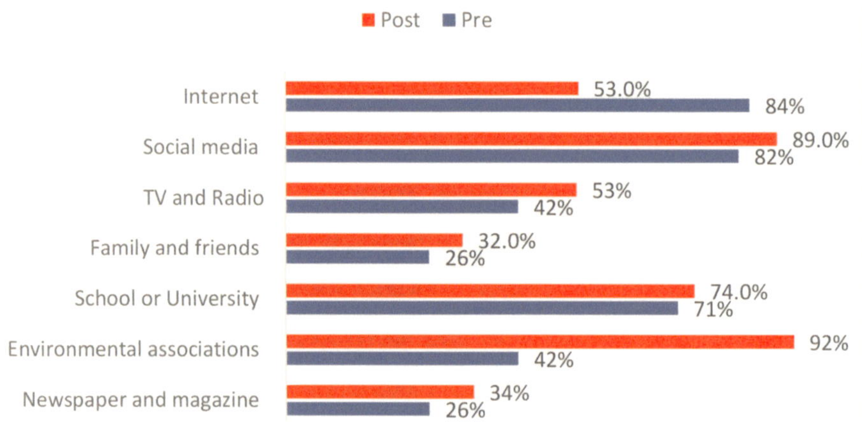

Figure 4.
Climate champions' responses regarding their source of information about climate change.

third with 71%, and environmental associations were rated with 42%. Family, friends, magazines and newspapers were viewed as the lowest percentage as a source of climate information (26%) in the pre-KAP questionnaire (**Figure 4**). On the other hand, the post-event KAP questionnaire showed that 92% of the respondents chose environmental associations as the main source for obtaining information related to climate.

In comparison, 89% of participants chose social media, making it the second choice. School or university was selected by 74% of the participants, and 53% of them selected the Internet as a source of information. Newspaper magazines were chosen by 34%, and family and friends were chosen by 32%.

Figure 5 illustrates the improvements in students' knowledge regarding climate change due to the enrichment units. All students recognised the Sustainable Development Goals (SDGs) in the post-questionnaire, compared to 66% in the pre-questionnaire and 63% new carbon footprint in both questionnaires. Only 5% of students recognised the three National Communication Reports (NCR) in the pre-questionnaire. In contrast, 21% of students knew the "NCR" by answering "Yes" in the post-questionnaire. Only 8% of students viewed the Nationally Determined

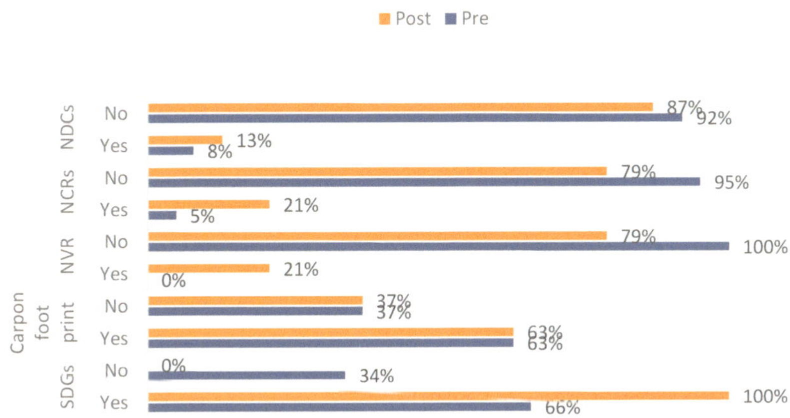

Figure 5.
Climate champions' knowledge about most reported climate change responses of Bahrain.

Contributions "NDCs" reports summited by Bahrain in the pre-event questionnaire and only 13% in the post. None knew VNR in the pre-questionnaire, and only 21% recognised it in the post-questionnaire.

Most respondents considered the climate change phenomenon as dangerous and that must be addressed (92% in the pre-event questionnaire and 97% in the post-event questionnaire). In contrast, few students thought that climate change deserves attention. Most respondents believed climate has changed in the past few years (89.55% in pre- and 75% in post-questionnaire). In comparison, some thought it was still going through a slow change phase (10% in the pre-and 23% in the post-questionnaire). Only one respondent believed that the climate did not change in the post-questionnaire. In addition, 92% of respondents knew that human factors are the reason for climate change. They also knew that the leading cause is the increase in GHGs; 73% in the pre-questionnaire increased to 81.5% in the post-questionnaire.

There was an enhancement in the respondents' knowledge; 95% figured that the recent concentration of carbon dioxide in the atmosphere was not 200 parts per million in the post-questionnaire compared to 16% in the pre-questionnaire. 53% of respondents knew that chlorofluorocarbons (CFCs) also belong to the GHGs. More than 30% knew that carbon dioxide (CO_2) and methane are GHGs. 29% of respondents only knew that water vapour (H_2O) is considered a GHGs, and few respondents did not realise that sulphur (S) and lead (Pb) are not GHGs even in the post-questionnaire.

In the post-questionnaire, sixty-six percent (66%) answered that the Paris Agreement calls for reducing emissions of GHGs for all countries of the world, whereas 16% believed the "Paris Agreement call" was only for developed countries.

In the post-questionnaire, the mean of climate change knowledge increased to 60% (acceptable), compared to the pre-test, where it was relatively low (46%). The training and lecture positively impacted the level of knowledge of climate change, even with limited training timing in the enrichment unit. These findings are consistent with what was mentioned in the study of educational groups in Bahrain's third national communication report.

A period sample t-test was performed to confirm the pre- and post-questionnaire variance in the *Knowledge* level regarding climate change. Results show that the

	Period sample t-test	T-value	Correlation	P-value
1	Pre-knowledge & post-knowledge	−7.603	0.456	0.000
2	Pre-attitude & post-attitude	−2.273	0.605	0.029
3	Pre-practice & post-practice	−3.016	0.363	0.005

Table 1.
Results of period sample t-test for pre- and post-KAP questionnaire.

p-value is $0.004 < 0.05$. A significant difference was observed between the levels of *Knowledge* before and after conducting the enrichment unit; correlation coefficient $r = 0.456$ (**Table 1**).

4.3 The level of *Attitude* regarding climate change

In the pre-and post-questionnaire, a period sample t-test was performed to measure the differences among *Attitudes* towards climate change. The mean increased from 4 to 4.2 (agree to strongly agree). The p-value for the period sample t-test for the pre-and post-questionnaire *Attitude* level was $0.00 < 0.05$; therefore, a significant difference was found between *Attitudes* in the pre and the post (**Table 2**).

The results show a positive *Attitude* towards climate change; for example, the majority (more than 90%) strongly agrees with replacing fossil fuels with renewable energy. In the future, many respondents would like to work for a company with an excellent environmental record and buy an electric car to reduce GHGs emissions. Half of the respondents support reducing water and electricity subsidies.

4.4 The level of *Practice* towards climate change

Students were asked to rate their *Practice* towards climate change. The overall mean of ratings of friendly environmental *Practices* changed from "sometimes" to "always". Period sample t-test has performed the p-value for the *Practice* level in the pre- and post-questionnaire. Accordingly, there was a significant difference between *Practice* levels in the pre-questionnaire (Mean = 2.25, SD = 0.612) and post-questionnaire (Mean = 2.48, SD = 0.580), conditions: $T(37) = -3.016$, P-value = 0.011.

In the pre-questionnaire, 26% of respondents (10) always use eco-friendly products, which increased to 42% (16) in the post-questionnaire; students may have better understood what eco-friendly products are after the enrichment unit and climate champions activity, which could have led some students who initially said they sometimes use eco-friendly products to change their answer to always. Or else, students may have started using more eco-friendly products after the enrichment unit. This could have been due to several factors, such as learning about new products or being more motivated to make environmentally friendly choices. Also, in the pre-questionnaire, most respondents never used public transportation or grew plants, but they changed their habits in the post-questionnaire to use public transportation some of the time and grow plants that require less water. Almost all respondents always turn off the air conditioner and lights before leaving the place, and they also rationally use water in both questionnaires. Students regularly participate in climate change campaigns. Most respondents were always interested in reading about climate change in the post, while they were sometimes spent only in pre.

#	Attitude towards climate change	Pre								Post							
		Strongly agree	Agree	Not sure	Disagree	Strongly disagree	Mean	SD	Agreement level	Strongly agree	Agree	Not sure	Disagree	Strongly disagree	Mean	SD	Agreement level
1	I would use public transportation to reduce emissions.	5	14	16	0	3	3.47	1.0060	Agree	13	15	8	1	1	4	0.958	Agree
2	I would rather use energy-saving light bulbs.	30	6	0	1	1	4.71	0.654	Strongly agree	26	9	3	0	0	4.61	0.638	Strongly agree
3	It is preferable to reduce water subsidies to reduce excessive consumption.	10	9	7	7	5	3.32	1.397	Not sure	12	10	6	4	6	3.47	1.447	Agree
4	It is preferable to reduce electricity subsidies to reduce excessive consumption.	10	12	6	6	4	3.47	1.330	Agree	15	8	6	4	5	3.631	1.441	Agree
5	It is better to replace fossil fuel-based energy with renewable energy.	23	9	6	0	0	4.45	0.760	Strongly agree	21	11	6	0	0	4.39	0.754	Strongly agree

#	Attitude towards climate change	Pre								Post							
		Strongly agree	Agree	Not sure	Disagree	Strongly disagree	Mean	SD	Agreement level	Strongly agree	Agree	Not sure	Disagree	Strongly disagree	Mean	SD	Agreement level
#	Attitude towards climate change	Pre								Post							
		Strongly agree	Agree	Not sure	Disagree	Strongly disagree	Mean	SD	Agreement level	Strongly agree	Agree	Not sure	Disagree	Strongly disagree	Mean	SD	Agreement level
6	I think decreased water use will reduce the fuel consumption for desalination.	11	14	13	0	0	3.95	0.804	Agree	16	13	7	2	0	4.13	0.905	Agree
7	Recycling organic waste reduces greenhouse gasses emissions.	21	9	6	0	0	4.39	0.755	Strongly agree	22	12	3	1	0	4.45	0.7604	Strongly agree
8	It is important to protect blue carbon areas (mangroves and seagrasses).	27	7	4	0	0	4.61	0.679	Strongly agree	25	11	2	0	0	4.61	0.594	Strongly agree
9	I do not encourage cutting vegetation (trees—shrubs—grass—flowers ...).	26	8	3	1	0	4.55	0.760	Strongly agree	26	9	2	0	1	4.55	0.828	Strongly agree
10	I do not think that climate change affects the spread of epidemics and diseases.	5	5	8	8	13	3.50	1.390	Agree	5	6	7	5	15	3.39	1.516	Agree

#	Attitude towards climate change	Pre								Post							
		Strongly agree	Agree	Not sure	Disagree	Strongly disagree	Mean	SD	Agreement level	Strongly agree	Agree	Not sure	Disagree	Strongly disagree	Mean	SD	Agreement level
11	I think that the spread of "Corona, COVID-19" has reduced greenhouse gasses emissions.	16	17	6	0	0	4.29	0.694	Strongly agree	16	14	5	0	2	4.08	1.075	Strongly agree
12	I'm thinking about working in an institution with a good environmental record in the future.	18	13	7	0	0	3.47	1.006	Agree	21	12	4	0	1	4.37	0.883	Strongly agree
13	I would like to buy an electric car in the future to reduce greenhouse gas emissions.	26	7	3	0	2	4.71	0.554	Strongly agree	23	10	4	1	0	4.45	0.795	Strongly agree
	Overall mean						4.06	0.9145	Agree						4.2	4.164	Strongly Agree

Table 2.
Attitude towards climate change in pre- and post-KAP questionnaire.

The study you are referring to found that students' ratings of their environmental practices changed from "sometimes" to "always" after completing a climate change education intervention, which suggests that the climate change education intervention effectively increased students' awareness of the importance of environmental practices and motivated them to make more environmentally friendly choices. The intervention may have also helped students to develop a better understanding of the causes and consequences of climate change.

Chi-square tests of independence were performed to measure the relationship between participant age group and *Knowledge, Attitude and Practice* (KAP). There were no significant relationships found between age group, *Knowledge* (Chi-square- = 9.182, df = 9, p-value =0.421) and *Attitude* (Chi-square = 2.867, df = 3, p-value =0.408) or *Practice* (Chi-square = 4.708, df = 3, value =0.189).

Chi-square tests of independence were performed to measure the relationship between nationality and *Knowledge*, *Attitude* and *Practice* (KAP). No significant relationship exists between nationality and *Knowledge* (Chi-square = 1.397, df = 3, p-values = 0.143) nor between nationality and *Practice* regarding climate change (Chi-square = 2.138, df = 1, p-value = 0.076). However, there was a significant relation between nationality and *Attitude* towards climate change (Chi-square = 8.478, df = 3, p-values = 0.037) for Bahrainis (69% agree, 21% strongly agree, 10% not sure). Bahrainis were likely more agree with positive *Attitudes* compared to non-Bahrainis.

Chi-square tests of independence were performed to measure the relationship between gender and the KAP. The results show that there was a relation between gender and *Attitude* (Chi-square = 14.198, df = 2, p-values = 0.027). There was no significant relation between gender and *Knowledge* (Chi-square = 0.310, df = 2, p-value =0.85) nor gender and *Practice* (Chi-square = 2.409, df = 6, p-values = 0.300). The female responses were strongly disagree =0%, not sure = 15.5%, agree = 69%, strongly agree = 15.5%, while the male participants' responses were strongly disagree =4%, not sure = 4%, agree = 56%, strongly agree = 36%.

Chi-square tests of independence were performed to measure the relationship between significance in secondary school and KAP (Chi-square = 33.56, df = 12, p-value = 0.001). There was a significant relation between the major in secondary school and *Attitude*. The unified track science respondents' 32 participants' responses were 33.5% strongly agree, 70% agree, 3.5% not sure.

The level of awareness regarding *Knowledge, Attitudes and Practice* (KAP) among undergraduate students was poor in the pre-KAP questionnaire but improved to be acceptable in the post. They were poor in the depth of scientific *Knowledge* regarding climate change, as shown in **Figure 6**. In the post-questionnaire, 63% got more than 60%, up from 13% in the pre-questionnaire in the knowledge section. As a result of the enrichment unit and activities in Climate Champions, students' *Attitudes* towards climate change have improved significantly, from "agree" to "strongly agree" and acting climate-friendly *Practices* from "sometimes" to "always".

The respondents' knowledge of climate change was evaluated as "poor" in the pre-event questionnaire (PRE). The respondents' evaluation improved to be acceptable in the post-event questionnaire (POV). The reason for improving respondents' knowledge could be the activities provided between the two events, PRE and POV. An intensive one-week program provided by the British Council in the form of the "active citizen program" has given its results. The youth respondents were fed with climate change knowledge and encouraged to provide awareness sessions to schools and multiple projects during six (6) months. One of these projects is the "Watergy" proposed by the students that implements an idea to collect the water produced by air conditioners in

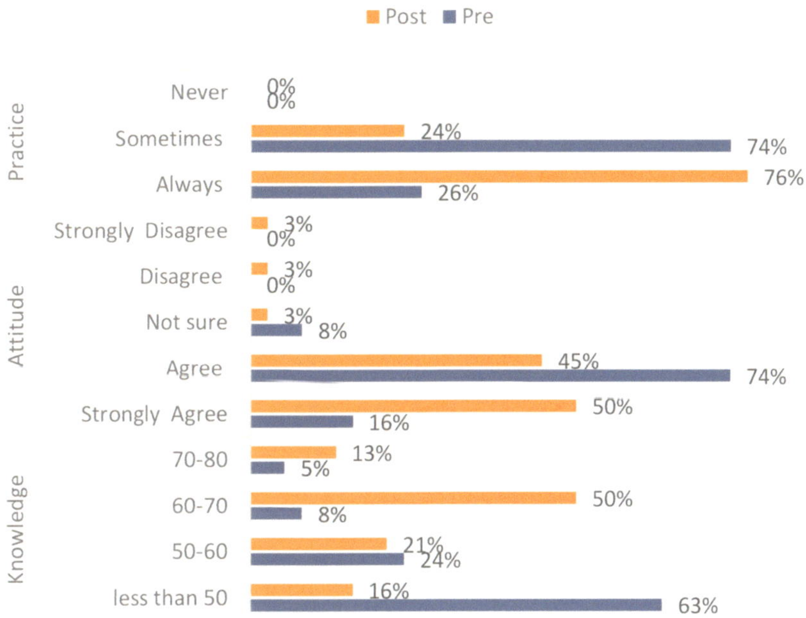

Figure 6.
Knowledge, attitude and practice levels in pre- and post-KAP questionnaires.

special tanks to be used using solar panels, to save water and produce energy, simultaneously. An environmental enrichment unit was provided at the end of the 6 months. As a result of the enrichment unit and activities in Climate Champions, students' *Attitudes* towards climate change have improved significantly, from "agree" to "strongly agree" and acting climate-friendly *Practices* from "sometimes" to "always".

According to the PRE questionnaire, respondents gained knowledge about climate change mainly from the internet, whereas they gained information primarily from environmental associations for the POE questionnaire. In other studies, the primary source of information was radio or TV [10, 15].

Bahrainis and non-Bahrainis reported similar percentages of attitudes towards climate change. This excellent result indicates that Bahrainis (69% agree, 21% strongly agree, 10% not sure) and non-Bahrainis (33% agree, 56% strongly agree, and 11% disagree) have a "responsible" attitude towards climate change. Researchers and climate practitioners/activists must exploit this fact to spread the responsibility towards climate change as a model for others to follow [3, 15].

No statistically significant differences were found between the "age group" and the level of knowledge regarding climate change issues or the attitude or behaviour (see results (above)). This could be because the majority, 86% of the student's age, is less than twenty-six (26). Therefore, the respondents' age group is very close. These results are reflected and found in other studies [15, 16].

Findings suggest that there is a shortage in the knowledge of university students. Their awareness level regarding climate change is rated "Poor". In comparison, their knowledge can be described as very poor. 63% of the surveyed students scored less than 60% in the knowledge section of the KAP. This might be due to the lack of climate change topics in the science and mathematics curricula in the governmental

	Practice towards climate change	Pre							Post						
		Always	Sometimes	Never	Mean	SD	%	Agreement	Always	Sometimes	Never	Mean	SD	%	Agreement Level
1	I use eco-friendly products.	10	27	1	2.24	0.490	75	Sometimes	16	21	1	2.39	0.547	80	Always
2	I rationalise water usage.	21	16	1	2.53	0.557	84	Always	23	15	0	2.61	0.495	87	Always
3	I grow plants that need less water.	10	18	10	2.00	0.735	67	Sometimes	22	15	1	2.55	0.555	85	Always
4	I make sure to turn off the Air Conditioner (AC) before leaving the room.	32	5	1	2.82	0.457	94	Always	32	6	0	2.84	0.370	97	Always
5	I make sure to turn off the lights before leaving the house.	33	5	0	2.87	0.343	96	Always	33	5	0	2.87	0.343	96	Always
6	I reduce red meat consumption.	15	18	5	2.26	0.685	75	Sometimes	18	16	4	2.37	0.675	79	Always
7	I use public transportation.	4	10	24	1.47	0.687	49	Never	9	13	16	1.82	0.801	61	Never
8	I share means of transportation with colleagues.	11	20	7	2.11	0.689	70	Sometimes	19	14	5	2.37	0.714	79	Always
#	**Practice towards climate change**	**Pre**							**Post**						
		Always	**Sometimes**	**Never**	**Mean**	**SD**	**%**	**Agreement Level**	**Always**	**Sometimes**	**Never**	**Mean**	**SD**	**%**	**Agreement Level**
9	I get my groceries from shops near my house.	26	12	0	2.68	0.471	89	Always	32	6	0	2.84	0.370	95	Always
10	I replace hard copies of documents with soft copies to save paper.	18	17	3	2.08	0.638	69	Sometimes	21	16	1	2.53	0.557	84	Always
11	I prefer distance learning to save the environment.	16	10	12	2.11	0.863	70	Sometimes	16	11	11	2.13	0.844	71	Sometimes
12	I sort my waste in the recycling bins.	8	26	7	2.03	0.636	68	Sometimes	19	14	5	2.42	0.642	81	Always
13	I do not encourage living by the sea because of the risks of sea-level rise due to climate change.	12	12	14	1.95	0.837	65	Sometimes	15	14	9	2.16	0.789	72	Sometimes

DOI: http://dx.doi.org/10.5772/intechopen.1002528

	Practice towards climate change	Pre							Post						
		Always	Sometimes	Never	Mean	SD	%	Agreement	Always	Sometimes	Never	Mean	SD	%	Agreement Level
14	I participate in climate change awareness campaigns.	21	15	2	2.50	0.604	83	Always	26	12	0	2.68	0.471	89	Always
17	I am interested in reading more about climate change to raise awareness and educate myself.	30	7	1	2.24	0.490	75	Sometimes	27	10	1	2.68	0.525	89	Always
	Overall mean				2.25	0.612		Sometimes				2.48	0.580		Always

Table 3.
Practice towards climate change in pre- and post-KAP questionnaire.

sector in the Kingdom of Bahrain. 74% of the participants chose "sometimes" when they were asked about their practices towards climate change, such as using eco-friendly products, reducing red meat consumption, sorting their waste in the recycle bins, replacing hard copies with soft copies and preferring distance learning to save the environment. 84% of them always turn off lights and air conditioners (AC) before leaving home and rationalise water usage (**Table 3**). Some *Attitudes* towards climate change were negative, such as unwillingness to reduce subsidies for water and electricity (73%). Other types of *Attitudes were positive*, including welling to protect blue carbon areas such as mangroves and seagrasses and preferring to work in an institution that acts towards the environment in the future (**Table 2**).

Most students (95%) agreed to adapt to the suggested *Practices* that support the environment in many ways. Therefore, the study sample had a Positive *Attitude* towards climate change, while 76% of them always apply environmental Practice in POST-KAP.

5. Conclusion

The enrichment unit and the climate champions' activity effectively increased students' knowledge of climate change, as evidenced by the increase in the percentage of students who scored more than 60% in the knowledge section of the KAP. The program also effectively changed students' attitudes towards climate change, as evidenced by the increase in the percentage of students who said they would "always" practice climate-friendly behaviours. These findings suggest that climate change awareness programs can effectively increase students' knowledge and change their attitudes and practices towards climate change.

Acknowledgements

The authors would like to acknowledge the British Council and the University of Bahrain (UOB) for their guidance and support during this research project.

Recommendation

The authors of the current research recommend paying more attention to the level of knowledge, awareness and Attitude towards climate change of youngsters. Awareness sessions have a favourable implication on the level of knowledge, particularly. Activities and awareness sessions are also beneficial in Attitude and Practice. This is also found in other studies in several domains [14, 17, 18].

List of abbreviations

ESD	education for sustainable development
GHGs	green house gases
KAP	knowledge, attitude and practice
SDGs	sustainable development goal
NCR	national communication report
NDCs	nationally determined contributions

Author details

Nahed Salman[1,2]*, Mariam Al-Mannai[3], Asma Abahussain[1] and Mahmood Saeed Mustafa Alalawi[4]

1 Environment and Sustainable Development Program, College of Science, University of Bahrain, Sakhir, Kingdom of Bahrain

2 Bahrain Polytechnic, Isa Town, Kingdom of Bahrain

3 Mathematics Division, College of Science, University of Bahrain, Sakhir, Kingdom of Bahrain

4 College of Information Technology, University of Bahrain, Sakhir, Kingdom of Bahrain

*Address all correspondence to: nahed.yusuf@gmail.com

IntechOpen

© 2023 The Author(s). Licensee IntechOpen. This chapter is distributed under the terms of the Creative Commons Attribution License (http://creativecommons.org/licenses/by/4.0), which permits unrestricted use, distribution, and reproduction in any medium, provided the original work is properly cited.

References

[1] World Bank. The World Bank Bahrain Data. World Bank Group. 2023. Available from: https://data.worldbank.org/country/BH [Accessed: January 26, 2022]

[2] Hosta S. Fact sheet: COP26—Children and climate change. UNICEF. 2021. Available from: https://www.unicef.org/turkey/en/press-releases/fact-sheet-cop26-children-and-climate-change [Accessed: February 04, 2022]

[3] Arab Youth Center. Arab Youth Council for Climate Change: Mobilising Youth for a Sustainable Future. 2021. Available from: https://arabyouthcenter.org/en/article/our-research/arab-youth-council-for-climate-change-mobilizing-youth-for-a-sustainable-future [Accessed: February 02, 2022]

[4] Lin MT. COP27, COP28 to see climate stocktaking, more focus on developing world, renewed fossil fuel debates. ISH Markit. 2022. https://cleanenergynews.ihsmarkit.com/research-analysis/cop27-cop28-to-see-climate-stocktaking-more-focus-on-developin.html

[5] Bahrain. Bahrain—High-level segment statement COP 26. In: Glasgow Climate Change Conference—October/November 2021. UNFCCC. 2021. pp. 1-2. Available from: https://unfccc.int/documents/309274 [Accessed: October 01, 2022]

[6] British Council. Bahrain Climate Champions: Supporting Young People to Create a Greener, more Sustainable World. Bahrain: British Council; 2021. Available from: https://www.britishcouncil.bh/en/programmes/education/climate-champions [Accessed: February 04, 2022]

[7] F.-R. RFD, J. Cruz, and F. Alshammari, "Knowledge of and attitudes toward climate change and its effects on health among nursing students: A multi-Arab country study," Nursing Forum, vol. 53, no. 2, pp. 179–189, 2018. Available from: http://www.uoh.edu.sa/en/Subgates/Faculties/Nursing/Documents/Knowledgeofandattitudestowardclimatechangeandits.pdf [Accessed: January 28, 2022]

[8] Kumar A, Bose DK, Jahanara RK. Awareness of university students about global warming in Allahabad district of Uttar Pradesh. Journal of Pharmacognosy and Phytochemistry. 2017;**6**(5):461-463

[9] Ibrahim AA, Fahmy HD, Mahmoud SR. Knowledge and attitude regarding global warming phenomenon among Assiut University students. Assiut Scientific Nursing Journal. 2018;**6**(18):13-21

[10] Saint Vincent and the Grenadines. Knowledge, Attitudes and Practice Study on Climate Change. 2017. Available from: https://www.adaptation-undp.org/sites/default/files/resources/knowledge_attitudes_and_practice_study_on_climate_change_saint_vincent_and_the_grenadines_final.pdf [Accessed: November 22, 2021]

[11] Al-Naqbi AK, Alshannnag Q. The status of education for sustainable development and sustainability knowledge, attitudes, and behaviors of UAE university students. International Journal of Sustainability in Higher Education. 2014;**15**(4):390-403

[12] Sanni A, Kaigama F, Salisu J, Dike O, Sadiq Y, Sharbatti SA. Knowledge, attitude and practices related to eco-friendliness among university students In Ajman, U.A.E. International Journal of Medical and Biomedical Studies. 2014;**3**(6):125-133. DOI: 10.32553/ijmbs.v3i6.299

[13] Freije AM, Hussain T, Salman EA. Global warming awareness among the University of Bahrain science students. Journal of the Association of Arab Universities for Basic and Applied Sciences. 2017;**22**:9-16. DOI: 10.1016/j.jaubas.2016.02.002

[14] Educational Group. A Summary of the Educational Group Report within the Committee of the Third National Statement-Report for the Kingdom of Bahrain in the UN Framework of Climate Change Agreement; UNFCCC. 2018

[15] Almheiri M. Climate Change Awareness in United Arab Emirates [Master's Thesis]. UAE: UAEU; 2022

[16] Fontenard T. UNDP-JCCCP in-Country Specific Campaign for Grenada Knowledge, Attitudes and Practices (KAP) Survey Report 2016. UNDP. 2016. pp. 1-37. Available from: https://www.undp.org/content/dam/barbados/docs/Publications/undp_bb_GRN_KAP_climate_change_kap_survey_report_grenada_0.pdf [Accessed: May 02, 2020]

[17] Smith GG, Besalti M, Nation M, Feldman A, Laux K. Teaching climate change science to high school students using computer games in an intermedia narrative. Eurasia Journal of Mathematics, Science and Technology Education. 2019;**15**(6):1-16. DOI: 10.29333/ejmste/103570

[18] SCE. Bahrain's Third National Communication: Climate Change 2020. UNFCCC. 2020. pp. 1-69. Available from: https://unfccc.int/files/national_reports/non-annex_i_parties/biennial_update_reports/application/pdf/tnc_report.pdf%0Ahttps://www.environment.gov.za/sites/default/files/reports/draftsouthafricas3rdnationalcommunication_unfccc2017.pdf

Chapter 10

Religion and Climate Change in Africa: A Gendered Perspective

Terence Mupangwa and Sophia Chirongoma

Abstract

This chapter reflects on the interface of religion, gender, and climate change in Zimbabwe. The writing of this chapter has been ignited by the ongoing climate change induced catastrophes such as cyclones and drought in Southern Africa. Cognizant of the pivotal role of religion in shaping people's worldviews not only in Southern Africa but the rest of the continent, the chapter discusses how religion shapes people's relationship with ecology. Informed by the African feminist theory, the chapter adopts a gendered perspective. It, therefore, discusses the intertwinement between African women's experiences of patriarchy and the earth's exposure to ecological degradation. Acknowledging the efficacy of fulfilling Sustainable Development Goals 4, 5, and 13, the chapter concludes by restating the importance for the church in Zimbabwe to adopt an agenda for gender empowerment and ecological stewardship.

Keywords: climate change, ecology, gender, religion, stewardship, sustainable development goals

1. Introduction

Climate change is the greatest challenge the world is facing today, and it is real in Zimbabwe. The devastation has been unleashed upon areas like Chimanimani and Chipinge, where lives have been lost, injuries sustained, and entire settlements swept away in the wake of Cyclone Idai which made landfall in March 2019. The consequences of climate change are evident through escalating global temperatures, rising sea levels, storm surges, recurrent floods, persistent droughts, heat waves, and the spread of infectious diseases. To combat this crisis, the pursuit of sustainable mitigation strategies has become imperative for both developing and developed nations. Human actions have played a significant role in driving these changes, resulting in profound effects on the environment as well as exacerbating the effects of climate change. This has prompted some to label this era as the Anthropocene epoch.

Given that climate change presents an existential threat, numerous calls have been made to address this issue from various quarters. However, discussions on climate change often overlook the role of religion and the specific impact on women in terms of mitigation and adaptation efforts. Women tend to be marginalized and relegated to a peripheral position when the intersection of religion and climate change is examined. In recent years, there has been increasing research and discussion surrounding the complex and multifaceted issue of how African women's experiences of patriarchy intertwine with ecological degradation. At its core, this issue revolves around the ways in which gendered power dynamics and environmental deterioration intersect and influence the experiences of African women. This includes their access to resources and their ability to mitigate the effects of climate change.

2. African feminism

This study employs the African feminist approach as a theoretical framework. African feminism falls within the category of global organizations dedicated to dismantling patriarchy in all its manifestations. However, they pursue this goal with an understanding that patriarchy varies across time and space, influenced by factors such as class, race, ethnicity, religion, and global power dynamics [1]. African feminism is an active and dynamic movement that recognizes the diversity of African cultures, the unique historical experiences of African women, and the specific social, cultural, political, and economic contexts they face [1].

African feminism is a discourse that seeks to comprehend the multiple challenges and obstacles faced by African women resulting from sexism, which is a product of patriarchy, poverty, and colonization [2]. This theoretical approach is particularly relevant to this study as it strives to educate, empower, and advance women, not in opposition to men, but alongside them. African feminism does not view African men as adversaries but rather calls upon them to recognize and address the specific forms of women's subjugation that exist alongside the overall oppression experienced by all Africans. It sees men as partners in the struggle against gender oppression [1]. Therefore, the discourse on African gender issues should involve both men and women in challenging oppressive African cultural norms. This characteristic makes this theory especially relevant to this study because, to eliminate the exclusion of women in climate change debates, it is necessary to engage men in the conversation. This theory deviates from conventional feminism that view men as enemies. Drawing from the principles of African feminism, our chapter affirms that involving men in this discourse will contribute to enhancing women's involvement in decision-making concerning the conservation of the environment.

African feminism is an activist movement and a set of ideas that advocate for positive social transformation. This study argues that while men and women may have biological differences, they deserve equal treatment, liberties, and rights, even in climate change issues. The approach acknowledges that gender inequality, discrimination, and oppression are realities in Africa.

3. The influence of religious teachings on gender roles and environmental ethics

Religious teachings play a significant role in shaping gender roles and hierarchies within societies, as well as influencing ethical perspectives on the environment. This section examines how religious teachings and interpretations can reinforce traditional gender roles and hierarchies while also shaping perspectives on stewardship, dominion, and the relationship between humans and the environment.

Many religious traditions have historically upheld patriarchal structures that assign specific roles and responsibilities to men and women. These teachings often emphasize male authority and female subordination within family and societal contexts [3]. For instance, some interpretations of Christian scriptures have been used to support the subordination of women, citing selected biblical passages such as Ephesians 5:22–24, which instruct wives to submit to their husbands. Similarly, certain interpretations of religious texts in other faith traditions have been used to justify gender inequalities, limiting women's participation and leadership roles within religious institutions [4].

Religious teachings also shape ethical perspectives on environmental stewardship, dominion, and the relationship between humans and the natural world. The concept of stewardship, as mentioned in Christianity, encourages believers to view themselves as caretakers of the Earth and emphasizes responsible care and protection of the environment [5]. However, the interpretation of dominion can vary. Some interpretations promote a responsible exercise of dominion, emphasizing the ethical responsibility to protect and preserve the Earth, while others may interpret dominion as a license for exploitation without considering environmental consequences [6].

The influence of religious teachings on gender roles and environmental ethics is a complex interplay that varies across different religious traditions, denominations, and cultural contexts [7]. It is important to recognize that interpretations of religious texts can be diverse and subject to evolution over time. Some religious communities are re-examining traditional teachings, promoting gender equality, and revisiting interpretations to foster more inclusive and environmentally conscious practices [8]. The intersectionality of gender and environmental ethics within religious contexts requires a nuanced understanding of how religious teachings and cultural norms intersect with ecological values and social justice concerns.

Religious teachings have a profound influence on both gender roles and environmental ethics. While certain interpretations can reinforce traditional gender hierarchies, religious perspectives on stewardship, dominion, and the relationship between humans and the environment provide a framework for understanding our ethical responsibilities toward the natural world. Recognizing the complexities of these influences allows for critical reflection, dialog, and the potential for transformative interpretations that promote gender equality, environmental sustainability, and social justice within religious contexts.

4. Patriarchal dynamics within church structures

Power structures within religious institutions in Africa have long influenced women's agency and participation in environmental decision-making processes. Religious institutions in Africa often reflect patriarchal norms and practices, with male-dominated power structures [4]. Some Pentecostal churches in Zimbabwe, like the Apostolic Faith Mission in Zimbabwe, are patriarchal and are led by a male bloated leadership [9]. Traditional interpretations of religious teachings can reinforce gender inequalities, limiting women's access to leadership roles, and diminishing their voices in environmental discussions. Women may face discrimination and exclusion from decision-making spaces due to societal and religious norms that prioritize male authority and control.

Recognizing and challenging patriarchal power structures within religious institutions is crucial for promoting women's agency and enhancing their involvement in environmental decision-making. Encouraging gender-sensitive interpretations of religious teachings can provide a basis for reimagining women's roles and empowering them to actively participate in environmental initiatives [9]. Fostering inclusive leadership and decision-making structures within religious institutions can create spaces for women's voices to be heard and influence environmental policies and practices [9]. Engaging religious leaders and communities in dialog and education on gender equality and environmental justice can help to challenge patriarchal norms and promote more equitable and sustainable approaches to environmental stewardship [9].

The lack of female representation in leadership positions within churches can limit women's influence and decision-making power in matters related to the environment [10]. Women are also endowed with indigenous knowledge which helps in preserving the environment and curbing some of the devastating effects of climate change. Women in Africa, just like their male counterparts, are equipped with the knowledge of how nature behaves and what predictions to make, using several observations on natural phenomena. More so, research shows that while the significance of indigenous knowledge has been recognized in the designing and implementation of sustainable development projects, not much has been done to include indigenous knowledge systems in formal climate change mitigation and adaptation strategies, especially in discussions with women, who are largely in touch with the environment in different ways. Regrettably, since very few, if any women sit in decision-making boards, their views are barely heard and their knowledge on ecological preservation is often underutilized [10].

Due to patriarchy, there is uneven access to resources, such as financial resources and education, which further contributes to gender disparities within churches. Limited access to resources can hinder women's capacity to engage in environmental initiatives, advocacy, and community-based projects. The unequal distribution of resources perpetuates gender inequalities and reinforces the exclusion of women from meaningful participation in environmental decision-making processes [10].

5. Gendered impacts of climate change on church-going women

Climate change has profound implications for the daily lives and well-being of church-going women, as they often have close connections to the natural environment and rely on it for their livelihoods. This section will explore the ways in which ecological degradation impacts the lives of church-going African women, considering their roles, responsibilities, and spiritual connections to the environment, with reference to the relevant studies and research.

Climate change manifestations vary across different regions of the world, and they are not globally uniform [11]. The impacts of climate change can be different significantly, with some regions experiencing increased rainfall levels while others face reduced levels, resulting in wetter or drier conditions. In areas with increased rainfall, there is a higher risk of flooding as the ground becomes saturated, rivers and tributaries exceed their carrying capacities, and dams may collapse or fail under excessive water pressure [11]. Zimbabwe serves as an example of a place facing diverse climate change challenges. It has witnessed the adverse effects of climate change including drought, erosion, and frequent flood disasters, impacting various parts of the country, and causing significant loss of life and property damage.

Climate change has significant implications for food security, and women in Zimbabwe are particularly vulnerable to its impacts. Climate change can disrupt agricultural systems, leading to decreased crop yields, shifts in rainfall patterns, increased frequency of extreme weather, and the spread of pests and diseases.

These changes directly affect the availability and accessibility of food, posing a threat to food security [12]. Climate change also has affected the agricultural practices of women in Chimanimani [13]. In Zimbabwe, women play a crucial role in agricultural production, particularly in small-scale farming and subsistence agriculture. They are involved in various activities, such as crop cultivation, and livestock rearing. However, climate change impacts such as prolonged droughts, erratic rainfall, and increased temperatures pose significant challenges to their farming practices. For example, the changing rainfall patterns can result in water scarcity, affecting the irrigation capabilities of women farmers. Reduced water availability hinders crop growth and productivity, leading to lower yields and limited food availability. Additionally, extreme weather events like floods or droughts can destroy crops, leading to food losses and compromising the food security for women and their families. Furthermore, women often have limited access to resources and technologies that can help them to adapt to climate change. They may lack access to improved seeds, irrigation systems, financial services, and information on climate-smart agricultural practices [12]. These limitations further exacerbate the vulnerability of women to climate change impacts on food security. The reduction in food production is a high threat to food security and this aggravates the problem of poverty already in existence. Women contribute 80% of the food in Africa, 60% of the food in Asia, and 40% of the food in Latin America [14]. The burden on women has increased because they are usually the ones who are responsible for finding other sources of food, which most of the time requires them to have some money to buy these alternative sources of food. Due to the ongoing economic challenges in Southern Africa, especially in Zimbabwe, this becomes a mammoth task. This is unlike in the good old days when they were assured of harvesting enough food from their fields.

Women in Africa often have strong spiritual connections to the natural environment, considering it sacred and imbued with spiritual meaning [15]. Ecological degradation has far-reaching consequences for the spiritual cultural practices deeply intertwined with the natural environment, significantly impacting the sense of identity, belonging, and spiritual fulfillment experienced by African women. The loss of biodiversity, destruction of sacred sites, and degradation of natural landscapes undermine the spiritual and cultural significance attributed to the environment, consequently affecting the well-being and resilience of women and their communities. Traditionally, African women have played essential roles as farmers, herbalists, and caretakers of water and fire, forging a profound connection with the land and its resources [16].

Climate change has also led to deforestation. Deforestation disrupts ecosystems and impacts women who heavily rely on forests for fuelwood, food, medicine, and income generation. The loss of forests threatens their livelihoods and exacerbates their vulnerability. Women's dependence on forest resources makes them more susceptible to the adverse effects of deforestation, leading to increased workloads, decreased access to resources, and limited income-generating opportunities. Deforestation also affects women's traditional knowledge and practices, as the destruction of forests erodes cultural and spiritual connections to the natural environment [17].

Women also become susceptible to dehumanization and exploitation because of the effects of climatic changes. For instance, in the aftermath of the Cyclone Idai carnage, some of the women who survived this ecological disaster in Chimanimani bemoaned how they were taken advantage of by some unscrupulous humanitarian officers who solicited for sex in exchange for the basic necessities such as food, tents, sanitary ware, which they were supposed to distribute free of charge [13].

An unprecedented increase in household chores is another gendered impact of climate change in Zimbabwe. In many African countries, Zimbabwe included, women are responsible for collecting firewood, fetching water, and growing crops, all of which are activities that are highly dependent on a healthy and sustainable environment [1]. When rains fail, not only do women find their workload doubled as they must walk longer distances to fetch water, but food security is also highly impacted, leading to ill-health and death [17]. Caregiving is another household chore for women, that increases because of climate change. This became apparent in the aftermath of the Cyclone Idai catastrophe which hit hardest in the Chimanimani and Chipinge districts of Zimbabwe. Many people were injured, and women had to take care of the sick and injured, increasing their workload. Women's role as caregivers and the primary providers of healthcare within their families means that environmental degradation and its associated health impacts disproportionately burden them [18].

Climate change has serious implications for human health. Drought also leads to water scarcity and a significant decrease in fresh water. This can compromise hygiene and increase the risk of diarrhea and other waterborne diseases such as typhoid fever, cholera, and river blindness. Floods contaminate freshwater supplies, heighten the risk of waterborne diseases, as well as creating a breeding ground for disease-carrying insects such as mosquitoes. Women are the ones who carry most of the burden for they are the ones who are supposed to maintain cleanliness in such difficult circumstances. Besides, they are the ones who also take care of the sick in the home [19]. Contaminated water sources, air pollution, and exposure to toxic substances can lead to increased health risks, including respiratory diseases, waterborne illnesses, and reproductive health issues. Food insecurity negatively impacts women's health and nutrition, leading to increased vulnerability to malnutrition and related health issues [20].

5.1 Resistance and transformation

Religious communities have got a crucial role to play, in addressing the environmental crisis and climate change. While religion alone may not be sufficient to tackle these challenges, it can serve as a vital partner, alongside fields such as economics, policy, science, and education. Religion has been recognized as a powerful force shaping human behavior and ideology concerning both mitigation and adaptation to environmental and climatic changes [21]. Unfortunately, decision-makers often overlook the religious and social dimensions when considering environmental conservation, climate change mitigation, and adaptation strategies.

Religion plays a pivotal role in shaping people's perceptions, attitudes, and actions toward various goals. It provides a framework through which values and orientations can be derived, potentially fostering world views and environmental ethics that promote the conservation of the environment and climate change adaptation [22]. By engaging with religious teachings and communities, it becomes possible to tap into these values and mobilize them to foster a deeper appreciation for the environment and drive sustainable actions. It is, therefore, apparent that recognizing and harnessing the potential of religion in climate change efforts is crucial. Collaboration between religious communities, policymakers, scientists, educators, and other stakeholders can lead to more holistic and effective approaches. By integrating religious perspectives, values, and practices into environmental initiatives, we can tap into a powerful force for change and promote a more sustainable and resilient future [10].

The church's widespread presence in the country grants it significant social and political influence, positioning it advantageously to make a greater contribution in mitigating, educating, and preventing climate change. By embracing this role, the church aligns with Christian ethics, which advocates for environmental stewardship, and the traditional African environmental philosophy, which emphasizes the interconnectedness between humans and nature. Leveraging its position and influence, the Church can effectively address climate change concerns and foster a sense of responsibility toward the environment among its followers.

By actively engaging in climate change initiatives, the church can fulfill its moral and ethical obligations while also leveraging its broad reach and influence to raise awareness, build capacity, and mobilize action. Emphasizing environmental stewardship and promoting sustainable practices, the Church can contribute to creating a more resilient and environmentally conscious society. Through its focus on women, youth, and local communities, the Church can address the specific vulnerabilities and challenges they face, ensuring a more inclusive and equitable approach to climate change mitigation and adaptation. Such efforts are crucial for addressing the urgent challenges posed by climate change and creating a more sustainable and just future for all [23].

The church has the potential to be a powerful space for resistance and transformation, challenging patriarchal norms and promoting ecological stewardship which will, in turn, benefit women in the long run. First, churches can actively work toward confronting and changing patriarchal traditions that perpetuate gender inequalities within religious communities. Such initiatives will promote gender equality, challenge traditional gender roles, and create spaces for women's leadership and participation [9]. For instance, the Church can start taking steps toward gender justice by ordaining women as priests and bishops and dismantling the male-bloated leadership structures within the church. By addressing gender inequalities within religious contexts and advocating for ecological stewardship, these initiatives contribute to a more holistic approach to justice, recognizing the interdependence of human well-being and the health of the planet. Church-based initiatives that prioritize gender equality and environmental justice contribute to a more inclusive and sustainable future, bridging the gap between religious teachings and social environmental action.

6. Overcoming barriers and empowering women

In sync with the principles of African feminism, which emphasizes women's empowerment as a cog for human development in Africa, the church can engage in a number of activities in empowering women to participate in climate change deliberations. First, education plays a pivotal role in empowering women within religious contexts. Providing access to quality education equips women with knowledge, skills and critical thinking abilities for challenging gender inequalities and promoting their agency [9]. Educational initiatives can focus on gender-responsive curricula, leadership development, and vocational training, enabling women to actively participate in decision-making processes and contribute to sustainable development efforts [9]. This dovetails with the UN Sustainable Development Goals 4 and 5, which are among the priority Sustainable Development Goals (SDGs) shaping Zimbabwe's development agenda. Sustainable Development Goal 4 is targeted toward ensuring inclusive and equitable quality education and promoting lifelong learning opportunities for all. On the other hand, Sustainable Development Goal 5 is tailored toward achieving gender equality as well as empowering all women and girls. Hence, by addressing these two SDGs, the church will be making an immense contribution toward the attainment of the overall goal of the 17 SDGs, which is to ensure that no one is left behind in the development agenda.

Theological re-interpretation can be a powerful tool in empowering church-going-women for them to participate in climate change debates. Re-examining religious texts and theological frameworks is crucial for promoting gender equality within religious institutions. Embracing the principles of African feminism and biblical interpretation helps to challenge patriarchal interpretations that reinforce gender disparities [24]. By highlighting women's agency, leadership roles, and contributions to religious narratives, theological reinterpretation promotes inclusive religious practices and fosters gender equality within churches. The main tenets of African women's theology can also provide a platform for feminist theological dialog and interpretation, facilitating transformative practices within religious communities [9].

More so, the church can engage in leadership and capacity building by supporting women's leadership development by creating opportunities for women to hold leadership positions in the church and environmental committees within the church. Capacity-building programs can be organized to enhance women's skills in key areas such as sustainable agriculture, natural resource management, and climate change [12].

Engaging women in community–based initiatives and grassroots activism is essential for empowering church-going women. Community participation also allows women to contribute their perspectives, knowledge, and skills to address social, economic, and environmental challenges [13]. Women's groups within religious communities can foster solidarity, organize advocacy campaigns, and implement sustainable projects that promote gender equality, environmental stewardship, and social justice [25]. Insights can be derived from the Green Anglicans movement, mentioned earlier in this chapter, which actively engages church communities in environmental activism and encourages women's leadership in ecological justice initiatives [5].

Furthermore, economic empowerment is crucial if ever women should participate fully in conserving the environment and preventing the drastic changes taking place in the climate. The church can facilitate income-generating activities for women that align with sustainable practices. For instance, women can be supported in starting eco-friendly businesses, such as organic farming, renewable energy enterprises, or sustainable crafts [26]. By promoting economic empowerment, the church can enable women to have a greater say in decision-making processes related to environmental conservation [12].

Advocacy and policy engagement is another strategy that the church can utilize in empowering women. The church can advocate for gender-responsive environmental policies and regulations that address the specific needs and priorities of women [27]. This can involve engaging with policymakers, participating in policy dialogs, and highlighting the importance of women's involvement in decision-making processes related to climate change and environmental issues [12]. Such an undertaking would feed into the main goals of Sustainable Development Goal 13 whose main target is to take urgent action to combat climate change and its impacts.

The church can also foster partnerships with other organizations, both within and outside the religious sector, as another way of amplifying its efforts in empowering women in the conservation of the environment. Collaborative initiatives can include joint projects, knowledge-sharing platforms, and resource mobilization for women-led environmental initiatives [12].

These strategies work synergistically to empower church-going Zimbabwean women, enabling them to challenge gender inequalities, promote sustainable practices, and contribute to social transformation within religious contexts. By implementing these strategies, the church can create an enabling environment for women to actively participate in environmental conservation and climate change prevention.

DOI: http://dx.doi.org/10.5772/intechopen.1002354

It can promote women's agency, leadership, and meaningful engagement in decision-making processes, ultimately contributing to sustainable development and a more equitable society.

7. Ways women can conserve the environment

There are several ways in which women can contribute toward the conservation of the environment. Advocating for environmental protection: Women can use their voices to advocate for environmental protection and raise awareness about the importance of conservation. For instance, Greta Thunberg, a young environmental activist, has been using her platform to raise awareness about climate change and inspire action to reduce carbon emissions [28]. The Chipko movement in India which is being led by women stopped the cutting of trees by physically surrounding the trees in the 1970s. Women around the world, through advocacy, continue to fight against climate change [29].

Women can support sustainable practices by using eco-friendly products, reducing waste, and choosing sustainable transportation options. A poignant example is the ecologically friendly practice by most rural women in the Democratic Republic of Congo, who use locally built energy-saving stoves for cooking. By doing, so they preserve the Virunga Park forests in Congo. In India, Vandana Shiva, an environmental activist and eco-feminist, has been advocating for sustainable agricultural practices that protect the environment and empower local communities [30].

Women can also engage in conservation efforts by volunteering for environmental organizations or participating in community-based conservation projects. The Nobel laureate and leading Kenyan environmentalist activist Wangari Maathai, founded the Green Belt Movement, which has planted over 50 million trees in Kenya and empowered women to participate in conservation efforts [31].

Women can also develop innovative solutions to environmental challenges. Aisa Mijeno, a Filipino engineer, invented a lamp that runs on saltwater, providing a sustainable and renewable source of light for communities without access to electricity [32]. In India, women also use homemade solar cookers to boil water. By so doing, they avoid trees, thus preserving the forests [29]. Other women from around the world can draw a leaf from these ecologically sensitive practices and domesticate such practices in their context. Together, men and women can work together to preserve and nurture Mother Earth.

8. Conclusion

This chapter has explored the intertwinement between church-going African women's experiences of patriarchy and ecological degradation, highlighting the significant impact that religious teachings and institutions have on gender roles, environmental ethics, and women's agency in environmental decision-making. The examination of patriarchal power structures in African religious institutions has revealed the ways in which women's participation and leadership are often limited, contributing to gender disparities in decision-making and resource access. Additionally, the analysis of specific environmental challenges faced by women, such as water scarcity, deforestation, and food insecurity, has underscored the need for gender-sensitive approaches to environmental conservation and sustainable development.

Despite these challenges, church-based initiatives have emerged as a space for resistance and transformation, challenging patriarchal norms and promoting ecological stewardship. Women's roles in environmental activism, conservation projects, and sustainable development efforts within religious contexts have been increasingly recognized, reflecting the potential for empowering women as agents of change.

To promote gender equality and women's agency within religious institutions, strategies such as education, theological reinterpretation, and community engagement are necessary. These strategies can foster sustainable practices that prioritize environmental stewardship and gender justice. Further research, interfaith dialog, and action are needed to continue advancing this important work.

In conclusion, addressing the intertwinement between patriarchy and ecological degradation within religious contexts requires a multifaceted approach that acknowledges the diverse experiences and perspectives of church-going African women. By working toward gender equality and ecological sustainability, we can create a more just and equitable world for all.

Author details

Terence Mupangwa[1] and Sophia Chirongoma[2,3]*

1 University of Cape Town, Cape Town, South Africa

2 Midlands State University, Gweru, Zimbabwe

3 Research Institute of Religion and Theology, University of South Africa, Pretoria, South Africa

*Address all correspondence to: sochirongoma@gmail.com

IntechOpen

© 2023 The Author(s). Licensee IntechOpen. This chapter is distributed under the terms of the Creative Commons Attribution License (http://creativecommons.org/licenses/by/4.0), which permits unrestricted use, distribution, and reproduction in any medium, provided the original work is properly cited.

References

[1] Mekgwe P. Theorizing African feminism (s). An African Journal of Philosophy/Revue Africaine de Philosophie. 2008;**10**:11-22

[2] Gatwiri GJ, Mclaren HJ. Discovering my own African feminism: Embarking on a journey to explore Kenyan women's oppression. Journal of International Women's Studies. 2016;**17**(4):263-273

[3] Smith J. Religion, gender and oppression. In: Code L, editor. Encyclopedia of Feminist Theories. New York: Routledge; 2000

[4] Ephesians 5:22-24, New International Version

[5] Chitando E. African Omen, religion, and health: A historical perspective. In: Archer-Straw, editor. The Wiley Blackwell Companion to African Religions. Chichester: Wiley-Blackwell; 2019

[6] Gottlieb RS. A Greener Faith: Religious Environmentalism and our planet's Future. New York: Oxford University Press; 2006

[7] Palmer M. Sacred Ecology: Traditional Ecological Knowledge and Resource Management. New York: Taylor and Francis; 1998

[8] Konig A. Religion, Gender and Culture in the Pre-Modern World. New York: Palgrave Macmillan; 2013

[9] Mupangwa T. The place of women in the leadership of the Apostolic Faith mission in Zimbabwe (AFMZ) [Doctoral Dissertation]

[10] Kilonzo SM. Women indigenous knowledge, systems, and climate change in Kenya. In: Chitando C, Kilonzo, editors. Perspectives on Religion and Climate Change. London: Routledge; 2022

[11] Nwagbara I, Iijoma J, Chima A. Climate change and the risk of vectorborne diseases: A study of malaria transmission dynamics in Nigeria. Research Journal of Applied Science and Technology. 2013;**6**(24):4623-4643

[12] Manjoro M, Makoni FS. Gender dimensions of climate change adaptation in rural farming communities of Zimbabwe. African Journal of Rural Development. 2019;**4**(3):646-654

[13] Chitando E. Who, me? Engaging religion and gender in environmental degradation and climate change. In: Religion, Gender, and the Environment. Geneva: World Council of Churches; 2014. pp. 3-16

[14] FAO. The State of Food and Agriculture 2001: Agricultural Trade and Poverty: Can Trade Work for the Poor? New York: Food and Agriculture Organization of the United Nations; 2005. Available from: http://www.fao.org/3/y5686e/y5686e00.htm#Contents

[15] Chitando E. Religion, spirituality, and the environment in Africa. In: Wiley Blackwell Companion to Religion and Ecology. Chichester: Wiley-Blackwell; 2018. pp. 321-332

[16] Dube MW. Mother earth, gender and biblical imagination. In: Mother Earth, Mother Africa and Biblical Studies, Bible in Africa Series (BIAS), University of Bamberg Press. Vol. 14. 2021. p. 234

[17] Chidumayo EN. Gender issues in African forestry: A review of key issues

and case studies. International Forestry Review. 2019;**18**(S2):25-42

[18] Gumbo D, Tumushime M, Chaminuka P. Impacts of climate change on food security and livelihoods in Africa: A review. Agriculture and Food Security. 2019;**8**(8):15

[19] Nabeela F, Azizulla A, Bibi R, Uzma S, Murad W, Shakir SK, et al. Microbial contamination of drinking water in Pakistan-a review. Environmental Science and Pollution Research. 2014;**21**:13929-13942

[20] FAO. The State of Food Security and Nutrition in the World. Rome: Food and Agriculture Organization of the United Nations; 2017. Available from: https://www.fao.org/3/I7787e/I7787e.pdf

[21] Bergmann S. Religion and Environmentalism: An Exploration of Environmental Engagement Among Religious Communities. New York: Routledge; 2016

[22] Hitzhusen GE, Tucker ME. Communicating climate change: The role of religious environmentalism in science, policy and practice. Journal for the Study of Religion, Nature and Culture. 2013;**7**(3):294-325

[23] Chirongoma S, Karanga-Shona. Rural Women's Agency in Dressing Mother Earth: A contribution towards an indigenous eco-feminist theology. Journal of Theology for Southern Africa (Essays in Honour of Steve de Gruchy). 2012;**142**:120-144

[24] Dube MW. Women bishops in southern Africa: Gendering Anglican ministry. Studia Historiae Ecclesiasticae. 2016;**42**(2):1-15

[25] Chirongoma S. Where earth and water meet: Exploring the impact of Tokwe-Mukosi dam in light of African spirituality and religion in Zimbabwe. In: Mendoza L, Zachariah G, editors. Decolonizing Ecotheology: Indigenous and Subaltern Challenges. Eugene: Wipf & Stock; 2022. pp. 146-161

[26] Chirongoma S, Chitando E. What did we do to our mountain?: African eco-feminist and indigenous responses to cyclone Idai in Chimanimani and Chipinge districts, Zimbabwe, in African. Journal of Religion and Gender. 2021;**27**(1):65-90

[27] Chirongoma S. Lament for the Chimanimani Community in Zimbabwe in the aftermath of cyclone Idai. In: Malcom H, editor. Words for a Dying World: Stories of Grief and Courage from the Gobal Church. London: SCM; 2020. pp. 92-97

[28] Thunberg G. No One Is Too Small to Make a Difference. Stockholm: Penguin Books; 2019

[29] Aragon J, Miller M. Global Women's Issues: Women in the World Today, Extended Version. United States department of State: Bureau of International Information Programs; 2012

[30] Shiva V. Earth Democracy: Justice, Sustainability, and Peace. London: Zed Books; 2016

[31] Maathai W. The Green Belt Movement: Sharing the Approach and the Experience. New York: Lantern Books; 2004

[32] Mijeno A. Sustainable lighting solutions for rural communities. In: Advances in Sustainable Energy and Fuels. New York: Springer; 2019. pp. 149-161

Chapter 11

Trends of Climate Variability over Two Different Eco-Regions of Ethiopia

Mohammed Gedefaw

Abstract

This paper investigated the trends of precipitation and temperature in two Eco-regions of Ethiopia. The climate trends were examined using MK, Sen's slope estimator test, and ITAM. The findings of the study revealed that the trends of precipitation showed a significant increasing trend in Gondar and Bahir Dar stations. However, the trend in Sekoru station showed slightly decreasing trend at highland eco-regions concerned. On the other hand, in lowland eco-regions a significant increasing trend was also observed in Gewane and Negele stations. However, the trend in Degahabur station showed a sharp decrease. As far as the trends of temperature are concerned, a statistically significant increasing trend was observed with Z > 4 in Gondar and Bahir Dar stations and a statistically sharp significant decreasing trend in Sekoru station of highland eco-regions. However, all stations of lowland eco-regions Gewane, Degahabur, and Negele show a statistically significant increasing trend with Z > 4. The consistency in precipitation and temperature trends over the two eco-regions confirms the robustness of the change in trends. The results of this study could help researchers, water resources managers, and decision-makers to understand the trends of climatic variables over the study eco-regions and become a base for further studies.

Keywords: trend analysis, precipitation, temperature, eco-regions, Ethiopia

1. Introduction

Increasing frequency and intensity of extreme climatic events have got widespread acceptance across the globe and are subject to variations associated with the climate change [1]. Changes in precipitation have a direct impact on floods, droughts, and water resources [2]. Thus, it is essential to study the variations of precipitation and temperature characteristics in the context of climate change. Predicting the trends of climatic variables are a prerequisite for water resources planning and management for sustainable development [3, 4].

Nowadays, the global hydrological cycle has been responding to observed global warming, which includes an increasing atmospheric water vapor content and changing precipitation pattern [5]. The precipitation and temperature vary with latitude, elevation, and physiography especially precipitation decreasing from south to north of the country [6].

The climate of Ethiopia is described by the statistical interpretation of precipitation and temperature data recorded over a long period of time. The regional and global changes in the weather systems and the topographic variation along with the seasonal cycles are responsible for the spatial variability of rainfall in the country [6–10]. The change in climatic variables such as precipitation and temperature can provide important geographic, environmental, resource, agronomic, economic, and sociological effects in the country [11].

Trend analysis of climatic variables is very essential to understand the climate system of a given area and become a vital research area for many researchers across the world [1, 8, 12–16]. For example, Mekasha et al. [17] analyzed the temperature and rainfall extremes for 11 stations and found a general tendency of increasing warm and decreasing cold extremes, whereas trends in precipitation extremes were much more variable and increasingly inconsistent among different stations. Bewket and Conway [18] also reported the temporal and spatial variability of precipitation for a relatively small portion of the country and found no consistent pattern in daily rainfall trends. Asfaw et al. [7] investigated the variability and time series trend analysis of rainfall and temperature in the Woleka sub-basin, Northcentral Ethiopia, and found a decreased annual rainfall, belg and kiremt rainfall, and [19] also found a significant increasing trend of annual and seasonal precipitation in Northwest China from 1960 to 2013. Furthermore, Refs. [20–23] reported the trends of spatial and temporal variability of climatic variables across different stations.

No study has been conducted so far on the trends of precipitation and temperature in two eco-environments of the country. Therefore, the output of this paper will provide insights for concerned body for future sustainable economic development, especially water resources management. The MK, Sen's slope estimator test, and innovative trend analysis (ITAM) methods were used to detect the trends of precipitation and temperature to confirm whether there is a change in climate.

This study aims to investigate the trends and changes in climatic variables of the two-eco-environments from 1980 to 2016. The objectives of the study are: (i) to analyze climatic trends, (ii) to investigate the temporal variation in precipitation and temperature, and (iii) to assess the correlation between precipitation and temperature.

2. Materials and methods

2.1 Study area

Ethiopia is found in East Central Horn of Africa lying between 3° and 15° North latitude, and 33°–48° East longitude. It has a total area of about 1.12 million km^2 and comprises 12 river basins with varying sizes and water resource potential [6] (**Figure 1**). The country is characterized by diversified climate due to its equatorial positioning and topography. Its climate is controlled by the seasonal migration of the intertropical convergence zone (ITCZ), atmospheric circulations, complex physiography, and the marked contrast in elevation [6]. The country is mainly divided into two eco-environments namely lowlands and highlands ranging from extreme heat at one of the lowest places the Danakil Depression to the highest places (4543 m) above sea level at Ras Dejen [24]. The rainfall also showed seasonal and inter-annual variability [6].

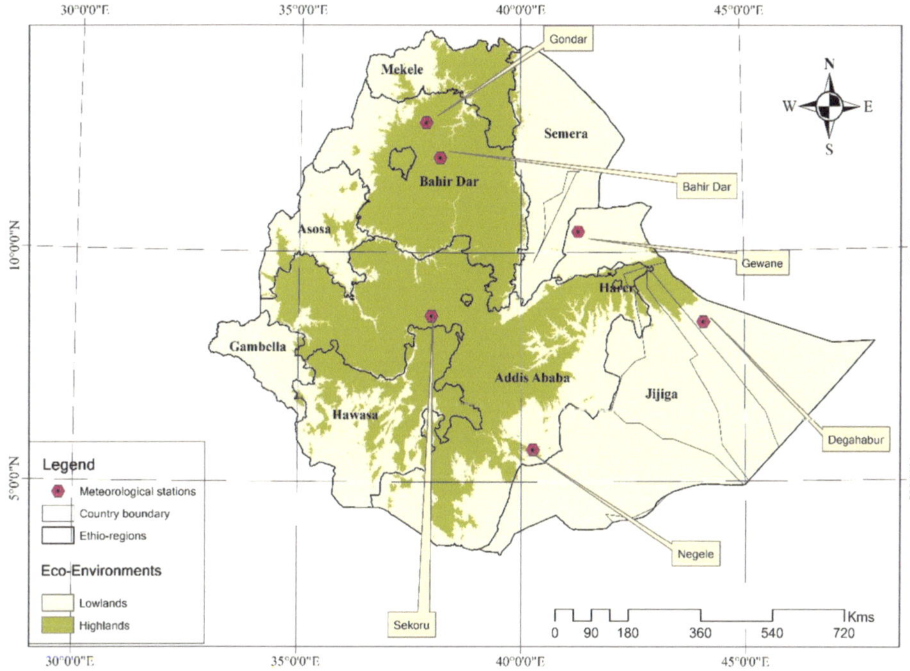

Figure 1.
Location map of the study area.

2.2 Data sources

Daily precipitation and temperature data were collected from 1980 to 2016 from the National Meteorological Services Agency of Ethiopia (NMSA). As the data series from 1980 to 2016 are completed, the observed precipitation and temperature data were selected as the basic analysis data in this study. We have selected six stations from two eco-environments for this study. By extending the individual missing data in this period with the interpolation method commonly used in hydrometeorology, the basic data was further reorganized to obtain the temperature and annual precipitation of each selected station. The stations were selected based on the length of a record period and the relative completeness of the data (**Table 1**).

2.3 Methods

The nonparametric Mann-Kendall test is used to detect the trends of hydro-climatic changes [25]. This study used 10%, 5%, and 1% significance levels to evaluate hydro-climatic changes.

2.3.1 Mann: Kendall trend detection (MK)

The Mann–Kendall (MK) test is used to detect the trends of hydro-climatic changes, which do not require the data series to be normally distributed [5, 20, 26]. The test statistic (S) is equated as follows:

Stations	Elevation (m)	Latitude (°N)	Longitude (°E)	Eco-environments
Gondar	1973	37.4319	12.5212	Highland
Bahir Dar	1827	37.322	11.6027	Highland
Sekoru	1928	37.4167	7.9167	Highland
Gewane	568	40.633	10.15	Lowland
Degahabur	1070	43.55	8.2167	Lowland
Negele	1544	39.5667	5.4167	Lowland

Table 1.
Basic information about meteorological stations.

$$S = \sum_{i=1}^{n-1} \sum_{j=i+1}^{n} sgn\left(X_j - X_i\right) \tag{1}$$

The trend test is applied to X_i data points (i = 1, 2, … , n−1) and X_j (j = i + 1, 2, … , n).

$$sgn\left(X_j - X_i\right) = \begin{cases} +1 \text{ if } \left(X_j - X_i\right) > 0 \\ 0 \text{ if } \left(X_j - X_i\right) = 0 \\ -1 \text{ if } \left(X_j - X_i\right) < 0 \end{cases} \tag{2}$$

where X_i and X_j are the values in periods i and j.

The variance Var(S) is calculated using the given equation below [27]:

$$Var(S) = \frac{n(n-1)(2n+5) - \sum_{k=1}^{m} t_k(t_k - 1)(2t_k + 5)}{18} \tag{3}$$

where m is the number of the tied groups in the time series, and t_k is the number of data points in the kth tied group. The test statistic Z is as follows:

$$Z = \begin{cases} \frac{S-1}{\delta} \text{ if } S > 0 \\ 0 \text{ if } S = 0 \\ \frac{S+1}{\delta} \text{ if } S < 0 \end{cases} \tag{4}$$

In time sequence, the statistics are defined independently:

$$UF_k = \frac{d_k - E(d_k)}{\sqrt{var(d_k)}} \ (K = 1, 2, \dots, n) \tag{5}$$

The time series is arranged in reverse order to equate it

$$UB_k = -UF_k \tag{6}$$

$$K = n + 1 - k \tag{7}$$

Finally, UB_k and UF_k are drawn as UB and UF curves [28].

2.3.2 Sen's slope estimator test

Slope estimator test is used to predict the magnitude of the trends as used by many researchers [29–31]. The slope (Q_i) between two data series is equated as follows:

$$Q_{i=} = \frac{X_j - X_k}{\mathrm{j} - \mathrm{k}}, for\ i = 1, 2, \dots, N \tag{8}$$

Then, the median of slope (β) is calculated as:

$$\beta = \begin{cases} Q[(N+1)/2] & \text{when N is odd} \\ Q[(N/2) + Q(N+2)/(2)/(2)] & \text{when N is even} \end{cases} \tag{9}$$

2.3.3 Innovative trend analysis method (ITAM)

The ITAM has been used by various climate scientists to investigate the trends across the globe [5, 20, 26]. In ITAM, the hydrometeorological observations were classified into two classes and then the data points were arranged independently in increasing order. Then, the two classified classes are placed on a coordinate system (X_i: i = 1, 2, 3, ..., n/2) on X-axis and (X_j: j = n/2 + 1, n/2 + 2, ..., n) on Y-axis.

This study also used ITAM to detect the trends using the following formula [5]:

$$\phi = \frac{1}{n} \sum_{i=1}^{n} \frac{10\left(X_j - X_i\right)}{\mu} \tag{10}$$

where, ϕ = trend indicator, n = number of observations in the subseries, X_i = data series in the first half subseries class, X_j = data series in the second half subseries class, and μ = mean of data series in the first half subseries class.

3. Results

3.1 Analysis of mean annual precipitation

From 1980 to 2016, the mean annual precipitation of the study area was found to be 834.97 mm, with a coefficient of variation (CV = 15%) and standard deviation of 122.27 mm. The minimum and maximum-recorded rainfalls were 509.93 and 1015.90 mm per year, respectively. An increase in precipitation was observed in 2000, 2005, 2007, 2010, and 2013 with a coefficient of determination for the trend line (R^2 = 0.01) and a sharp decreasing trend in 1992 (**Figure 2**). The areas with mean annual precipitation of more than 650 mm were located at the highland eco-environments of the country (Gondar, Bahir Dar, and Sekoru), which accounts for about 20.30% of lowland eco-environments (Gewane, Degahabur, and Negele).

3.2 Analysis of trend of precipitation

The MK curve annual precipitation (UF and UB = Changing Parameters) shows the trends of precipitation in highland and lowland eco-environments of the study area. A significant increasing trend was observed in Gondar from 1999 to 2012 (Z = 1.69), a

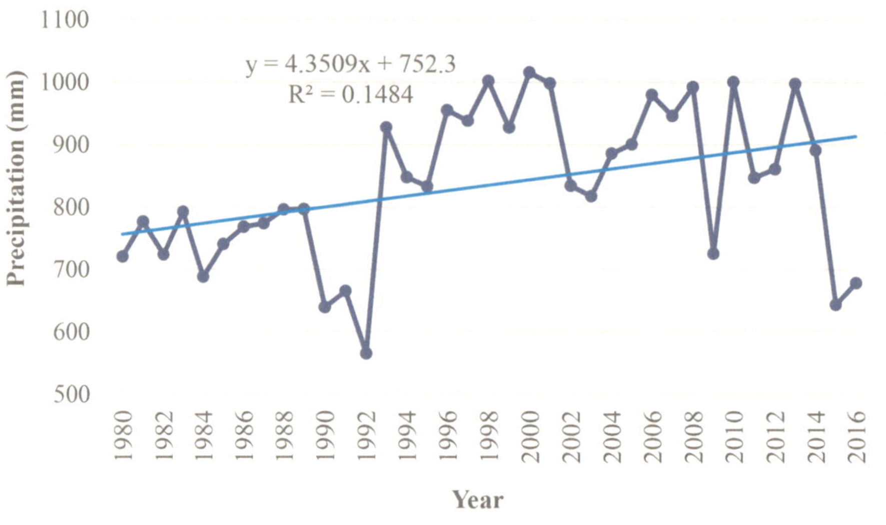

Figure 2.
Mean annual precipitation of the study area.

significant increasing trend in Bahir Dar from 1994 to 2014 (Z = 0.72), and a slightly decreasing trend was observed in Sekoru from 1983 to 2016 (Z = 0.45) stations. On the other hand, in lowland eco-environments, a significant increasing trend was observed from 1994 to 2006 and 2006 to 2015 in Gewane (Z = 0.80) and Negele (Z = 0.72) stations, respectively. However, a sharp decreasing trend was observed in Degahabur station (Z = 0.30) (**Figure 3**).

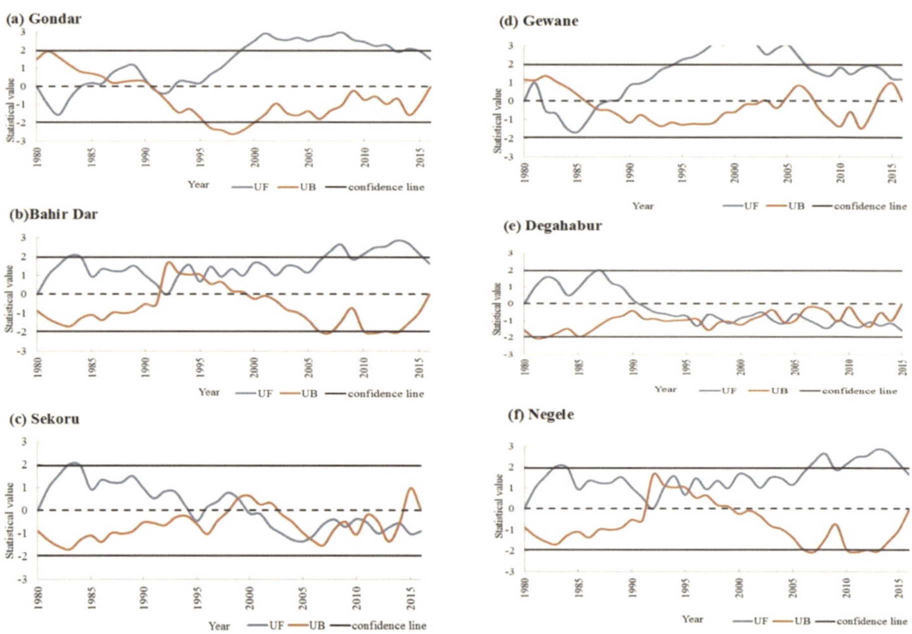

Figure 3.
Trends of annual precipitation across stations (note: UB = −UF).

No.	Stations	Z	φ	β
1	Gondar	1.69**	0.54	1.84**
2	Bahir Dar	−0.07*	−23.51	1.80*
3	Sekoru	1.37	0.21	0.01
4	Gewane	5.59**	0.69	0.10**
5	Degahabur	0.30	−0.56	4.13
6	Negele	0.72**	−0.03	23.40**

Trends at 0.1 level. Trends at 0.05 level.*

Table 2.
Results of MK, ITAM, and Sen's slope estimator test for precipitation.

The trend analysis of precipitation in two eco-environments using the MK test, ITAM, and Sen's slope estimator test results are presented in **Table 2**. Significance levels at $\alpha = 0.05$, $\alpha = 0.1$ were taken to detect the trends in all stations.

3.3 Analysis of mean annual temperature

The maximum and minimum mean annual temperature of the study was found to be 30.35 and 27.92°C, respectively. Whereas, the annual mean temperature was 29.16°C from 1980 to 2016. The temperature showed a dramatic increasing trend in 2010 and 2015 with a coefficient of determination for the trend line ($R^2 = 0.67$) and a decreasing trend in 1989 (**Figure 4**). The highest temperature was recorded at the lowland eco-environments (Gewane, Degahabur and Negele). Whereas, a slightly lower temperature was observed in highland eco-environments (Gondar, Bahir Dar, and Sekoru).

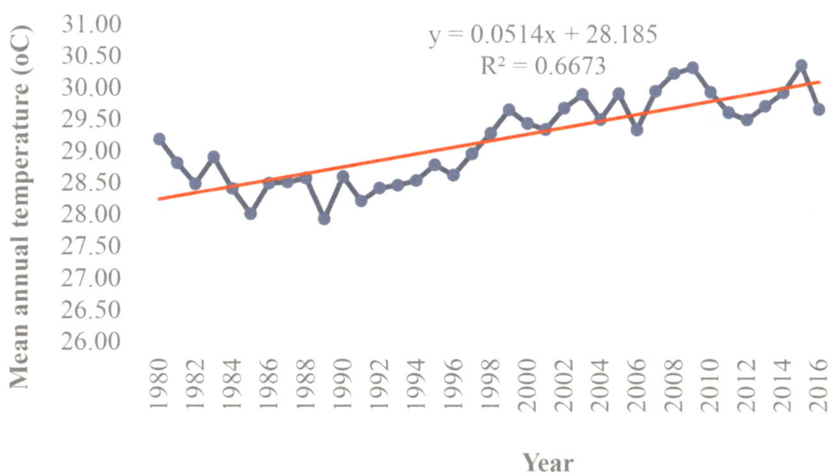

Figure 4.
Mean annual temperature of the study area.

3.4 Trend analysis of temperature

The statistical test result of this study showed that the trends of temperature in Gondar (Z = 5.68) and Bahir Dar (Z = 7.59) stations showed a statistically significant increasing trend and a statistically sharp significant decreasing trend in Sekoru (Z = 1.37) station of highland eco-environments. However, all stations of lowland eco-environments (Gewane (Z = 5.59), Degahabur (Z = 4.78), and Negele (Z = 8.01)) showed a statistically significant increasing trend (**Figure 5**). The trend test results are presented in **Table 3**.

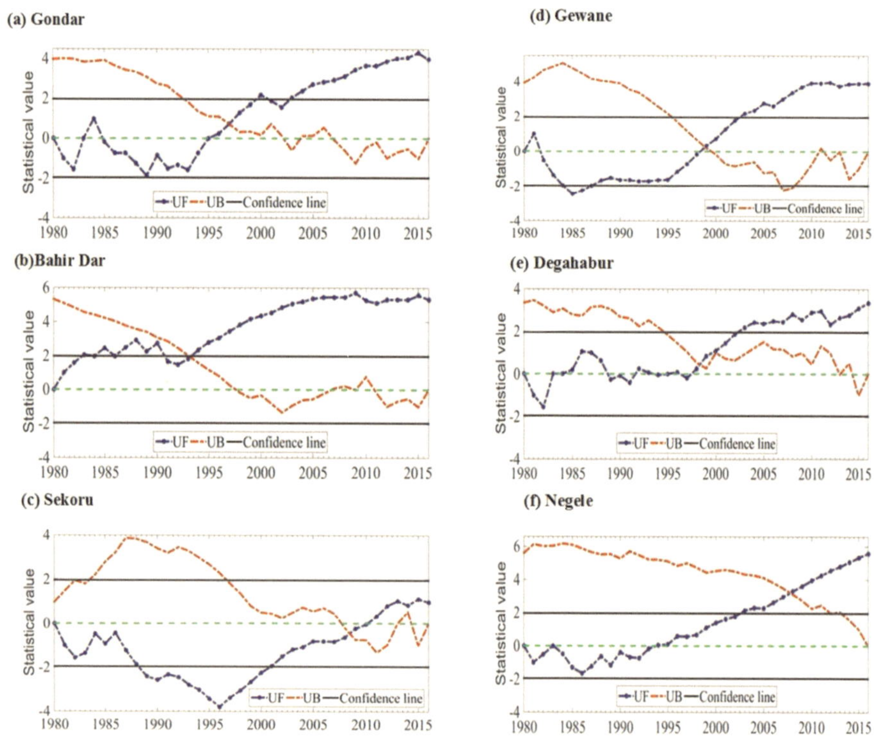

Figure 5.
Trends of annual temperature across stations (note: UB = −UF).

No.	Stations	Z	Φ	β
1	Gondar	5.68**	0.35	0.04**
2	Bahir Dar	7.59**	0.62	0.08**
3	Sekoru	1.37**	0.21	0.01**
4	Gewane	5.59**	0.69	0.10**
5	Degahabur	4.78*	0.18	0.03*
6	Negele	8.01*	0.48	0.07*

Trends at 0.1 level. Trends at 0.05 level.*

Table 3.
Results of MK, ITAM, and Sen's slope estimator test for temperature.

3.5 Correlation analysis between precipitation and temperature

The high correlations were expected, as one can reasonably assume a cause-and-effect relationship between precipitation and temperature when considering the average annual values. The correlation between precipitation and temperature was found to be very high (R^2 = 0.99) in this study. The correlation between precipitation and temperature is shown in **Figure 6**.

4. Discussion

Ethiopia is characterized by inter-annual variability of rainfall and temperature. The rainfall variability affects crop production in most parts of the country. There is also significant spatial variation in the mean range of temperature. This change may express the diurnal variation and demand atmospheric moisture. The findings of this study showed that the frequency and intensity of warm temperature extremes have increased, while cold temperature extremes decreased in most stations.

The mean annual precipitation and temperature of the study eco-regions showed a consistently increasing trend. We have noticed that the correlation between precipitation and temperature was very strong in this study. It was confirmed that precipitation is mainly caused by cold summer, and thus correlates to a large extent with temperature in the study area.

The studies eco-regions are characterized by maximum precipitation in kiremt (June to August) season. There is fluctuation of precipitation and temperature over the stations in each eco-region. The trend detection methods showed that decreasing and increasing trends of precipitation and temperature were observed across the two eco-regions. The results of this study are consistent with [5–8, 17, 20, 32, 33]. The change of these climatic variables across the stations during the study period (1980–2016) could be associated with human activities and climate change.

The analysis of the correlations between precipitation and temperature shows a coherent pattern of relationship. The precipitation is marked with blue and the temperature with red colors as shown in (**Figure 7**).

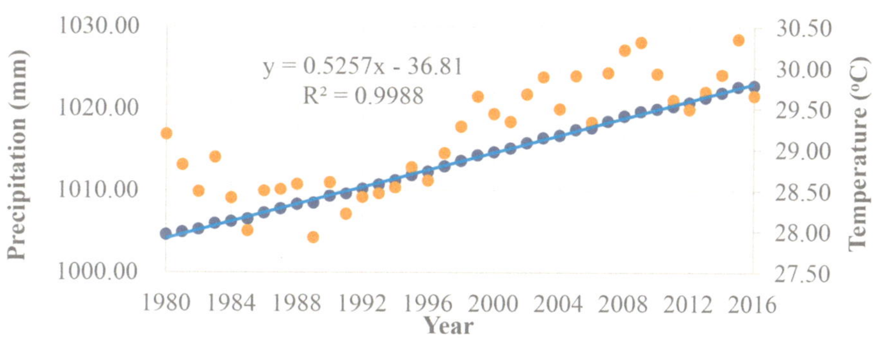

Figure 6.
Correlation coefficient between precipitation and temperature.

15

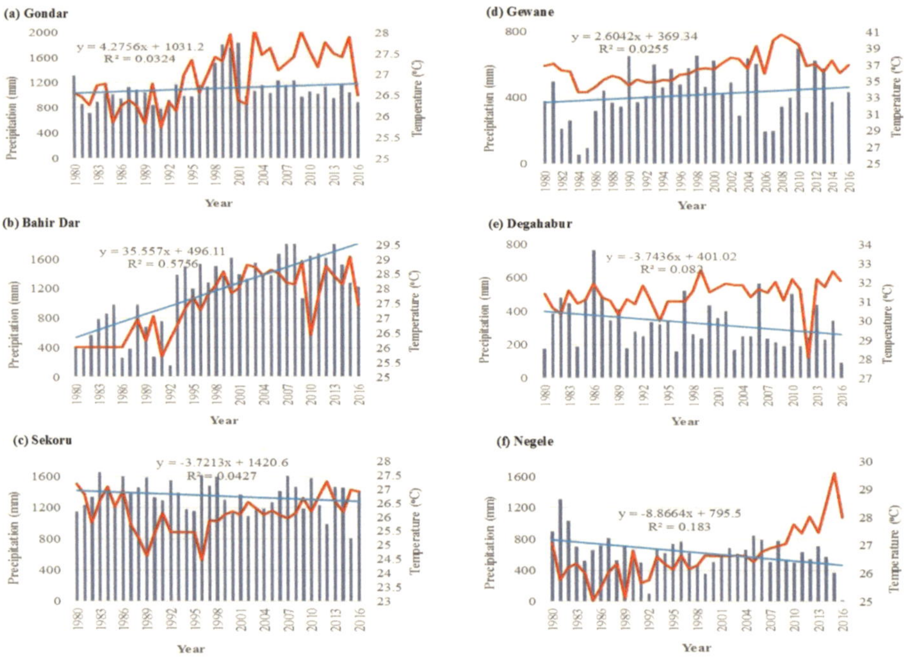

Figure 7.
Long-term relation between precipitation and temperature.

The changes in water vapor divergence for the rainy season (June to September) exhibited a regional increasing drying trend in the country from 1980 to 2016, which was consistent with the variation in consecutive dry days.

5. Conclusions

This paper investigated the trends of precipitation and temperature in two eco-environments of Ethiopia from 1980 to 2016. The temporal variability of precipitation and temperature was also investigated in all stations. A significant increasing trend of precipitation was observed in highland eco-environments (Gondar and Bahir Dar stations), whereas a slight decreasing trend was detected in Sekoru stations.

As far as temperature trends are concerned, statistically significant increasing trends were detected in two stations of highland eco-environments. Whereas, in lowland eco-environments, all stations show a statistically significant increasing trend from 1980 to 2016.

The increasing and decreasing of both precipitation and temperature across all stations during the study period is probably due to human activities and periodic drought. Further studies should be conducted to confirm the change of climate over the two eco-regions by increasing the sample meteorological stations. Other studies on indices such as rainfall intensity and frequency of wet and hot days are also recommended. This study only provides the trends of precipitation and temperature over the study region. This could be very essential to predict the climate condition of the study area and sustainable development.

6. Recommendations

Further study should be conducted by taking more meteorological stations to clearly see the climate conditions of the two eco-regions in Ethiopia.

Acknowledgements

The authors would like to thank the National Meteorological Service Agency of Ethiopia for providing the raw meteorological data and the National Key Research and Development Project (Grant No. 2016YFA0601503) for financing this work.

Conflict of interest

The authors declare no conflict of interest.

Author details

Mohammed Gedefaw
College of Agriculture and Environmental Sciences, University of Gondar, Gondar, Ethiopia

*Address all correspondence to: mohammedgedefaw@gmail.com

IntechOpen

© 2023 The Author(s). Licensee IntechOpen. This chapter is distributed under the terms of the Creative Commons Attribution License (http://creativecommons.org/licenses/by/4.0), which permits unrestricted use, distribution, and reproduction in any medium, provided the original work is properly cited.

References

[1] Pingale SM, Khare D, Jat MK, Adamowski J. Spatial and temporal trends of mean and extreme rainfall and temperature for the 33 urban centers of the arid and semi-arid state of Rajasthan, India. Atmospheric Research. 2014;**138**: 73-90

[2] Wen X, Wu X, Gao M. Spatiotemporal variability of temperature and precipitation in Gansu Province (Northwest China) during 1951–2015. Atmospheric Research. 2017; **197**:132-149

[3] Buchori I et al. A predictive model to assess spatial planning in addressing hydro-meteorological hazards: A case study of Semarang City, Indonesia. International Journal of Disaster Risk Reduction. 2018;**27**:415-426

[4] Wang H et al. Hydro-climatic trends in the last 50 years in the lower reach of the Shiyang River Basin, NW China. Catena. 2012;**95**:33-41

[5] Gedefaw M et al. Trend analysis of climatic and hydrological variables in the awash river basin, Ethiopia. Water. 2018;**10**(11):1-14. DOI: 10.3390/w10111554

[6] Seleshi Y, Zanke U. Recent changes in rainfall and rainy days in Ethiopia. International Journal of Climatology. 2004;**24**:973-983

[7] Asfaw A, Simane B, Hassen A, Bantider A. Variability and time series trend analysis of rainfall and temperature in northcentral Ethiopia: A case study in Woleka sub-basin. Weather and Climate Extremes. 2017;**19**: 29-41. DOI: 10.1016/j.wace.2017.12.002

[8] Tsidu GM. High-resolution monthly rainfall database for Ethiopia: Homogenization, reconstruction, and gridding. Journal of Climate. 2012;**25**: 8422-8443

[9] Yenehun A, Walraevens K, Batelaan O. Spatial and temporal variability of groundwater recharge in geba basin, Northern Ethiopia. Journal of African Earth Sciences. 2017;**134**:198-212. DOI: 10.1016/j.jafrearsci.2017.06.006

[10] Behailu S, Melesse AM, Bhat MG, Mcclain ME. Catena assessment of water resources availability and demand in the Mara River Basin. Catena. 2014;**115**: 104-114

[11] Gong B et al. Variation of hydrothermal conditions under climate change in Naqu Prefecture, Tibet Plateau, China. International Journal of Environmental Research and Public Health. 2018;**15**:2271

[12] Lyon SW, King K, Polpanich O, Lacombe G. Journal of hydrology: Regional studies assessing hydrologic changes across the Lower Mekong Basin. Journal of Hydrology: Regional Studies. 2017;**12**:303-314

[13] Fathian F, Aliyari H. Temporal trends in precipitation using spatial techniques in GIS over Urmia Lake Basin, Iran Farshad Fathian * and Hamed Aliyari Ercan Kahya Zohreh Dehghan. International Journal of Hydrology Science and Technology. 2016;**6**:62-81

[14] Hamisi J. Study of rainfall trends and variability over tanzania a research project submitted in partial fulfilment of the requirements for the postgraduate diploma in meteorology university of nairobi. Research Project. 2013

[15] Girma E, Tino J, Wayessa G. Rainfall trend and variability analysis in Setema-Gatira area of Jimma, Southwestern Ethiopia. African Journal of Agricultural Research. 2016;**11**:3037-3045

[16] Panthi J et al. Spatial and temporal variability of rainfall in the gandaki river basin of nepal Himalaya. Climate. 2015;**3**: 210-226

[17] Mekasha A, Tesfaye K, Duncan AJ. Trends in daily observed temperature and precipitation extremes over three Ethiopian eco-environments. International Journal of Climatology. 2014;**34**:1990-1999

[18] Bewket W. Soil and water conservation intervention with conventional technologies in northwestern highlands of Ethiopia: Acceptance and adoption by farmers. Land Use Policy. 2007;**24**:404-416

[19] Yang P, Xia J, Zhang Y, Hong S. Temporal and spatial variations of precipitation in Northwest China during 1960–2013. Atmospheric Research. 2017; **183**:283-295

[20] Gedefaw M, Yan D, Wang H, Qin T, Girma A, Abiyu A, et al. Innovative trend analysis of annual and seasonal rainfall variability in amhara regional state, Ethiopia. Atmosphere. 2018;**9**:326

[21] Toride K, Cawthorne DL, Ishida K, Kavvas ML, Anderson ML. Science of the Total Environment Long-term trend analysis on total and extreme precipitation over Shasta Dam watershed. Science of the Total Environment. 2018;**626**:244-254

[22] Theobald A, Mcgowan H, Speirs J. Trends in synoptic circulation and precipitation in the Snowy Mountains region, Australia, in the period 1958–2012. Atmospheric Research. 2016;**169**: 434-448

[23] Mbungu W, Ntegeka V, Kahimba FC, Taye M, Willems P. Temporal and spatial variations in hydro-climatic extremes in the Lake Victoria basin. Physics and Chemistry of the Earth. 2012;**50–52**:24-33

[24] Wondie M, Schneider W, Melesse AM, Teketay D. Spatial and temporal land cover changes in the simen mountains national park, a world heritage Site in northwestern Ethiopia. Remote Sensing. 2011;**3**:752-766

[25] Fersi W, Lã©zine AM, Bassinot F. Hydro-climate changes over southwestern Arabia and the Horn of Africa during the last glacial–interglacial transition: A pollen record from the Gulf of Aden. Review of Palaeobotany and Palynology. 2016;**233**:176-185

[26] Basin LB et al. Observed trends of climate and river discharge in mongolia' s selenga sub-basin of the lake Baikal basin. Water. 2018;**10**(10):1-18. DOI: 10.3390/w10101436

[27] Ma X, He Y, Xu J, Noordwijk M, Van – Lu X. Catena Spatial an temporal variation in rainfall erosivity in a Himalayan watershed. Catena. 2014;**121**: 248-259

[28] Zhang Q, Sun P, Singh VP, Chen X. Spatial-temporal precipitation changes (1956-2000) and their implications for agriculture in China. Global and Planetary Change. 2012;**82–83**:86-95

[29] Martinez CJ, Maleski JJ, Miller MF. Trends in precipitation and temperature in Florida, USA. Journal of Hydrology. 2012;**452–453**:259-281

[30] Gu X, Zhang Q, Singh VP, Shi P. Changes in magnitude and frequency of

heavy precipitation across China and its potential links to summer temperature. Journal of Hydrology. 2017;**547**:718-731

[31] Laurent JS, Mazumder A. Influence of seasonal and inter-annual hydro-meteorological variability on surface water fecal coliform concentration under varying land-use composition. Water Research. 2013;**48**:170-178

[32] Nkiaka E, Nawaz NR, Lovett JC. Analysis of rainfall variability in the Logone catchment, Lake Chad basin. International Journal of Climatology. 2017;**37**:3553-3564. DOI: 10.1002/joc.4936

[33] Tekleab S, Mohamed Y, Uhlenbrook S. Hydro-climatic trends in the Abay/Upper Blue Nile basin, Ethiopia. Physics and Chemistry of the Earth. 2013;**61–62**:32-42

Section 4

Climate Change: Food Production and Environmental Factors

Chapter 12

Nature and Extent of Air Pollution and Climate Change Related Stresses on Cocoyam Production in Nigeria

Dennis Mark Onuigbo, NwaJesus Anthony Onyekuru, Anthonia Ifeyinwa Achike, Chinasa Onyenekwe and Eric Eboh

Abstract

There has been a dramatic decline in cocoyam production in Nigeria in recent years due to climate related stressors. We investigated the nature and extent of these impacts and the resilience building strategies used by cocoyam farmers in Nigeria. Data were collected from the farmers in 2010 and 2017 and from FAO statistical database. Results show a yield decline from 8mt/ha in 2005 to 4mt/ha in 2016, due to acid rain, causing bleaching of cocoyam leaves, die back, smaller tubers, early decay of the tubers and decline in the quality and taste of the tubers. Kinds of climate change impact include unusual early rains followed by weeks of dryness, delay onset of rain, long period of dry season, higher temperature, drought and high rate of disease incidence. Resilient building strategies adopted by the farmers were planting deeper or shallower, processing tubers, intensive manure application, multiple cropping and mixed farming.

Keywords: cocoyam, acid rain, climate change, farmers, abiotic stress, adaptation

1. Introduction

Cocoyam refers to two members of the Araceae family (Colocasia esculentum (L.) Schott) and Xanthosoma sagittifolium (L.) Schott) which are staple foods for many people in developing countries in Africa, Asia and the Pacific and the other parts of tropical regions of the world [1, 2]. It is mainly grown for its starchy corms although the leaves are also consumed as vegetables or medicine [3]. They are stem tubers that are widely used for human food, providing food for over 400 million people around the world [4]. Cocoyam has been reported to be the third most important staple root/tuber crop after yam and cassava in Nigeria, second to cassava in Cameroon and first in Ghana [5, 6]. In terms of volume of production, the global production in 2019 stood at 10.54 million tonnes; Nigeria is the largest producer of cocoyam globally, accounting for about 27.14% of the total production with Cameroun, China and Ghana ranked second, third, and fourth [7, 8]. FAO statistical data [7] show that total output of cocoyam in Nigeria for 2019 is close to about 2.86 million tonnes and yield of about 2.88tons/ha, thus underscoring the potential of the crop in meeting the Nigeria food security gap. Cocoyam is commonly grown among small scale farmers who operate within the subsistence economy in most parts of West and Central Africa [1, 9]. It is a widely cultivated crop in both western and eastern region of Nigeria in terms of the area devoted to it and the number of farmers growing it. Indeed, it is a food security crop as almost every household in these regions grows the crop.

In terms of its agronomy, cocoyam thrives well in warm temperatures above 21°C and low temperatures not less than 10°C. It has high moisture requirement and adapts to tropical regions with adequate rainfall distribution. In many countries, cocoyam is often planted under permanent plantations like banana, coconut, citrus, oil palm, cocoa and other agroforestry systems in order to meet the temperature and rainfall requirements [10]. Cocoyam thrives better on a well-drained sandy loamy soil. It produces optimally when planted in fertile soil with a good water retention capacity. Compared to grains, cocoyam is more tolerant in low soil fertility and more resistant to drought, pests and diseases and its roots are storable in the ground for months after they mature [11]. In the traditional farming system where labour is a very limiting factor, cocoyam has endeared itself in the heart of farmers, This is because, in comparison with other root crops cocoyam require the least amount of labour input, thrive better in marginal soils, is shade tolorant and therefore most preffered for interplanting with trees. Thus, has become part of the traditional agroforestry system.

In spite of the many potentials and advantages of cocoyam production, less attention is paid to it when compared to cassava and yam as root crops. Skott et al. [12] and Talwana et al. [13] have identified that research on cocoyam has trailed behind that of other staples in Nigeria and has not received enough deliberate attention to address its research and development in all regional agricultural research centres and therefore, its contribution to food security and economy is underestimated. Authors such as Bello et al. [14], Eboh et al. [15] and Enibe et al. [16] have written on impact of climate change on agriculture, barriers to climate change adaptation in Nigeria, and cocoyam marketing. However, there is paucity of information on the nature and extent of climate change impact on cocoyam production. Chemura et al. [10] assessed the impacts and adaptation strategies to climate change in Zimbabwe. This study will be among the earliest attempts to investigate climate change impact on cocoyam production in Nigeria. The need for such information has become very important due to recent development in cocoyam productions; there is a dramatic declining trend in cocoyam production as well as a shortage of its supply in domestic markets as in recent years [7].

While this decline can be attributed to several causes such as poor cultural practices, inability to adopt new technologies or even pest and diseases; the nature of changes noticed by the farmers as will be discussed below tends to rule out some of the afore-mentioned causes of decline in productivity and suggests stresses occasioned by air pollution and climate change. This is because; cocoyam production is predominantly confined to the forest zones due to its high moisture requirements for optimum yield [17]. According to Taylor et al. [18] the agro-ecological range of cocoyam will be altered under climate change because the crop thrives best under wet, humid environments with little or no ability to adapt to drought. This assertion was collaborated by a modelling study by Kodis et al. [19] who concluded that the potential for cocoyam production will be greatly reduced in Hawaii, especially

DOI: http://dx.doi.org/10.5772/intechopen.1002244

under high-emission scenarios. Thus, changes in the cocoyam output may occur given a change in the climate; reduced or delayed rainfall, which is frequently noticed in recent time. In any given scenario, a larger, faster or more radical change in the weather element may well result in vegetation stress, rapid plant loss (in this case fall in cocoyam output) [20]. However, a mild increase in precipitation and warmth will bring about an improved plant growth and thus cocoyam output and the subsequent sequestration of airborne CO_2. It was on this premise that this study was borne; to further investigate the nature and extent of these stresses on cocoyam production and resilience building strategies among cocoyam farmers in Nigeria.

2. Methodology

2.1 Study area

The study was conducted in Nigeria. Nsukka area of southeastern Nigeria (**Figure 1**) was specifically selected because of the high concentration of cocoyam production and high level of its utilization in the area. Nsukka has a total population of 309,633 people [21], and is pojected to be 4832,56 by 2023 [22]. Primary data were collected with the aid of structure questionnaire from the cocoyam farmers. The prominent climatic seasons in the area include rainy season, lasting from April to October and the dry season lasting from November to March, the average rainfall is 1579 mm per annum, whereas the average temperature is 24.9°C [23]. The vegetation of the area is derived savannah. The primary occupation of the people in the area is farming; and agricultural production constitutes a major source of income for a greater percentage of the population.

Cocoyam in Nsukka and in many areas of South Eastern Nigeria, is known to be the most important staple crop with the most varied local uses of all the root crops. This is because, apart from meeting immediate food need, it is always cooked, sliced and dried (called '*Echicha*') and preserved to last throughout the year till the next harvesting season and used to blend several other local food stuffs into different kinds of delicacies, to bridge the gap between time of abundance and scarcity of other food sources, thus it is refered to as a food security crop.

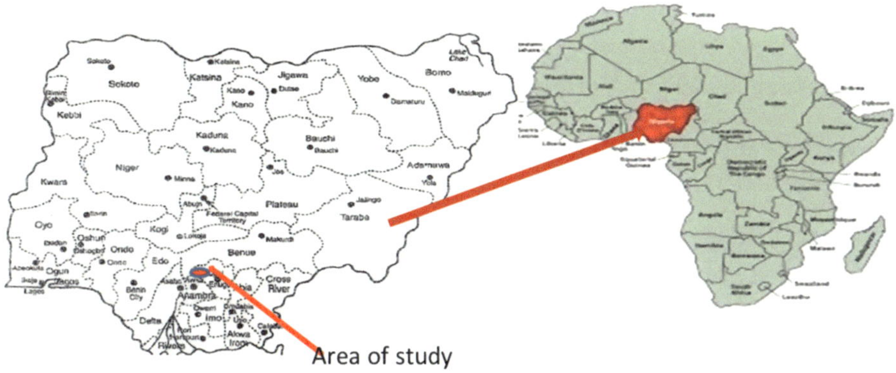

Figure 1.
Showing area of study in Nigeria.

2.2 Sampling procedure

Random sampling was adopted in the selection of respondents. The method involves random selection of three towns from the area. From the towns, 20 cocoyam farmers were randomly selected from a frame of cocoyam farmers prepared with the help of the local farmers' organizations in the towns, making a total of 60 farmers in 2010 and 60 in 2017.

2.3 Data collection

Data were collected with the aid of pre-tested questionnaire focusing on the socio-economic characteristics of cocoyam farmers, vulnerability of cocoyam production to climate change in the area, effects on productivity, adaptation strategies in use and impediments to adaptation. Secondary data of historical cocoyam yield per hectare in Nigeria from 1961 to 2016 were collected from FAO statistical data base.

2.4 Method of data analysis

Data were analyzed using descriptive statistics; tables, pie charts, bar chart, graphs and mean score (Likert-type scale rating). A Likert scale is a psychometric scale in survey research. When responding to a Likert questionnaire item, respondents specify their levels of agreement with respect to the extent of the impact of the different kinds of phenomena on their farming activities. The average extent agreed upon is determined by the summation of the different values of a 5-point Likert-type scale in the case of this study (To a very great extent = 5, To a great extent = 4, To some extent = 3, To a little extent = 2, and To no extent = 1). The mean score (MS) of the respondents was computed as: $(5 + 4 + 3 + 2 + 1)/5 = 3$ cut off point. Based on this, any score below 3 ($MS < 3$) is taken as a weak factor and may not be considered, while those with mean score above 3 ($MS > 3$) is taken as strong factor and considered to have impact on the farmers.

3. Results

This section presents the results of the study in different subsections as shown below.

3.1 Trends in cocoyam production

An analysis of the trends in cocoyam production is presented below. For the years 2007 to 2009 in the study area **Figure 2** shows a declining trend in output of cocoyam from 2007.

In 2009, output was 73tonnes less than that of year 2008 and about 105 tonnes lesser compared to year 2007. This result is in resonance with the trend analysis of cocoyam yield from FAO statistical database (**Figure 3**).

The result showed that the yield of cocoyam per hectare rose to an unprecedented pick in 2005 and started a gradual drop which continues to as low as 4mt ha^{-1}. More revealing is the fact that this result agrees with the result of the field study on the time the farmers started noticing the problems with cocoyam production, as is shown in **Figure 4**.

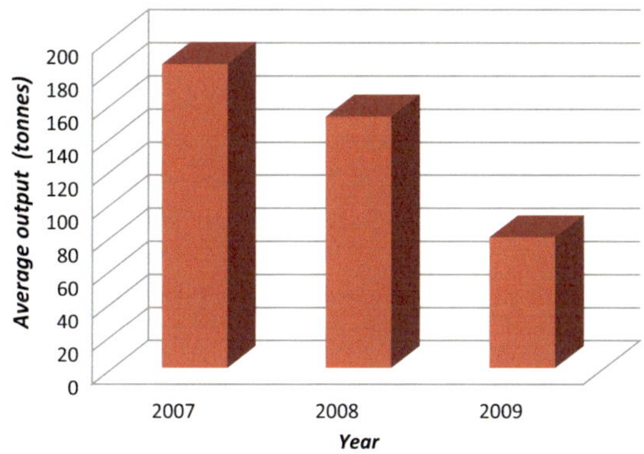

Figure 2.
Average output of cocoyam (tonnes) per farmer for years 2007, 2008, 2009. Source: Field survey, 2010.

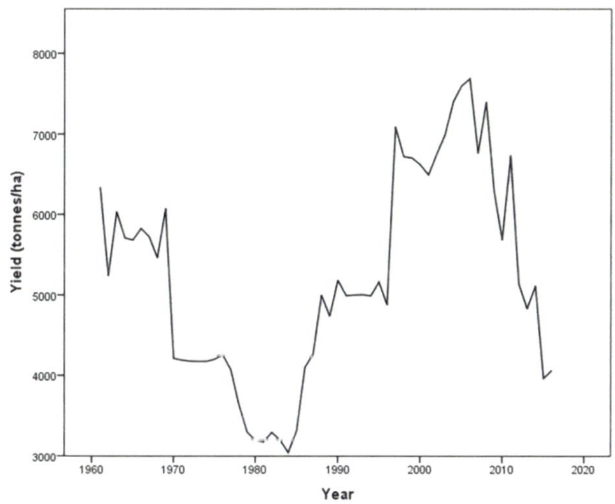

Figure 3.
Trend analysis of the yield of cocoyam in Nigeria (1960–2016).

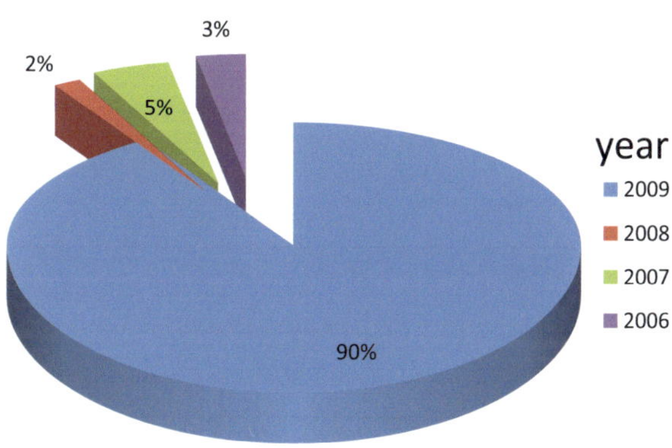

Figure 4.
Distribution of respondents according to year sudden decline started. Source: Field survey, 2010.

The result in **Figure 4** indicates that about 90% of the respondents indicated that the sudden decline in cocoyam output occurred in the year 2009. This finding resonates with the result from the trend analysis of the FAO data (**Figure 3**) and agrees with the period the problem of acid rain was reported in the country and therefore in agreement with farmers' attribution of the problem to acid rain. This finding is in agreement with those of Ume et al. [24] and Ukonze [25].

3.2 Nature of pollution and climate change impact on cocoyam

Analysis of the nature of pollution and climate change impact on cocoyam are as presented in **Table 1** and **Figures 5–7**.

Table 1 shows that about 75% of cocoyam farmer in the area said the unusual effect occurred during the period acid rain was reported in 2009, they noticed that leaves of cocoyam appeared as if hot water was poured on them after which they turned brown/yellow and gradually died back (**Figure 5**).

As a result, most of the cocoyam plants die prematurely in the field before they are matured for harvesting. This finding is not far different from that also found in more detailed analysis in 2017 about the nature of impact on the leaves and tubers of cocoyam, as in **Figures 6** and 7.

The results reinforce earlier findings indicating the issue of bleaching of the leaves of the cocoyam plant which results in the poor yield and small tubers (**Figure** 7).

3.3 Extent of climate change impact on cocoyam production

To verify if the afore-mentioned phenomena are associated with climate change, a detailed assessment of the different forms of climate change impacts was done in the study area and the result is presented in **Table 2**.

The result shows that the farmers in the area agreed that climate change has impacted on their production in different forms; including unusual early rains that are followed by weeks of dryness, delay in the unset of rain, long period dry season, less rainfall, higher temperature, drought, increase in pests' problems, high rate of disease incidence and so on. Furthermore, the farmers were asked to indicate whether these forms of climate change impact were decreasing, increasing or has not changed. This result is presented in **Table 3**.

How it occurred	Percentage (Frequency)
Noticed leaves of cocoyam appeared as if hot water was poured on it after which they turn brown/yellow and gradually eaten up from either the middle or from edges of the leaves until there was no leave left on the plant	75 (45)
Do not know	23.3 (14)
Evil spirits	1.7 (1)
Total	**100**

Source: Field Survey, 2010.

Table 1.
Distribution of respondents according to how unusual effect occurred.

(a)

(b)

Figure 5.
(a) Showing the effect of acid rain on cocoyam leaves (picture taken on 5th may 2010 in Nsukka, Nigeria). (b) Cocoyam tubers.

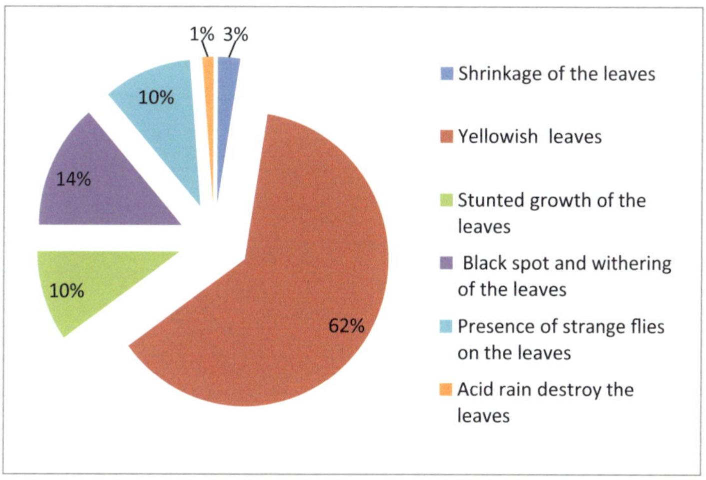

Figure 6.
Nature of effects on cocoyam leaves. Source: Field data 2017.

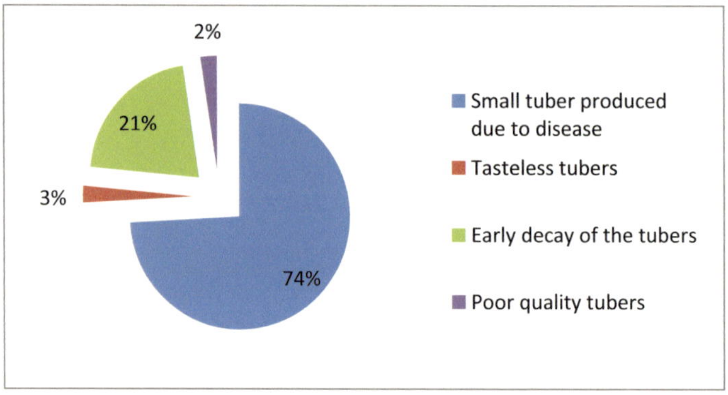

Figure 7.
Nature of climate change impacts on cocoyam tubers. Source: Field survey 2017.

Average extent of impact	Decision rule	
= (1 + 2 + 3 + 4 + 5)/5=3	If average of individual element is > = 3, then the people agreed that climate change has impact on the particular element in the area. If not, they disagree.	
Variable	**Mean score**	**Decision**
Unusual early rains that are followed by weeks of dryness	3.45	Agreed
Erratic rainfall pattern	2.93	
Delay in the unset of rain	3.25	Agreed
Long period dry season	3	Agreed
Heavy and long period of rainfall	2.47	
Less rainfall	3.05	Agreed
No or reduced harmattan	2.63	
Long period of harmattan	2.55	
Higher temperature	3.87	Agreed
Thunderstorms	1.88	
heavy winds	2.17	
floods and erosion	2.64	
Drought	3.18	Agreed
heat waves	2.77	
heavy rainfall	2.32	
desertification or loss of forest resources	2.64	
Increase in pests' problems	3.12	Agreed
High rate of disease incidence	3.65	Agreed
Increase weed infestation	2.43	
Loss of soil fertility	2.77	
Drying up of streams/rivers	2.31	
Overflowing of streams/rivers	1.23	

Source: Field Survey, 2010 and 2017.

Table 2.
Farmers' perception of different forms of climate change impact and their extent on cocoyam production.

Climate change impact	Decreasing	Percent Increasing	No change
Unusual early rains that are followed by weeks of dryness	0	100	0
Erratic rainfall pattern	0	95	5
Delay in the unset of rain	1.7	98.3	0
Long period dry season	1.7	98.3	0
Heavy and long period of rainfall	78.3	21.7	0
Less rainfall	10	81.7	8.3
No or reduced harmattan	8.3	43.3	48.3
Long period of harmattan	33.3	26.7	40
Higher temperature	1.7	98.3	0
Thunderstorms	43.3	5	51.7
heavy winds	48.3	21.7	30
floods and erosion	30	45	25
Drought	0	100	0
heat waves	25	61.7	13.3
heavy rainfall	90	6.7	3.3
desertification or loss of forest resources	6.7	45	48.3
Increase in pests' problems	6.7	93.3	0
High rate of disease incidence	0	100	0
Increase weed infestation	36.7	60	3.3
Loss of soil fertility	0	100	0
Drying up of streams/rivers	15	85	0
Overflowing of streams/rivers	98.3	0	1.7

Source: Field Survey, 2010 and 2017.

Table 3.
Directions of the impact of the different forms of climate change.

Results show that climate change impact on the underlined variables which include: unusual early rains that are followed by weeks of dryness, Erratic rainfall pattern, Delay in the unset of rain, Long period dry season, Less rainfall, Higher temperature, Drought, Increase in pests problems, High rate of disease incidence, Loss of soil fertility, Drying up of streams/rivers are on the increase in the area.

3.4 Resilient building strategies to the impacts of climate change

In this section the various resilient building strategies employed by the farmers to cope with climate change effect on cocoyam are as illustrated in the **Table 4**.

Results indicate that the most prominent strategies adopted by the farmers are planting deeper or shallower than the usual planting depth, processing crops to minimize post-harvest losses, intensive manure application, erosion control, multiple cropping, move to a different site, mixed farming and intercropping – main crops planted with subsidiaries at low densities.

Strategies	Used	Percent
		Unused
Purchase of water for irrigation (liters)	0	100
Mulching (kg)	23.3	76.7
Use of wetlands/river valleys (e.g. Fadama) (cost of use) (hectares)	21.7	78.3
Contour cropping across hill slopes (hectares)	3.3	96.7
Planting deeper or shallower than the usual planting depth (hectares)	88.3	11.7
Construction of drainage or dam within the farm/household (Area)	3.3	96.7
Afforestation: planting of trees (number)	23.3	76.7
Use of resistant varieties (number)	23.3	76.7
Processing crops to minimize post-harvest losses (Kg)	95	5
Expansion of cultivated land area (ha)	23.3	76.7
Increased use fertilizers, seeds (kg)	20	80
Cultivation on marginal lands	3.3	96.7
Intensive manure application (kg)	83.3	16.7
Increased weeding (ha)	20	80
Use of chemicals: herbicides, pesticides etc. (kg)	43.3	56.7
Erosion control (ha)	100	0
Move to a different site (ha)	55	45
Soil conservation practice (ha)	1.7	98.3
Shading and shelter (ha)	10	90
Change in the timing of land preparation activities	38.3	61.7
Changes in planting dates	38.3	61.7
Changes in harvesting dates	13.3	86.7
Multiple cropping	93.3	6.7
Mixed farming (crop and animal production)	70	30
Relay cropping – planting and harvesting in succession	10	90
Use of different varieties	18.3	81.7
Use of different planting dates	30	70
Intercropping – main crops planted with subsidiaries at low densities	63.3	36.7
Reducing your animal stock	18.3	81.7
Culling of infected animals	8.3	91.7
Crop replacement	43.3	56.7
Change from crop production to animal rearing	5	95
Change from animal production to crop production	11.7	88.3
Agro-forestry practice	16.7	83.3
Change from production to marketing of agricultural products	10	90
Total change from farming to other occupations	10	90
Replanting	43.3	56.7
Prayers	80	20

Source: Field survey, 2010 and 2017.

Table 4.
Resilient building strategies to the impacts of climate change among cocoyam farmers.

DOI: http://dx.doi.org/10.5772/intechopen.1002244

3.5 Factors affecting the adoption of different resilience building practices

This section presents the factors that affect the adoption of adaptation practices in the area. The mean score of each constraint was determined using a five-point Likert scale with the following ratings; To a very great extent = 5, = 4, To some extent = 3, To a little extent = 2, and To no extent = 1), giving a mean of 3, as shown in **Table 5**.

Results show that the major constraints to adoption of adaptation practices in the area high cost of farm labour having an average of 3.08 and limited income scores an average of 3.05. The farmers complained that getting farm labour is quite expensive and they have limited income to pay for such.

Average	Decision rule	
3	If the average of the variable is 3 and above, it means that such variable is an important constraint to adaptation in the area, but if lower it is not an important constraint	
Strategy	**Mean score**	**Decision**
Limited availability of land for farming	2.63	
High cost of farmland	2.27	
Inherited system of land ownership	2.75	
Communal system of land ownership	2.1	
Poor access to information sources	2.33	
Non-availability of credit facilities	2.52	
High cost of irrigation facilities	2.48	
Non-availability of farm inputs	2.75	
High cost of fertilizers and other inputs	2.38	
Inadequate knowledge of how to cope	2.53	
Non-availability of improved varieties	2.2	
High cost of improved varieties	2.05	
Non-availability of farm labour	2.2	
High cost of farm labour	3.08	Agreed
Lack of access to weather forecasts	2.26	
Poor response to crises related to climate change by the government's agencies and interest groups	2.42	
Non-availability of storage facilities	2.35	
Limited income	3.05	Agreed
Non-availability of processing facilities	2.83	
High-cost processing facilities	2.62	
Traditional beliefs/practices e.g. on the commencement of farming season etc.	1.25	

Source: Field Survey, 2010 and 2017.

Table 5.
Factors affecting adoption of different adaptation practices.

4. Discussion

The results of this study show that the progressive decline in the production of cocoyam since the mid 2000 are associated to different kinds of abiotic stress on the crop. A detailed analysis of the trend in the production of the crop reveals that there has been a disturbing trend in its output. Both results from field interview and that of the trend analysis of FAO data set agree, indicating a declining trend in output of cocoyam from 2007 (**Figures 2** and **3**). The FAO statistics shows that the yield declined from an unprecedented height of about 8mt ha^{-1} in 2005 progressively to as low as 4mt ha^{-1} in 2016 and this trend has continued to date. This disturbing trend also resonates with the findings of Talwana et al. [13] about a declining trend in the output of cocoyam in Uganda; which coincides with the same period of this decline in Nigeria. This finding is also supported by the position of experts and stakeholders in Cocoyam value chain in 2017, who lamented the declining trend in cocoyam output over the years [26]. They also called for improved technology and practices that would mitigate the observed adverse effects of climate change on the crop. This position was made in a workshop targeted at assisting cocoyam farmers to sustainably adapt to the severe effects of climate change in Nigeria, The workshop was sponsored by the Association of Commonwealth Universities, African Academy of Sciences and UK Aid, under the Climate Impact Research and Leadership Enhancement programme, focused on: "Climate Change Adaptation and Constraints Faced By Cocoyam Farmers in Southeast Nigeria" and held at the National Root Crop Research Institute (NRCRI), Umudike, Abia State, with participants drawn from universities, agriculture research institutions and farmers groups, among others.

There is therefore evidence that something is wrong with cocoyam production in the tropics. This trend has brought untold hardship on the households in the cocoyam production areas of Nigeria. This is because the crop has formed a part of daily staple for the households, especially in the rural areas. Therefore, the shortfall has brought about unnecessary food security crisis among the households. It was for this reason that this stud was initiated to investigate the nature and extend of the problem.

Further analysis showed that this phenomenon became very prominent in the later part of 2000. Specifically, they believed it was in the year 2009, which is also in agreement with the FAO data. Coincidentally, these findings agree with the period the problem of acid rain was reported in different parts of the country [27–29], and therefore in line with farmers' attribution of the problem to acid rain. The severity of the frustration occasioned by this problem is depicted in the words of Hassan, thus.

> *"As the call to battle hunger in the world continues to grow, another dimension to the battle against climatic forces has emerged in Nigeria in the form of acid rain threat. Acid rain presents yet another challenge to smallholder farmers especially, because it has the potential of destroying crops in the farms thereby affecting the total yield at the end of the farming season" [28].*

Also describing the nature of the problem, most of the farmers indicate that cocoyam crops in the field die before they are matured for harvesting. In their words;

> *"cocoyam leaves appear as if they are bleached with hot water, after which they turn brown/yellow and gradually eaten up from either the middle or from edges of the leaves until there is no leave left on the plants".*

And this happens usually after rainfall and is the case in most parts of eastern Nigeria, where cocoyam is predominantly utilized. The farmers linked this scenario to acid rain that has been reported in Nigeria. Acid rain results from air pollution and climate change. Acid rain come about due to sulfur dioxide (SO_2) and nitrogen oxide (NO_x) gases and their particulate matter derivatives from industrial and automobile emissions, they react with falling water to make it acidic. Acid rain does not usually kill plants directly. Instead, it weakens the plants by damaging their leaves, limiting the nutrients available to them, or exposing them to toxic substances slowly released from the soil. Quite often, injury or death of trees is a result of these effects of acid rain in combination with one or more additional threats [30]. As a result of this scenario output declined from about 1.8 tonnes to about 0.6 tonnes per household from 2007 to 2009 respectively.

Further analysis in 2017 also corroborates the 2010 findings with more detailed description of the effects on the leaves and tubers of cocoyam; bleaching of cocoyam leaves, die back, smaller tubers, early decay of the tubers and decline in the quality and taste of the tubers. Because of these problems most of the cocoyam plants die prematurely in the field before they are matured for harvesting.

Analysis of the experiences of the farmers concerning the problem shows that the farmer highlight some climate change related factors that affect their cocoyam production, The different forms; that affect their production include unusual early rains that are followed by weeks of dryness, delay in the unset of rain, long period dry season, less rainfall, higher temperature, drought, increase in pests problems and high rate of disease incidence. These phenomena were shown to have mean scores above 3, which is the cut-off point that shows the factor which important from the mean scores' analysis.

Particularly, the farmers complained of the problems of delayed unset of rain and early unset of rainfall that is followed by long period of dryness that have become rampant in recent times. These conditions are not favorable for cocoyam production, as the crop require high amount of moisture for optimum yield. More so, some local farmers who planted their cocoyam seedlings after the first and second send rains when asked, complained of unexpected rise in insolation and temperature of the soil which resulted either in the cooking of the tubers in the soil, destruction of some seedlings newly planted or delayed germination and/green leaf formation. Thus, the issue of poor adoption of improved technology or poor cultural practices, pest and disease become secondary among the causes of decreased output as has been suggested b some scholars. These findings therefore indicate that climate change interacts with other abiotic factors to predispose the cocoyam and other crops in the country to stress as has also been reported by other scholars [31].

With respect to the resilient building strategies adopted by the farmers to cope with some of these challenges, the study shows that the farmers have tried different approaches like planting deeper or shallower than the usual planting depth, depending on the crop variety to reduce the effect of scorching sun or excessive flooding. To prevent decay of the tubers, the farmers are forced to process them into more durable forms which are stored for future use. To boost crop yield, the farmers practice intensive manure application, this because the felt that the cause of the poor yield was due to poor soil fertility. They also practice multiple cropping/mixed cropping as a form of crop insurance, so that if the cocoyam crop fails they can harvest the other crops. This is practices as a traditional way of ensuring against total crop failure. The also abandon the fields where their crops fail and, move to a different sit, with the hope that the new sites would be free from the disease that killed their crop or are more fertile than the previous sites. The practice mixed farming and whereby the also keep animals as a means of insurance.

On the other hand, the farmers pointed out some of the challenges that militate against their ability to effectively cope with the challenges as high cost of farm labour and limited income. They believe farm labour is expensive, and they have limited income to pay for labour and the required farm inputs that they require in the production process, like fertilizers and other agrochemicals. The trio; farm labour, credit and inputs are the major limiting factors in agricultural production in Nigeria aside from the problems of climate change. They are often not available, where they are; they are limited, expensive or require very stringent bureaucratic bottlenecks to access, in the case of credit. So, farmers resort to using family labour to do the much they can. The buy the fertilizers and chemicals they can afford and therefore do not use them in the recommended rate for optimum results. Thus, they continue in the vicious cycle of low productivity and poverty.

5. Conclusion and recommendation

The study investigated the problem of the declining trend in cocoyam production since the mid-2000. Preliminary investigations show that farmers were experiencing poor yield and were gradually moving away from the cultivation of cocoyam to other crops. The results indicate that air pollution, which results in the production of acid rain and climate change combine to limit the productivity of cocoyam in the study area.

This problem has resulted in low output of the crop and reduced welfare of the households who depend on the crop for their livelihood. Though the farmers are making genuine efforts to cope with the situation, but their efforts are limited by access to the required labour, capital, inputs and other necessary support that will assist them in building resilience to the problem.

We therefore posit that the threat to cocoyam has assumed a frightening dimension. Thus, there is need for more detailed integrated research to unravel and solve the problem of cocoam production in Nigeria. Concerted efforts is also needed to help the farmers copped with the present problems which are not the making of theirs. If nothing is doneurgentl to stem this negative tide cocoyam is heading towards extinction. Policies that target input provision, reduction in greenhouse gas emission and increase in farmers' share/income must be encouraged as these will help build their resilience to the impact of climate change.

Acknowledgements

We wish to thank friends and family for their patients and kind supports during this research.

Funding

This research did not receive any specific grant from funding agencies in the public, commercial, or not-for-profit sectors.

Conflict of interest

None.

Author details

Dennis Mark Onuigbo[1], NwaJesus Anthony Onyekuru[2*], Anthonia Ifeyinwa Achike[2], Chinasa Onyenekwe[2] and Eric Eboh[2]

1 University of Tasmania, Hobart, Australia

2 Resource and Environmental Policy Research Centre, EfD, University of Nigeria, Nsukka, Nigeria

*Address all correspondence to: anthony.onyekuru@unn.edu.ng

IntechOpen

© 2023 The Author(s). Licensee IntechOpen. This chapter is distributed under the terms of the Creative Commons Attribution License (http://creativecommons.org/licenses/by/4.0), which permits unrestricted use, distribution, and reproduction in any medium, provided the original work is properly cited.

References

[1] Onyeka J. Status of cocoyam (Colocasia Esculenta and Xanthosoma Spp) in West and Central Africa: Production, household importance and the threat from leaf blight. CGIAR Research Program on Roots, Tubers and Bananas. 2014. Available from: www.rtb.cgiar.org

[2] Si H, Zhang N, Tang X, Yang J, Wen Y, Wang L, et al. Transgenic research in tuber and root crops. Genetic Engineering of Horticultural Crops. 2018;**2018**:225-248. DOI: 10.1016/B978-0-12-810439-2.00011-8

[3] Akwee P, Netondo G, Palapala VA. A critical review of the role of taro Colocasia esculenta L. (Schott) to food security: A comparative analysis of Kenya and Pacific Island taro germplasm. Scientia Agriculturae. 2015;**9**(2):101-108

[4] Onokpise O, Wutoh J, Ndzana X, Tambong J, Meboka M, Sama A, et al. Evaluation of macabo cocoyam germplasm in Cameroon. Perspectives on New Crops and New Uses. 1999;**631**:1991-1993. DOI: 10.1007/s00590-012-1063-3

[5] Echebiri R. Socio-economic factors and resource allocation in cocoyam production in Abia State, Nigeria: A case study. Journal of Sustainable Tropical Agricultural Research. 2004;**9**:69

[6] Knipscheer A, Wilson U. Importance of cocoyam production in Nigeria. In: Proceeding of National Workshop on Cocoyam. 2010

[7] FAOSTAT. Food and Agriculture Organisation of the United Nations Statistical Database; Statistical Division. Rome, Italy: FAO; 2021. Available from: http://www.fao.org/statistics/en/

[8] Otekunrin OA, Sawicka B, Adeyonu AG, Otekunrin OA, Racho'n, L. Cocoyam [*Colocasia esculenta* (L.) Schott]: Exploring the production, health and trade potentials in sub-Saharan Africa. Sustainability. 2021;**13**:4483. DOI: 10.3390/su13084483

[9] Oshunsanya SO. Quantification of soil loss due to white cocoyam (*Colocasia esculentus*) and red cocoyam (*Xanthosoma sagittifolium*) harvesting in traditional farming system. Catena (Amst). 2016;**137**:134-143

[10] Chemura A, Kutywayo D, Hikwa D, Gornott C. Climate change and cocoyam (*Colocasia esculenta* (L.) Schott) production: Assessing impacts and potential adaptation strategies in Zimbabwe. Mitigation and Adaptation Strategies for Global Change. 2022;**27**: 42.ol.:(0123456789). DOI: 10.1007/s11027-022-10014-9

[11] Oguniyi L. Profit efficiency among cocoyam producers in Osun state, Nigeria. International Journal of Agriculture and Rural Development. 2008;**1**:38-46

[12] Skott GJ, Best R, Rosegrant M, Bokanga M. Root and tubers in the global food system: A vision statement of the year 2020. 2000

[13] Talwana HL, Tumuhimbise R, Osiru D. Comparative performance of wetland taro grown in upland production system as influenced by different plant densities and seedbed preparation in Uganda. Journal of Root Crops. 2010;**36**:65-71

[14] Bello OB, Ganiyu OT, Wahab MKA, Afolabi MS, Oluleye F, Ig SA, et al. Evidence of climate change impacts on agriculture and food security in Nigeria. International Journal of Agriculture and

Forestry. 2012;**2**(2):49-55. DOI: 10.5923/j.ijaf.20120202.08

[15] Eboh EC, Ujah O, Garforth CJ. Barriers to climate change adaptation among farming households of southern Nigeria. Journal of Agricultural Extension. 2010;**14**(1):114-124

[16] Enibe DO, Nwobodo E, Nworji MJ, Okonkwo CA. Economic analysis of cocoyam Marketing in Anambra Agricultural Zone of Anambra State, Nigeria. Asian Journal of Agricultural Extension, Economics & Sociology. 2019;**29**(3):1-10. DOI: 10.9734/AJAEES/2019/46476

[17] Regina S. Climate Change and Root Crop Production in Ghana. Accra: Environmental Protection Agency (EPA) Report; 2006

[18] Taylor M, Lebot V, McGregor A, Redden RJ. Sustainable production of roots and tuber crops for food security under climate change. In: Yadav SS, Redden RJ, Hatfield JL, Ebert AW, editors. Danny Hunter Food Security and Climate Change. 2019

[19] Kodis M, Galante P, Sterling EJ, Blair ME. Ecological niche modeling for a cultivated plant species: A case study on taro (Colocasia esculenta) in Hawaii. Ecological Applications. 2018;**28**(4):967-977

[20] Bachelet D, Neilson RP, Lenihan JM, Drapek RJ. Climate change effects on vegetation distribution and carbon budget in the United States. Ecosystems. 2001;**4**:164-185. DOI: 10.1007/s10021

[21] National Population Commission. Nigeria Population 200. 2006

[22] Cityfacts. Nsukka, Enugu Nigeria. 2023. Available from: https://www.city-facts.com/nsukka/population

[23] ClimateData. No Title [WWW Document]. 2018. Available from: www.en.climate-data.org/location/404904/

[24] Ume S, Ike EC, Okeke CC. Effect of Climate Variability on Cocoyam Production by Farmers in Afikpo South Local Government Area of Ebonyi State, Nigeria. 2018. DOI: 10.22192/iajmr.2018.4.2.4

[25] Ukonze IA. Impact of Climate Change on Cocoyam Production in South Eastern Nigeri. 2012. Available from: https://www.researchgate.net/publication/249007700_Impact_of_Climate_Change_on_Cocoyam_Production_in_South_Eastern_Nigeria

[26] Udeajah G. Experts Lament Decline in Cocoyam Output, State Panacea. The Guardian. 2017. Available from: https://guardian.ng/features/health/experts-lament-decline-in-cocoyam-output-state-panacea/

[27] Efe SI. Spatial variation of acid rain and its ecological effect in Nigeria. COLERM Proceedings. 2011;**234**:381-396

[28] Hassan TA. Nigeria: Impact of Acid Rain on Crops, Food Availability. Nigeria: Dly Trust; 2010

[29] Mustapha S. Nigeria: Acid Rains, the Current Environment Controversy. 2010

[30] U.S. Environmental Protection Agency. Effects of Acid Rain. 2023. Available from: https://www.epa.gov/acidrain/effects-acid-rain

[31] Onyekuru N, Marchant R. Climate change impact and adaptation pathways for forest dependent livelihood systems in Nigeria. African Journal of Agricultural Research. 2014;**9**:1819-1832. DOI: 10.5897/AJAR2013.8315

Chapter 13

Impact of Climate Change on the Dairy Production in Fiji and the Pacific Island Countries and Territories: An Insight for Adaptation Planning

Royford Bundi Magiri, Phillip Sagero, Abubakar Danmaigoro, Razia Rashid, Wati Mocevakaca, Shivani Singh, Walter Okello and Paul A. Iji

Abstract

Climate change affects weather patterns, leading to changes in average temperatures, increased frequency, variability, and intensity of extreme weather events, especially in the Pacific Island countries. Climate change poses the greatest threats to the sustainability of smallholder dairy farming in Fiji, with the farmers being highly vulnerable, yet their adaptive capacity is low. Additionally, the Pacific's current and future sustainable livestock development will heavily depend on its ability to cope with climate variability and adapt to future climate changes. Available data indicate that there is high spatial and temporal variability of rainfall over Fiji Island with the mean annual rainfall ranging from 1600 to 3600 mm, with Rotuma station receiving the highest rainfall over Fiji Island. Rainfall in Fiji has shown an increasing and decreasing trend, where both minimum and maximum temperatures have shown an increasing trend. This will have a great impact on the smallholder dairy farmers who consist of over 95% of the existing farmers. Using available information and drawing from other contexts or countries where data or information is unavailable, we provide an overview of dairy production in Fiji as a prototype to other Pacific Island Countries and Territories (PICTs), highlighting smallholder dairy systems in the Fijian dairy sector, challenges, and opportunities of the dairy sector in the PICTs. We conclude that climate change significantly impacts dairy production in Fiji and the Pacific.

Keywords: climate change, adaptation, dairy, Pacific, planning, climate resilience

1. Introduction

Global human population is expected to increase from 7.2 to 9.6 billion by 2050 representing a population increase of 33% [1]; hence, the demand for global animal-source protein such as dairy product is expected to increase due to increased human population and associated rise of the middle class. To meet the demand for animal-source protein, global livestock production needs to increase by 70%, putting pressure on agricultural land since about one-third of the global cereal harvest will be used for livestock feeds. However, climate change is a threat to livestock industry such as dairy production because of the impact on quality of feed crops and forages, water availability, diseases, animal reproduction, and biodiversity at a time when it is most needed. Therefore, the global challenge is to maintain a balance between dairy production, food security, and environmental preservation.

The impact of climate change on food systems and livelihoods is becoming more apparent globally. These impacts are more severe in low-middle income countries like the Pacific Island countries and territories (PICTs) that continuously experience extreme weather events such as droughts, cyclones, and heat stress. Additionally, Pacific Island Countries and Territories (PICTs) have variable and unique geographical distributions and face similar challenges such as recurrent natural disasters, limited availability of fertile land, susceptibility to biosecurity threats, small economies, overreliance on tourism and changes in global markets, and small human and livestock populations [2, 3]. Furthermore, human population in PICTs is predicted to increase in several countries [4–6]. This will result in rapid expansion of the agricultural subsectors [7]. In dairy production, extreme weather events culminate in decreased animal health and productivity, decreased animal welfare, and reduced income to the affected communities and countries. Adapting dairy production practices can be useful in assisting dairy farmers in the PICTs adapt to climate changes. However, apart from climate change, there are other factors that may affect dairy production and these need to be considered to have a holistic view on challenges affecting dairy production in the PICTs and country's ability to adapt and mitigate climate change. The expected expansion of the livestock sector including dairy production offers the PICTs an opportunity to increase their contribution to human diets. The impact of climate change to the livestock sector including dairy production in the Pacific is primarily due to rising temperatures, alterations in atmospheric CO_2 levels, changes in precipitation that results in floods and droughts, and a mix of the climate components that results in recurrent cyclones [8, 9]. The combination of these factors leads to poor quality and quantities of feed crops and forages, water availability for livestock, animal growth productivity, emergence of diseases, mortalities, and reduced reproduction [10, 11].

Factors such as climate change make the dairy systems vulnerable thus requiring adaptation to sustainable systems. Although livestock sector including dairy is unquestionably an essential component of food and socioecological systems [12–14], there are few studies on how it can be improved to become sustainable in the PICTs under climate change [15, 16]. Therefore, to enable PICTs transition to sustainable dairy production systems under climate change, it is necessary to: (a) understand potential challenges affecting sustainability, profitability, and resilience of dairy production systems in the PICTs and (b) provide strategies that could promote adaptation for dairy farmers in the PICTs.

We conclude that dairy production in the PICTs faces a myriad of challenges, and there is an opportunity to sustainably develop the sector given the increasing demand for dairy products. However, factors such as land conservation and management, effective disease control, and public financing of the dairy sector need to be considered to enhance dairy management practices in the PICTs. This study is an important starting point in developing sustainable and climate resilient dairy farming systems in the PICTs.

2. Fiji geographical location and topography

Fiji's Island is unique in terms of land mass and diversity in topography. The nation consists of over 300 islands with a total of around 18,300 km^2 of land in the South Pacific region. The country lies between longitudes 175° E and 178° W, and latitudes 15° S and 22° S. Viti Levu, the country's largest island (**Figure 1**), is characterized by relatively high topographical variation (**Figure 1**). Vanua Levu and Kadavu are located to the northeast and south of Viti Levu, respectively. The larger island is characterized by mountainous topography, which is known to influence rainfall over the country, with windward having high rainfall compared to the leeward side. This explains why orographic rainfall is the dominant form of precipitation in Fiji [17]. Furthermore, Fiji and the Pacific have been identified as particularly vulnerable to climatic change and variability, because of low adaptive capacity of the communities [18]. Below is the map of Fiji with its elevation values (**Figure 1**).

3. Overview of rainfall patterns over Fiji

Rainfall in Fiji Island is highly variable in space and time, influenced by the islands topography, and prevailing southeast trade winds. The mountains in Viti Levu create wet climatic zones on their windward side and dry climatic zones on their leeward side [19]. The South Pacific Convergence Zone (SPCZ) controls the seasonal rainfall distribution over Fiji which is the main rainfall-producing synoptic system over the region. From November to April, the Fiji Islands are also frequented by tropical cyclones originating in the Pacific Ocean, which results in

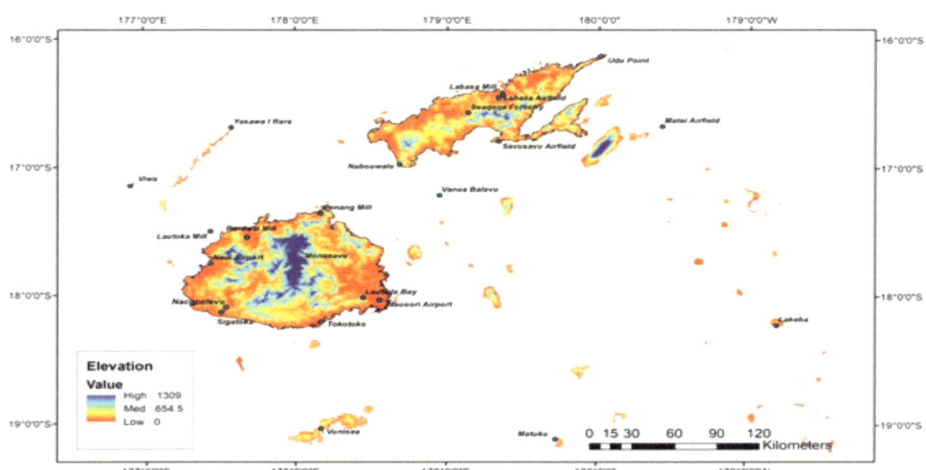

Figure 1.
Map of Fiji showing the islands and their elevation.

prolonged heavy rainfall and flooding of low-lying coastal areas [20]. The Pacific Ocean also plays central roles in the interannual variability of rainfall in Fiji. It is strongly associated with El Niño and La Niña Southern Oscillation (ENSO) [21]. La Niña is associated with above normal rainfall and a strong cyclone system, which results in floods. On the other hand, El Niño is associated with severe droughts that adversely affect agriculture and, in turn, cause food shortages [22]. Therefore, Fiji's current and future sustainable socioeconomic development will heavily depend on its ability to cope with climate variability and adapt to future climate changes. The country experiences a unimodal rainfall pattern during the austral summer months (November to April), the months of May to October, that are generally dry (**Figure 2**). March appears to be the month with the highest rainfall amounts (wettest), while July is the driest. The average annual rainfall over Fiji is 2500 mm, which varies from year to year and place to place.

The annual rainfall in Fiji's western region ranges from 3000 to 4900 mm in the higher mountains, while it averages approximately 1900 mm in the center region (**Table 1**). It is evident from the fact that the yearly rainfall at Nadi (on the leeward side) is only approximately 60% of what is observed at Suva (on the windward side), and that orography has a significant effect in the spatial distribution of rainfall over Viti Levu. Summer rainfall in Nadi (western division) can reach 76% of the yearly total, while Suva (central division) receives 63% of the annual total (**Table 1**). Western is where the wet and dry seasons are more clearly visible, whereas Eastern and Central are wetter on both seasonal and annual average. The western division typically experiences bright or partly overcast skies, plenty of sunshine, a strong west to northwesterly sea breeze during the day, and generally cold temperatures at night. Summertime thunderstorms are frequent and are to blame for short bursts of heavy rain [19]. However, the relationship between rainfall patterns and dairy production has not well understood in the Pacific and requires investigation.

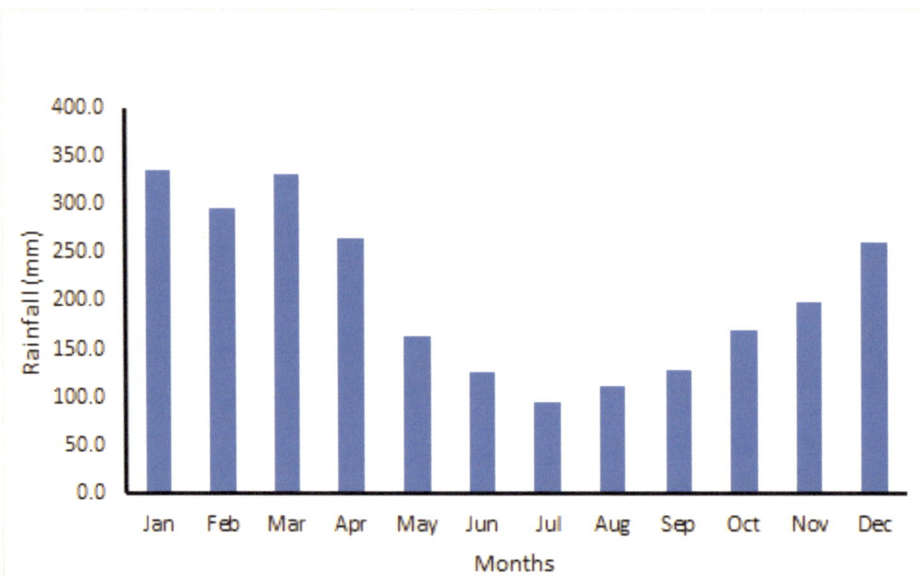

Figure 2.
Mean monthly rainfall over Fiji Island for the period 1971–2020.

	Station	Lat	Lon	Annual	Summer Rainfall	Winter Rainfall	% contribution of summer Rain	% contribution of Winter Rain
1	ono-i-Lau	−20.67	178.72	1489.7	936.1	559.4	62.8	37.5
2	Yasawa-i-Rara	−16.7	177.58	1554.5	1131.1	426.4	72.8	27.4
3	Nacocolevu	−18.1	177.54	1738.6	1232.3	503.6	70.9	29.0
4	Matuku	−19.13	179.73	1663.7	1048.9	612.0	63.0	36.8
5	Nadi Airport	−17.76	177.44	1937.7	1502.8	437.1	77.6	22.6
6	Lakeba	−18.23	−178.8	1898.9	1281.0	623.5	67.5	32.8
7	Lautoka	−17.55	177.44	1970 2	1522.7	447.4	77.3	22.7
8	Savusavu	−16.81	179.34	2160.9	1381.7	774.8	63.9	35.9
9	Labasa Airfield	−16.47	179.34	2212.2	1732.6	468.7	78.3	21.2
10	Vunisea	−19.05	178.17	2174.2	1425.7	762.2	65.6	35.1
11	Penang Mill	−17.37	178.17	2298.4	1747.2	550.8	76.0	24.0
12	Seaqaqa	−16.59	179.14	2326.1	1829.0	491.1	78.6	21.1
13	Nabouwalu	−16.99	178.69	2404.1	1670.2	731.3	69.5	30.4
14	Udu Poin	−16.14	−179.99	2419.5	1681.6	733.6	69.5	30.3
15	Metei Airfield	−16.69	−179.58	2495.8	1645.2	862.9	65.9	34.6
16	Nausori Airport	−18.05	178.56	2963.1	1912.5	1059.2	64.5	35.7
17	Laucala	−18.03	178.45	3043.8	1919.5	1132.8	63.1	37.2
18	Rotuma	−12.5	177.05	3369.5	1868.0	1493.9	55.4	44.3
19	Tokotoko_Navua	−18.22	178.17	3232.8	1897.6	1290.4	58.7	39.9
20	monasavu	−17.75	178.05	4942.3	3249.9	1628.2	65.8	32.9

Table 1.
Fiji weather stations and rainfall distributions.

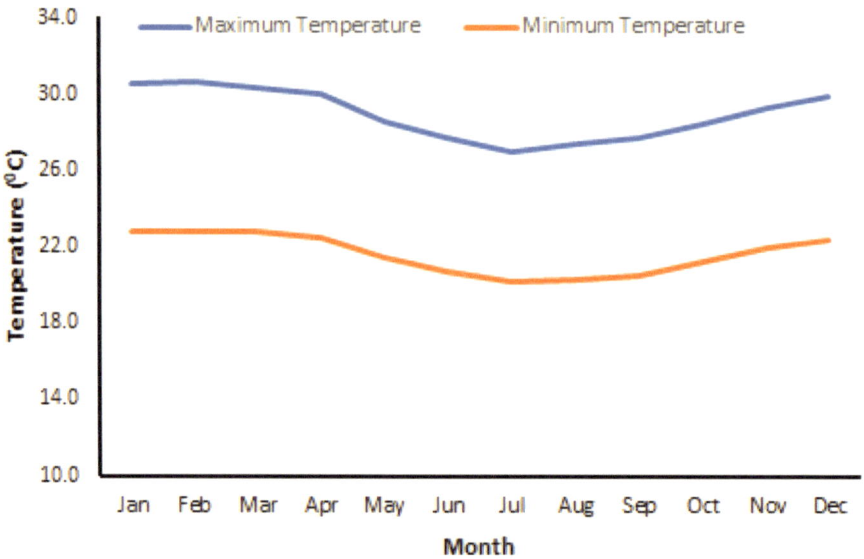

Figure 3.
Annual temperature cycle Fiji based on dataset 1970–2020.

4. Overview of temperature pattern in Fiji

Fiji's climate is tropical and surface air temperatures in summer (between November and April) are, in general, only slightly higher than in the winter months (between May and October) such that the seasonal variations are not large (**Figure 3**). Fiji being an Island country, the diurnal temperature range is generally (8.0°C) low, and this is due to contrasts in heat capacity and moisture supply between land and ocean. The mean temperature has low temporal variability, with a difference of 3.61°C between the warmest month (February) and the coolest month (July) [23]. The cool season, the months of May to October, is generally dry. The western side of Viti Levu is a relatively dry zone and is subject to large seasonal variations in both maximum and minimum temperatures relative to the eastern wet side.

5. Rainfall and temperature trends

Both minimum and maximum temperatures over Fiji have increased overtime (**Figure 4**), Temperature maximum has the greatest increase as compared with the minimum temperature. Ongoma et al. found out that the temperature change is 0.1°C/decade for maximum temperature, 0.04°C per decade for minimum temperature, and 0.07°C/decade for mean temperature [23]. This change in temperature in Fiji is slightly lower than the global of 0.24°C/decade [24].

On the other hand, rainfall in Fiji has a higher year-to-year variation than the variation due to the long-term trend (**Figure 5**). There are stations that show decreasing trends, while others show increasing trends. The increasing trends are observed in stations in the western part of Fiji, while those in the central and Easter show a decreasing trend, although the trends are statistically insignificant. However, the effect of rainfall and temperature changes on dairy production is not investigated in the Pacific.

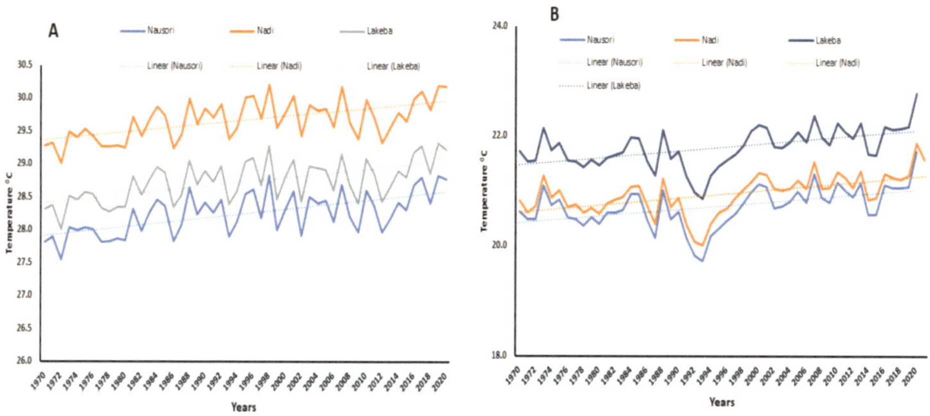

Figure 4.
Temperature trend for Fiji (a) maximum temperature and (b) minimum temperature.

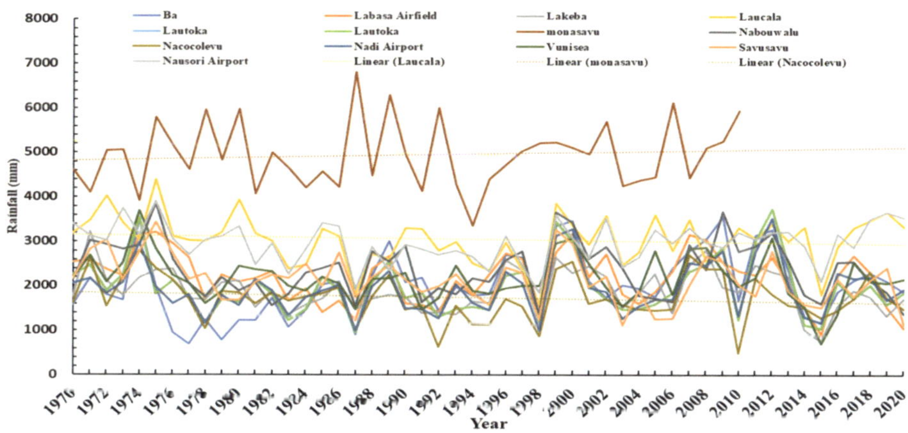

Figure 5.
Annual rainfall time series with linear trend line for various meteorological stations in Fiji.

6. Dairy farming systems in the Pacific

The rural population in the Pacific Island Countries (PICs) benefits greatly from raising ruminant livestock such as beef and dairy cattle. Dairy cattle produce milk that can be converted into cash to cater for significant costs like higher education [25]. The production system is largely determined by land area, land demand, culture, tradition, number of cattle, and feed availability with little variation between countries. However, these grazing systems, including tethering and free-range smallholder units, are found in every Pacific Island country. Low-quality feeds are supplemented by shrub leaves for maintenance, weight gain, and other physiological processes. In Fiji, the cornerstones of dairy feeding systems are tethering with limited grazing, along with some cut-and-carry to the cattle feeding zone. There is usually lack of feedlots for livestock although dairy animals are frequently fed supplementary feeds [26]. This alone makes the systems vulnerable to climate change. According to Nube and Voortman [27] previously, the South Pacific region relied mostly on crop agriculture for sustenance, with animals kept as a backup in case of crop failure; as a result, the ruminant livestock sector is not as well-established. The increasing demand for meat, dairy, and other animal products has increased livestock production among farmers. The overall underdevelopment of the livestock sector including dairy production in the region may be attributed to a few factors, including a shortage of feed, negative social attitudes, insufficient technology for disease management, marketing, and extension services.

There is a dry and a rainy season in Fiji whereby the country receives the majority of its yearly precipitation during the rainy season (November–April), which is characterized by intense, transient local showers. On the major islands, annual precipitation varies between 2000 and 4500 millimeters (mm) in the highlands and between 1600 and 3000 mm (mm) in low-lying areas and along the shore. It is common for smaller islands in Fiji to get rainfall ranging from 1500 to 3500 mm less precipitation than the main island [28]. Livestock is gaining significance among smallholders in cyclone-prone countries, such as Fiji, due to the frequent occurrence of cyclones during the wet season. This is attributed to the greater resilience of livestock as compared to crops in the face of cyclonic events [25].

6.1 Dairy production in Fiji

According to the dairy milk, production sector in Fiji has not been characterized by a well-structured farming system [29]. Most dairy farms in the region are classified as semi-commercial based on the quantity of milk delivered to the processor. The number of smallholder farmers has increased over the past decade; however, dairy farming has had a significant decline due to large farms closure that has greatly impacted rural household lifestyles and income, although over 95% are smallholder dairy farmers (**Figure 6**). The dairy farming industry has had a favorable impact on augmenting the livelihoods and earnings of rural households, which has resulted in a rise in the number of smallholder farmers over the past decade, as reported (**Figure 6**) [29–31].

Fiji's agricultural sector comprises three primary farming systems, namely subsistence, semi-commercial, and commercial [32]. Many farms are owned by the

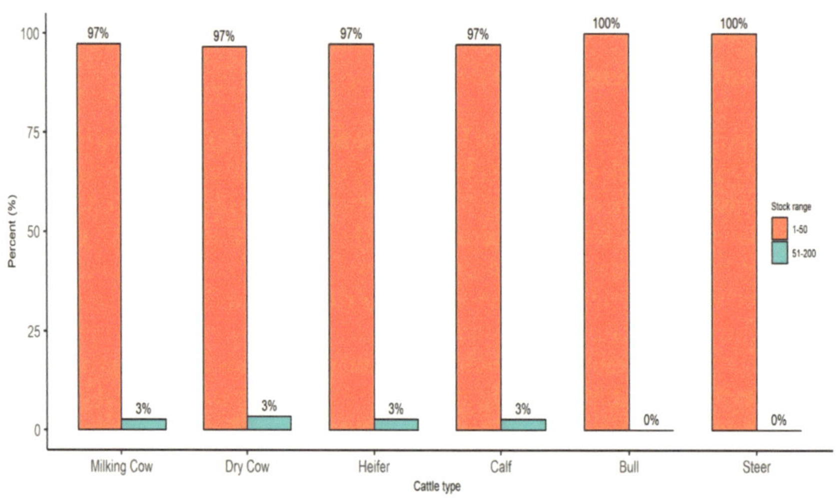

Figure 6.
Dynamics and distribution of diary cattle in the 2023 registered farms by FCDCL in Fiji.

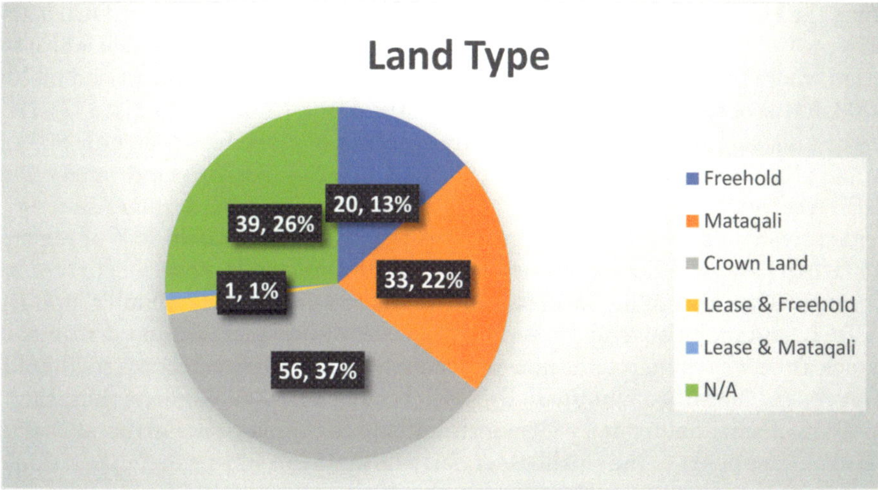

Figure 7.
Characterized the land type in Fiji dairy farms. Most of the farms belong to crown land with a total of 37% (n = 56) followed by Mataqali land with a total of 22% (n = 33). About 13% (n = 20) of the farms are freehold land with 1% of the farms under Lease & Mataqali and lease and freehold.

indigenous population, also known as iTaukei. Specifically, 65.4% of farms are owned through the traditional mataqali system, while 17% are leased from the iTaukei Land Trust Board (iTLTB). A mere 7.7% of farms are classified as freehold (**Figure 7**). Approximately 90% of all farmers are classified as smallholder subsistence to semi-commercial farmers (**Figure 6**). The remaining 10% of the population comprises commercial farmers who play a significant role in generating export earnings and providing employment opportunities for Pacific Islanders [33]. The agricultural sector in Fiji is predominantly reliant on smallholder farmers, which poses a challenge to the improvement of farm profitability.

As per the findings of the 2020 Fiji Agriculture Census, Fiji currently has a collective count of 6144 farms and an excess of 49,650 dairy cattle, constituting 26.7% of the overall livestock population. Additionally, it has been reported that more than 300 farms are registered under Fiji Dairy Corporate Company Limited (FCDCL), as per Group, W. B. [31] findings. With respect to the objective of dairy production, the majority of the milk produced, specifically 79.5%, was sold, while 17.1% was utilized for domestic consumption, 2.3% was distributed as gifts, and 1.1% was deemed unsuitable for use. Furthermore, the Central Division accounted for 71.2% of the total milk production at the divisional level, followed by the Western Division at 24.3%, the Northern Division at 4.5%, and the Eastern Division at 0.03% [30]. Annually, the nation's milk consumption amounts to approximately 80 million liters, while the domestic milk production amounts to a total of 20 million liters, with approximately half of it being contributed by the commercial sector and the other half by the unofficial sector. Consequently, the local demand for milk is met by importing approximately 60 million liters of milk annually [29].

6.2 Contribution of smallholder dairy systems to the Fijian dairy industry

The primary enterprise responsible for milk processing in Fiji is the Rewa Co-operative Dairy Company (RCDC), which underwent a name change in 2010 to become the Fiji Co-operative Dairy Company Limited (FCDCL). The FCDCL is situated in the Central Division of Viti Levu, the primary island of Fiji, which is also the location where the majority of dairy farmers operate [34]. During the period of 2005 to 2008, it was observed that the Fiji Co-operative Dairy Company Limited (FCDCL) received milk supply from a total of 260 dairy milk suppliers. However, only 80% of these suppliers were estimated to have provided milk to FCDCL. As indicated earlier, FCDCL received milk from a total of 206 farmers, while 35 other farmers directly provided fresh milk, ghee, and cream to consumers. Approximately 80% of registered dairy farmers are smallholder farmers who operate on a smaller scale due to the size of their farms. The smallholder farmers contributed to 44% of the fresh milk supplied to FCDCL, and they also tend to keep other species of livestock. Although ruminant livestock are not as numerous as non-ruminant livestock, they can boost smallholders' revenues. The bulk of ruminant livestock is owned by smallholders; thus, the adaptation of smallholder dairy farmers to climate change is vital. Further, PICs governments support to the smallholder dairy farmers can substitute importation of dairy products, create self-sufficiency for animal protein, and generating revenue for farmers [25].

7. Potential challenges affecting sustainability and resilience of dairy production in Fiji and the PICTs to climate change

7.1 Small and varying dairy cattle populations

Profitable dairy production depends on productive lactating cattle and replacement stock at the farm level. In the PICTs, although dairy cows' population data are limited, available data for Fiji and Tonga indicate a varying degree in dairy population growth. The data indicate that in Fiji dairy cattle increased from 22,551 in 2009 to 49,650 in 2019 representing an increase in 54.5% [32].

7.2 Global market shocks and importation of processed dairy products

There is limited data on imported dairy products in PICTs. There are few countries with reliable but dispersed data like Fiji which reveal that various dairy products are imported into PICTs [35]. In Fiji, the milk imported in 2020 was 2077.5 tons and 2740.6 tons in 2021 representing an increase in 24.2%, while other dairy products were 3795.6 and 4023.2 tons in 2020 and 2021, respectively, which represent an increase of 5.7%.

7.3 Occurrence of infectious cattle diseases and poor husbandry practices

A search for published articles on cattle diseases in the PICTs revealed limited information. Data from the few studies on dairy cattle showed that leptospirosis, brucellosis, tuberculosis, and helminths are the major dairy cattle diseases [36–38].

Animal husbandry practices are poor mainly due to insufficient technical knowledge, skills, and methods by the existing extension officers [39]. The government on the other hand continues to support and increases the number of trainings on dairy husbandry specialist areas, e.g., disease management, through their policies and programs but the university curriculums might need to be revised [39]. Farmers who have no formal dairy training try to improve their management and husbandry skills by going online for husbandry information, some of which could be found from the dairy officers within the government offices.

7.4 Limited livestock genetic resources

In the Pacific region, there exists a limited quality of genetic materials for breeding and production. This is worsened by limited access to improved genetic material from the neighboring developed economies such as Australia and New Zealand; when accessed they are usually expensive for the local farmers. Some of the factors contributing to the limited improvement of genetic resources in PICTs include lack of incentives for breeding and production, unknown diversity of animal genetic resources, and the conservation of specific animal genetic resources for future utilization. Available local breeds that are not high producers have resorted to inbreeding that gives rise to poor production.

7.5 High cost and lack of quality animal feeds

The high cost of livestock feed is one of the major constraints in the dairy industry. The main reason for the high cost of stock feed is due to its importation in the PICTs. As a result of the local stock feed shortage, imported feed is essential to cater for the dairy population. High costs of compounded or balanced feed (most imported) result in high cost of local products, and local producers cannot compete with imported products. Expensive animal feeds lead to higher production costs, thus contributing to higher prices for local dairy products supply, which further leads to cheaper consumer alternatives found on the imported products. There is limited availability (seasonal) and capacity to incorporate locally available feedstuffs in feed rations. Research on cheaper, more readily available alternatives for animal feed, will be very helpful for the dairy industries in the Pacific countries.

7.6 Inadequate livestock investment policies

In many Pacific Island countries, agriculture is a top governmental priority to encourage employment creation, food and nutrition security, and economic growth. Many of the existing policies are geared toward crop production with little attention to livestock such as dairy sector. Despite the existence of codified plans and strategies for agricultural development, these objectives have generally not been adequately funded by the government, and there has also been a dearth of implementation of funding to the livestock sector. To encourage and defend private sector investment and ensure the sector's viability, countries in the Pacific region must enact policies, provide high-level assistance, enforce rules, invest in infrastructure, and promote institutional innovations. However, potential policy pathways to improve access to and affordability of quality inputs and productive assets for feed and livestock production are possibly poorly understood by policymakers and decision-makers or are understudied. These aspects should be a future direction for research.

8. Climate change and its effect on dairy production in Fiji

The agricultural industry is highly susceptible to climatic changes, and the potential consequences of climate change could significantly jeopardize the sector's sustained productivity in the long term [40]. Heat stress is one of the significant impacts of climate change on dairy productivity. Heat stress lowers the organic and inorganic milk components produced by dairy cows, causing substantial financial burdens on the farmers [41]. It affects milk production and animal health and reduces reproduction efficiency [42]. The hot climate impacts cattle and causes direct and indirect heat stress on livestock output [43]. The reproduction efficiency of livestock is also highly vulnerable to climate change [44]. In cows, increasing temperature and high heat radiation load can negatively impact the reproductive rhythm *via* the hypothalamic-hypophyseal-ovarian axis [45]. Heat stress also results in reduced length and intensity of the estrous period, adversely affecting their conception due to reduced estradiol secretion [46, 47]. In addition, during pregnancy, heat stress can slow down embryonic development, resulting in reduced fetal growth and subsequently small calf size [48]. In some instances, heat stress in dairy livestock has caused early embryonic deaths [49].

Cyclones, droughts, floods, heat waves, and other extreme weather phenomena like storm surges frequently seen in the Pacific region can cause reduced livestock production as well as injury and mortality, particularly in more intense commercial operations with pigs, chickens, and dairy cattle. Flooding can restrict access to usable pasture in the short term, and if it continues, it will eventually harm or kill inundated grass. In addition to causing damage to roads, fences, wells, feeding stalls, and other management infrastructure, high winds and flooding can also have a negative impact on livestock productivity. Floods can also endanger cattle and humans by spreading water-borne illnesses [50] as well as infections that are transmitted by aquatic-stage vectors [51]. Annually, milk production is the lowest in the summer when rejection of milk is common due to high curd content. It is at this time that rainfall and temperature are highest, flood, and cyclones threats are the greatest in the Pacific. A 2% decrease in production level will occur for every degree that the ambient temperature rises [52]. Fiji milk production has been declining in the last decade due to extreme climatic changes and emergence of livestock diseases.

9. Recommended strategies to support climate change adaptation for sustainable dairy production in the PICTs

Different species of dairy cattle, dairy production systems and climate conditions are observed in the Pacific. This requires dairy production in the Pacific to adjust towards a sustainability climate change adaptation and mitigation systems [53]. Additionally, it is important to consider the environmental impact due to emission and pollution that may occur due to these dairy mitigation strategies in the Pacific [54]. Specifically, methane emissions by enteric fermentation are reported to be increased with greenhouse gas emission, with nitrous oxide reduction causing the leaching of nitrate and ammonia in the mixed farming system [55]. However, this is not well understood in the Pacific and requires further investigation. Mitigation mostly occurs directly or indirectly by improving dairy production system efficiency [56, 57]. For sustainable livestock production in the Pacific region, policies that encourage climate change adaptations, mitigations, greater productivity, and broader markets for animal products are required [58]. The next section discusses several potential strategies that can support ecosystem-based adaptation measures to boost dairy population, productivity, reduce dependency on imported processed dairy products, and mitigate impacts of climate change.

9.1 Development of appropriate and efficient dairy production systems

Globally, intensive livestock production systems have become key significant drivers in promoting sustainable livestock production. However, enforcing biosecurity, environmental regulations, and zoning will be required for disease prevention and management [59]. Approximately 90% of the cattle produced on the Pacific Island nations today are raised extensively. This production system remains within limits that are sustainable to the environment even if it is expanded to more than three times the current production levels. Many modern dairy facilities, however, are not well-regulated and are situated close to water coasts, rendering them susceptible to flooding and frequent cyclone storms in the Pacific. Additionally, they follow limited sound dairy husbandry principles. Ignoring good animal husbandry practices while setting up and managing livestock production operations is damaging to the environment and could also disrupt investments and decrease profitability.

In the Pacific region, governments play a vital role in controlling the growth and operation of the dairy sector. Modern dairy production growth depends on sustainable dairy intensification [60]. However, small- and medium-scale farmers can expand their operations and increase production with correct zoning, site selection for livestock, resource efficiency enhancements, energy-efficient water waste systems, recirculating animal waste systems, and backyard farming. These home farming methods can be used in many locations, even those close to cities and in climates that are unsuited for raising large numbers of livestock. Small-scale farming in cities, which is anticipated to take place because of population expansion, and the production of fresh goods closer to cities can enhance urban nutrition security.

9.2 Livestock genetic resources improvement and management

National resources for breeding and strain improvement in livestock are limited and insufficient. Because livestock strains in the PICTs yield less than optimum compared to improved strains elsewhere, access to superior animal breeds would increase output. Disease resistance and feed efficiency standards should be included in livestock breeding because most livestock in the PICTs are not selected. However, genetically modified livestock encounter opposition in many nations, which is motivated by worries they would threaten the survival of native livestock breeds. To understand how well-performing native and non-genetically modified livestock breeds performed in various management and agro-ecological systems, it will be necessary to gather enough benchmarking data, expand breed performance trials, conduct risk and economic assessments of alternative breeds, and increase the capacity and readiness of national strategies and the private sector to manage the modified livestock breeds in the PICTs.

Breed selection is currently employed to improve dairy production efficiency, and to improve and upscale massive production rate in the livestock sectors [61]. However, some selected breeds of species with improved productivity elicit higher metabolic heat production which exposes them to heat stress [62, 63]. Many livestock breed in the PICT remain uncharacterized. Thus, there is a great need for selection of genetically improved breed with adaptive mechanism against the effect of climate change toward improving productivity and efficiency [64]. Genetic selection of improved breed with heritability trails toward mitigating climate change effects as seen in the selection of *Bos indicus* cattle in Africa could be beneficial for the PICTs [64, 65].

9.3 Animal disease management practices to improve livestock health and production

In order to increase animal health and welfare, lower the financial burden of animal diseases, increase food safety, and lessen the possibility of antibiotic resistance, it is also crucial to invest in veterinary services and animal disease surveillance [66]. Animal disease prevention can stop the transmission of deadly zoonotic diseases at their source, where it is most cost-effective to act, as well as the spread of animal viruses to humans. Good livestock management improves the "One Health" plan, which strives to improve the health of humans, animals, and the environment.

Significant output losses are brought on by infectious diseases. In this context, several PICTs have stopped importing livestock inputs and products from nations with suspected transboundary animal diseases, significantly reducing supply and productivity. However, long-term solutions necessitate ongoing disease monitoring and surveillance both within and across borders, rapid diagnosis, and strengthening biosecurity at the farm and breeding locations. Unfortunately, currently in the region, there are hardly any specialists and reference laboratories specializing in livestock health management.

9.4 Innovation for affordable and quality animal feeds

A few large-scale feed companies exist (e.g., Pacific Feeds). For local feed producers, importing raw materials and processing equipment are expensive. Research on cheaper, more readily available alternatives for animal feed will be very helpful for the livestock industries in the Pacific countries. Recent innovations can increase livestock productivity, generate environmental benefits through efficient use of feed waste, and contribute to a circular economy. These new innovations utilize food that has been abandoned along the food supply chain as well as underutilized local products as a carbon source, including cassava peels, rice bran, and maize bran.

About 60% of the global biomass produced worldwide enters the livestock subsystem as feed or bedding material [67]. Greenhouse gas emissions from feed production represent 80% of the emission coming from eggs, chicken, and pigs and 35–45% of the milk and beef sectors [68]. The application of manure as fertilizer for crops and the deposition of manure on pastures generate a substantial amount of nitrous oxide emissions representing about half of these emissions [69]. Although livestock feed production often involves large applications of nitrogen to the soils, good manure management can reduce the need for manufactured fertilizers to be added to the soil during farming [70, 71].

9.5 Altering grazing intensity and/or manure use to enhance carbon sequestration

Overgrazing by livestock and other herbivores is associated with decreased plant growth, vegetation density, and biomass that lead to carbon input to the soil system [72]. Carbon sequestration is linked to excessive grazing, which is often detrimental to plant communities and soil carbon stocks [73]. The grazing system is fundamentally related to soil carbon stock [73]. The change in grazing system to intensive system depletes the soil carbon stock and grassland ecosystem with associated impact to sequestration of atmospheric CO_2 [74]. The grassland ecosystem takes up atmospheric CO_2 and mineral nutrients which are all converted into organic products to enhance forage growth. However, in grassland ecosystem, carbon stock on the soil is assimilated directly toward fiber, forage production, and growth conditions [75]. The ecosystem of the PICTs could greatly benefit from such integration. It is well established that ecosystems are a major source of biogenic greenhouse gas, CO_2, nitrous oxide, and methane gas [76]. However, when altered the carbon balance changes resulting in loss of respiration from photosynthesis even in matured old grasses [77, 78]. Notably, excessive forage consumption by livestock leads to substantial loss in carbon from soil and grasses [79]. Biomass in grassland is herbaceous with a small transient carbon pool, which mostly determines the carbon stock [80]. Intensive livestock production systems increase livestock forage production, which may contribute to soil carbon stock resulting in sequestration of atmospheric carbon in the soils [74]. Changes in the management system augment this effect leading to improvement in the favorable forage grasses and legumes. This management system includes soil fertilization, irrigation, and integrated grazing management system in the PICTs [81]. A rotation farming system with grass, hay, or pasture results to large impact on soil carbon stocks, with manure adding to soil organic matter build-up on the farmland [79]. Additionally, the use of seeded grasses for cover cropping increases carbon input to the soil by increasing the time required to by plants to atmospheric CO_2 in cropland systems [82]. Altering the management of grazing system with may reverse carbon sequestration produced by bushfire management and fertilization [83].

9.6 Improving grazing management

The changes in soil health improve the dairy farmer's potential to adapt to and mitigate the impact of climate change, which is the greatest factor that affects dairy cattle grazing management [84]. Grazing and grassland conversation has gained a lot attention toward dairy cattle grazing management [85]. Approximately, 25–30% of the earth surface can be used for grazing, and proper grazing management will increase production in the dairy production system [85, 86]. However, the livelihood of large populations of individuals is affected by climate change's effect on the grazing land [87]. Grazing land provides a significant ecosystem service by reducing runoff and erosion, protecting water quality, and providing wildlife habitat and recreational areas, thereby supporting habitat biodiversity [88]. It is reported that management of grassland together with grazing management system services contributes to climate change mitigation and adaptation. However, the lichening act of overgrazing and erosion results in changes in the soil texture and impacts crop production. Identifying the best management system in grazing and ecosystem services will go a long way in protecting and improving the grazing management toward providing adequate feed to the livestock in the PICTs [89].

Due to several uncertainties in the PICTs, several approaches to improve grazing management have earlier been proposed [84]. Integrated grazing and pastures could be intensively managed in a diverse cropping system with extended resting of lands and ecosystem services to improve grazing management systems [84]. This will have a great impact on the tradeoffs related to ecosystem, livelihoods, and socioeconomic factors in the PICTs [84]. Ecosystem service is a once clear factor that majorly affects the grazing land management system with climate change resilience serving as a critical point for increase rainfall variability in an area such as PICTs [90]. However, improving the effective water management in soils to capture more rainfall and make its available to plants during drier times is recommended [90]. In addition, soil water content is known to be linked to soil carbon and soil organic matter content that could further improve both crop and grazing land management's [91]. Hydro-properties associated with porosity and plant available water are linked to agricultural systems [92]. Continuous grazing strategy is a more complex strategy in grazing management toward improving full grass-based systems [92]. Changing stock densities is reported by several studies to be another basic strategy for improving grazing management systems in mitigating the effect of climate change [89]. Grazing system increases infiltration rates, which is basic mechanisms in grazing management strategy [84].

Sustainability of land and soil management systems is important in improving grazing management to enhance grass yield production for livestock [93]. The nature of the soil represents the major tools for climatic change mitigation, by altering the management process in which soil provides a wide range of services in sustainability of the crop production [94]. The grassland management system, soil properties, and quality support dairy farmers in protecting the negative impact of climate change on livestock production.

9.7 Management of land resources

The geomorphological parameters such as hill-slope structure, runoff pathways, topographic wetness, and sediment sinks determine the potential for water to be transported through a dairy production system [95]. The soil and landscape processes together form the connectivity of the landscape [95]. The process of understanding this makes it possible to predict how the dairy system reacts to changes in the drivers in terms of water and sediment connectivity. The design of these landscapes is mostly based on the cascade of processes that work together with nature to manage land and water resources [95]. The assessment can be linked to all parameters and processes that influence the structural and functional of dairy production systems in the PICTs. For example, mulching promotes plot scale infiltrations, reduces the runoff, and increases water availability for agriculture [96]. These management strategies can mitigate risks like flooding, agricultural droughts, and extreme soil erosion events to enhance biodiversity of pastures for dairy production [96, 97].

9.8 Provision of subsidized extensions services to livestock farmers to enhance their knowledge

The shortage of skilled staff is a recurring issue for livestock enterprises in the Pacific region. A lack of cutting-edge knowledge and agribusiness skills, as well as poor record keeping, hygiene, stocking, feeding, and water management procedures, hinders the production and revenues of small-scale agricultural businesses. Governments of Pacific Island countries must invest in boosting the skills of livestock extension agents and livestock-farmer groups in order to encourage training and disseminate sound farming techniques. Practical animal husbandry training programs in universities and technical institutions, in addition to government extension initiatives, would provide professionals with crucial knowledge [39]. Additionally, information and communication technology, such as radio, television, and smartphone apps, provide information and help to farmers at a reasonable cost and can be used for data collection to monitor and improve productivity.

9.9 Adjustment of public investment policy and support to farmers

In many Pacific Island countries, agriculture is a top governmental priority in order to encourage employment creation, food and nutrition security, and economic growth. Despite the existence of codified plans and strategies for agricultural development, these objectives have generally not been adequately funded by the government, and there has also been a dearth of implementation. To encourage and defend private sector investment and ensure the sector's viability, countries in the Pacific region must enact policies, provide high-level assistance, enforce rules, invest in infrastructure, and promote institutional innovations. Simpler and more efficient business practices reduced taxes to promote the expansion of agricultural businesses and lower import duties on items like machinery needed for domestic production of cattle and animal feeds are just a few examples of enabling policies. However, potential policy pathways to improve access to and affordability of quality inputs and productive assets for feed and livestock production are possibly poorly understood by policymakers and decision-makers or are understudied. These aspects should be a future direction for research.

Due to high transportation costs and unpredictable energy, the Pacific has a difficult time attracting and maintaining private sector investment and livestock value chains. Additionally, livestock, especially poultry, are highly susceptible to mortality, and livestock products can spoil while being produced, distributed, and transported. The building and maintenance of roads, transit systems, and electricity to cold power chains therefore demand investment. For instance, FCDCL's initiative to construct chilling milk plants in Fiji has increased dairy farmers' income and reduced milk losses. In some Pacific countries, current efforts to save infrastructure and logistical expenses include developing clusters or hubs of livestock farms and satellite farming that can be combined with service packages to minimize costs. Intensive livestock farmers are increasingly interested in solar energy as it becomes necessary for more intense, highly productive livestock companies.

Dairy farming has recently become riskier because of animal diseases like bovine tuberculosis, illustrated by the forced closure of some farms in Fiji with high prevalence. Due to this, private sector investment in this area has stagnated. In the short and medium term, government, development financial institutions, impact investors, and innovative financing may result to interest-free loans to farmers to initiate fresh waves of economic growth in the dairy subsector. In order to increase livestock output, correct zoning and location in acceptable and suitable locations are required. To prevent negative environmental effects and disruptions in investments and activities related to livestock, good practices must be enforced.

Data-driven and evidence-based policy reform and decision-making by stakeholders will be made possible by improving the quality of data collection, monitoring systems, and assessment for livestock food systems from the economic, social, and environmental dimensions and utilizing low-cost digital technologies, creative crowdsourcing, and public-private partnerships. Currently, few accurate production statistics available, and data from institutions such as the UN Food and Agriculture Organization, do not always align with other sources such as industry production estimates. Overall, there are not many thorough impact analyses of livestock rising in the Pacific.

10. Conclusion

Although livestock farmers in the PICT face several challenges such as low livestock productivity, livestock diseases, overreliance on imported livestock products, and impacts of climate change, there is an opportunity to improve dairy production systems to cater for the growing population. Ecosystem-based adaptation practices, such as land conservation and management, nutrient recycling through use of animal manure, and utilization of genetic diversity, and climate change mitigation measures such as efforts to reduce greenhouse gas emissions, will be required to help conserve the ability of the agro-ecological systems to sustain dairy production, profitability, and ecosystem services. However, broader policies on dairy production such as disease control, animal feed processing, and public financing of the livestock sector will be required to aid transitioning to resilient dairy production in the PICTs.

Acknowledgements

The study was funded by the Australian Centre for International Agricultural Research (ACIAR), project number LS/2019/119.

Author contributions

RM conceived and drafted the article. AD reviewed the first draft and wrote sections of the manuscript. WM, SG, and RR summarized the data from animal census reports. PS and WO carried out analysis and interpretation of results and wrote part of the manuscript. PI revised the manuscript critically for important intellectual content. All authors read and approved the manuscript.

Disclosure statement

The authors declare no conflict of interest.

Ethics and consent

Not applicable.

Data availability statement

The data that support the findings of this study are openly available in Ministry of Agriculture; National Agriculture census report published online and Fiji Meteorological Service (FMS).

Author details

Royford Bundi Magiri[1]*, Phillip Sagero[2], Abubakar Danmaigoro[1], Razia Rashid[1], Wati Mocevakaca[1], Shivani Singh[1], Walter Okello[3] and Paul A. Iji[1]

1 Department of Veterinary Science, College of Agriculture, Fisheries and Forestry, Fiji National University, Nausori, Fiji

2 Department of Geography, Earth Science and Environment, University of South Pacific, Suva, Fiji

3 Commonwealth Scientific and Industrial Research Organization, Black Mountain Science and Innovation Park, Australia

*Address all correspondence to: royford.magiri@fnu.ac.fj

IntechOpen

© 2023 The Author(s). Licensee IntechOpen. This chapter is distributed under the terms of the Creative Commons Attribution License (http://creativecommons.org/licenses/by/4.0), which permits unrestricted use, distribution, and reproduction in any medium, provided the original work is properly cited.

References

[1] Nations U. World Population Projected to Reach 9.6 Billion by 2050. USA: Department of Economic and Social Affairs, Online; 2013

[2] Scandurra G et al. On the vulnerability of small island developing states: A dynamic analysis. Ecological Indicators. 2018;**84**:382-392

[3] Rosegrant MW et al. Food Security in a World of Natural Resource Scarcity: The Role of Agricultural Technologies. Washington, D.C., USA: The International Food Policy Research Institute; 2014

[4] Thornton PK et al. The impacts of climate change on livestock and livestock systems in developing countries: A review of what we know and what we need to know. N.W. Washington, D.C., U.S.A: Agricultural Systems. 2009;**101**(3):113-127

[5] Dearie C et al. Trends in mortality and life expectancy in Fiji over 20 years. BMC Public Health. 2021;**21**(1):1-14

[6] Lal S et al. Projecting populations for major Pacific island countries with and without COVID-19: Pro-active insights for population policy. Journal of Population Research. 2022;**39**(2):257-277

[7] Barrett CB et al. Agri-food value chain revolutions in low-and middle-income countries. Journal of Economic Literature. 2020;**58**:1-67

[8] Bakare AG et al. Impact of climate change on sustainable livestock production and existence of wildlife and marine species in the South Pacific island countries: A review. International Journal of Biometeorology. 2020;**64**(8):1409-1421

[9] Sánchez Mendoza B et al. Causes and consequences of climate change in livestock production and animal health. Review. Revista Mexicana de Ciencias Pecuarias. 2020;**11**:126-145

[10] Tinsley TL et al. Managing cow herd dynamics in environments of limited forage productivity and livestock marketing channels: An application to semi-arid Pacific island beef production using system dynamics. Agricultural Systems. 2019;**173**:78-93

[11] Sisifa A et al. Pacific communities, agriculture and climate change. In: Vulnerability of Pacific Island Agriculture and Forestry to Climate Change. Australia: University of Sunshine Coast; 2016. pp. 5-45

[12] Le QB. Food System Resilience: Defining the Concept. United States: Global Food Security; 2016

[13] Lawrence MA et al. Sustainable, resilient food systems for healthy diets: The transformation agenda. Public Health Nutrition. 2019;**22**(16):2916-2920

[14] Medina Hidalgo D et al. Adaptation, sustainable food systems and healthy diets: An analysis of climate policy integration in Fiji and Vanuatu. Climate Policy. 2022;**22**:1-16

[15] Claudine H, Hart T. Livestock production and livelihood of people in Australia. Journal of Agriculture. 2020;**4**(1):53-59

[16] Herrero M et al. Livestock, livelihoods and the environment: Understanding the trade-offs. Current Opinion in Environmental Sustainability. 2009;**1**(2):111-120

[17] Chattopadhyay M, Katzfey J. Simulating the climate of South Pacific islands using a high resolution model. International Journal of Climatology. 2015;**35**(6):1157-1171

[18] Secretariat I. IPPC Procedure Manual for Implementation and Capacity Development. Rome: IPPC; 2022

[19] Mataki M, Koshy KC, Lal M. Baseline climatology of Viti Levu (Fiji) and current climatic trends. Pacific Science. 2006;**60**(1):49-68

[20] Rhee J, Yang H. Drought prediction for areas with sparse monitoring networks: A case study for Fiji. Water. 2018;**10**(6):788

[21] Kumar R, Stephens M, Weir T. Rainfall trends in Fiji. International Journal of Climatology. 2014;**34**(5): 1501-1510

[22] Glantz MH. Currents of Change: Impacts of El Niño and La Niña on Climate and Society. United Kingdom: Cambridge University Press; 2001

[23] Ongoma V et al. Variability of diurnal temperature range over Pacific island countries, a case study of Fiji. Meteorology and Atmospheric Physics. 2021;**133**:85-95

[24] Thorne P et al. Reassessing changes in diurnal temperature range: Intercomparison and evaluation of existing global data set estimates. Journal of Geophysical Research: Atmospheres. 2016;**121**(10):5138-5158

[25] Aregheore EM. Feeds and Forages in Pacific Islands Farming Systems. Fiji. Available from: http://www.fao.org/ag/AGP/AGPC/doc/Newpub/feeds_forages/feeds_forgaes.htm: Animal Science Department Alafua Campus, the University of the South Pacific; 2005

[26] Aregheore E, Perera D. Effects of Erythrina variegata, Gliricidia sepium and Leucaena leucocephala on dry matter intake and nutrient digestibility of maize stover, before and after spraying with molasses. Animal Feed Science and Technology. 2004;**111**(1-4):191-201

[27] Nube M, Voortman R. Simultaneously Addressing Micronutrient Deficiencies in Soils, Crops, Animal and Human Nutrition: Opportunities for Higher Yields and Better Health. Center for World Food Studies. Amsterdam, Netherlands: Staff Working Paper; 2006

[28] Bonell M, Callaghan J, Connor G. 11 synoptic and mesoscale rain producing systems in the humid tropics. In: Forests, Water and People in the Humid Tropics: Past, Present and Future Hydrological Research for Integrated Land and Water Management. Cambridge: Cambridge University Press; This page intentionally left blank; 2004. p. 194

[29] Kumar S, Reddy M. Fiji's dairy industry: A cost and profitability analysis. Pacific Economic Bulletin. 2019;**23**:23-28

[30] Ministry of Agriculture, 2020. Fiji Agriculture Census. Suva: Fiji Government; 2020

[31] Group WB. Global financial development report 2014: Financial inclusion. Vol. 2. USA: World Bank Publications; 2013

[32] Bacolod ED, Tabunakawai N, Natasiwai T. Fiji 2020 agriculture sector policy agenda. Fiji Islands: Ministry of Agriculture; 2020

[33] Iese V et al. Impacts of COVID-19 on agriculture and food systems in Pacific island countries (PICs): Evidence from communities in Fiji and Solomon islands. Agricultural Systems. 2021;**190**:103099

[34] Igbal MR. Bovine mastitis in Fiji: Economic implications and management—A review. Journal of Agricultural Science. 2021;**13**(10):162

[35] Snowdon W et al. Processed foods available in the Pacific islands. Globalization and Health. 2013;**9**(1):1-7

[36] Brioudes A et al. Diseases of livestock in the Pacific islands region: Setting priorities for food animal biosecurity. Acta Tropica. 2015;**143**:66-76

[37] Tukana A, Hedlefs R, Gummow B. Brucella abortus surveillance of cattle in Fiji, Papua New Guinea, Vanuatu, the Solomon Islands and a case for active disease surveillance as a training tool. Tropical Animal Health and Production. 2016;**48**(7):1471-1481

[38] Borja E et al. A retrospective study on bovine tuberculosis in cattle on Fiji: Study findings and stakeholder responses. Frontiers in Veterinary Science. 2018;**5**:270

[39] Magiri R et al. The role of agricultural institutions in providing support towards sustainable rural development in South Pacific Island countries. Journal of Agricultural Science. 2022;**14**(2)

[40] Iese V et al. Agriculture under a changing climate. In: Climate Change and Impacts in the Pacific. Switzerland: Springer Nature; 2020. pp. 323-357

[41] Summer A et al. Impact of heat stress on milk and meat production. Animal Frontiers. 2019;**9**(1):39-46

[42] Abdurehman A, Ameha N. Prospects of climate change on livestock production. Journal of Scientific and Innovative Research. 2018;7(4):100-105

[43] Lacetera N. Impact of climate change on animal health and welfare. Animal Frontiers. 2019;**9**(1):26-31

[44] Hansen P. Exploitation of genetic and physiological determinants of embryonic resistance to elevated temperature to improve embryonic survival in dairy cattle during heat stress. Theriogenology. 2007;**68**:S242-S249

[45] Sheikh AA et al. Effect of climate change on reproduction and milk production performance of livestock: A review. Journal of Pharmacognosy and Phytochemistry. 2017;**6**(6):2062-2064

[46] Naqvi S, Sejian V. Global climate change: Role of livestock. Asian Journal of Agricultural Sciences. 2011;**3**(1):19-25

[47] Baumgard LH, Rhoads RP, Rhoads ML, Gabler NK, Ross JW, Keating AF, et al. Impact of Climate Change on Livestock production. Environmental stress and amelioration in livestock production; 2012. pp. 413-468

[48] Das S. Impact of climate change on livestock, various adaptive and mitigative measures for sustainable livestock production. Approaches in Poultry, Dairy and Veterinary Sciences. 2017;**1**:33

[49] Skliarov P et al. Impaired reproductive performance of dairy cows under heat stress. Agriculturae Conspectus Scientificus. 2022;**87**(2):85-92

[50] Hoberg E et al. Pathogens of domestic and free-ranging ungulates: Global climate change in temperate to boreal latitudes across North America. Revue Scientifique et Technique. 2008;**27**(2):511-528

[51] Taylor M, McGregor A, Dawson B. Vulnerability of Pacific Agriculture and

Forestry to Climate Change. Suva, Fiji: Secretariat of the Pacific Community; 2016

[52] Makanjuola BO et al. Effect of recent and ancient inbreeding on production and fertility traits in Canadian Holsteins. BMC Genomics. 2020;**21**(1):1-15

[53] Koop SH, van Leeuwen CJ. The challenges of water, waste and climate change in cities. Environment, Development and Sustainability. 2017;**19**(2):385-418

[54] Mcleod E et al. Lessons from the Pacific islands–adapting to climate change by supporting social and ecological resilience. Frontiers in Marine Science. 2019;**6**:289

[55] Eckard R, Clark H. Potential solutions to the major greenhouse-gas issues facing Australasian dairy farming. Animal Production Science. 2018;**60**(1):10-16

[56] Malhi GS, Kaur M, Kaushik P. Impact of climate change on agriculture and its mitigation strategies: A review. Sustainability. 2021;**13**(3):1318

[57] Fawzy S et al. Strategies for mitigation of climate change: A review. Environmental Chemistry Letters. 2020;**18**:2069-2094

[58] Michalk DL et al. Sustainability and future food security—A global perspective for livestock production. Land Degradation & Development. 2019;**30**(5):561-573

[59] Ragasa C et al. Sustainable aquaculture development in sub-Saharan Africa. Nature Food. 2022;**3**(2):92-94

[60] Campbell BM et al. Sustainable intensification: What is its role in climate smart agriculture? Current Opinion in Environmental Sustainability. 2014;**8**:39-43

[61] Faverdin P et al. Animal board invited review: Specialising and intensifying cattle production for better efficiency and less global warming: Contrasting results for milk and meat co-production at different scales. Animal. 2022;**16**(1):100431

[62] Osei-Amponsah R, Asem EK, Obese FY. Cattle crossbreeding for sustainable milk production in the tropics. International Journal of Livestock Production. 2020;**11**(4):108-113

[63] Osei-Amponsah R et al. Heat stress impacts on lactating cows grazing Australian summer pastures on an automatic robotic dairy. Animals. 2020;**10**(5):869

[64] Palanivel H, Shah S. Unlocking the inherent potential of plant genetic resources: Food security and climate adaptation strategy in Fiji and the Pacific. Environment, Development and Sustainability. 2021;**23**(10):14264-14323

[65] Aliloo H et al. Ancestral haplotype mapping for GWAS and detection of signatures of selection in admixed dairy cattle of Kenya. Frontiers in Genetics. 2020;**11**:544

[66] Otte M, Chilonda P. Animal Health Economics: An Introduction. Rome, Italy: Animal Production and Healthy Division (AGA), FAO; 2000. p. 12

[67] Meisterling KW. Climate Implications of Biomass Appropriation: Integrating Bioenergy and Animal Feeding Systems. USA: Carnegie Mellon University; 2011

[68] Barthelmie RJ. Impact of dietary meat and animal products on GHG

footprints: The UK and the US. Climate. 2022;**10**(3):43

[69] Sommer SG, Hansen MN. Incorporation of deep litter reduce ammonia volatilization: Effect of incorporation depth and timing. Atmospheric Environment: X. 2022;**13**:100148

[70] Tian M et al. Pollution by antibiotics and antimicrobial resistance in livestock and poultry manure in China, and countermeasures. Antibiotics. 2021;**10**(5):539

[71] Rayne N, Aula L. Livestock manure and the impacts on soil health: A review. Soil Systems. 2020;**4**(4):64

[72] Patón D. Climatic and biological factors related with goat grazing management in the arid grassland of the Coquimbo region (Northern Chile). Ecologies. 2021;**2**(4):345-365

[73] Madigan AP et al. Full inversion tillage (FIT) during pasture renewal as a potential management strategy for enhanced carbon sequestration and storage in Irish grassland soils. Science of the Total Environment. 2022;**805**:150342

[74] Abdalla K et al. Grassland rehabilitation significantly increases soil carbon stocks by reducing net soil CO_2 emissions. Soil Use and Management. 2022;**38**(2):1250-1265

[75] Padbhushan R et al. Delineate soil characteristics and carbon pools in grassland compared to native forestland of India: A meta-analysis. Agronomy. 2020;**10**(12):1969

[76] Dabi T, Khanna V. Effect of climate change on rice. Agrotechnology. 2018;7(2):2-7

[77] Leakey AD et al. Genomic basis for stimulated respiration by plants growing under elevated carbon dioxide. Proceedings of the National Academy of Sciences. 2009;**106**(9):3597-3602

[78] Dusenge ME, Duarte AG, Way DA. Plant carbon metabolism and climate change: Elevated CO_2 and temperature impacts on photosynthesis, photorespiration and respiration. New Phytologist. 2019;**221**(1):32-49

[79] Brown JN et al. Diagnosing the weather and climate features that influence pasture growth in Northern Australia. Climate Risk Management. 2019;**24**:1-12

[80] Yang W-S et al. SOC changes were more sensitive in alpine grasslands than in temperate grasslands during grassland transformation in China: A meta-analysis. Journal of Cleaner Production. 2021;**308**:127430

[81] Berça AS et al. Advances in pasture management and animal nutrition to optimize beef cattle production in grazing systems. In: Animal Feed Science and Nutrition-Production, Health and Environment. London, UK: IntechOpen; 2021

[82] Porwollik V et al. The role of cover crops for cropland soil carbon, nitrogen leaching, and agricultural yields–a global simulation study with LPJmL (V. 5.0-tillage-cc). Biogeosciences. 2022;**19**(3):957-977

[83] Xu Z, Zhao Z, Lu C. The contribution of land use and land cover on carbon storage in the North Tibet plateau, China. JAPS: Journal of Animal & Plant Sciences. 2021;**31**(6):1598-1609

[84] Teague R, Kreuter U. Managing grazing to restore soil health, ecosystem function, and ecosystem services. Frontiers in Sustainable Food Systems. 2020;**4**:534187

[85] Wezel A et al. Good pastures, good meadows: Mountain farmers' assessment, perceptions on ecosystem services, and proposals for biodiversity management. Sustainability. 2021;**13**(10):5609

[86] Schmitt TM et al. Ecosystem services from (pre-) alpine grasslands: Matches and mismatches between citizens' perceived suitability and farmers' management considerations. Ecosystem Services. 2021;**49**:101284

[87] Paudel B et al. Climate change and its impacts on farmer's livelihood in different physiographic regions of the trans-boundary koshi river basin, central Himalayas. International Journal of Environmental Research and Public Health. 2021;**18**(13):7142

[88] Yang Y. Evolution of habitat quality and association with land-use changes in mountainous areas: A case study of the Taihang Mountains in Hebei Province, China. Ecological Indicators. 2021;**129**:107967

[89] Baronti S et al. Rotational pasture management to increase the sustainability of mountain livestock farms in the Alpine region. Regional Environmental Change. 2022;**22**(2):50

[90] Oyebola OO et al. Potential adaptation strategies for climate change impact among flood-prone fish farmers in climate hotspot Uganda. Environment, Development and Sustainability. 2021;**23**:12761-12790

[91] Verma P et al. Strategies to improve agriculture sustainability, soil fertility and enhancement of farmers income for the economic development. In: Soil Fertility Management for Sustainable Development. Switzerland: Springer; 2019. pp. 43-70

[92] Eze S et al. Impacts of conservation agriculture on soil structure and hydraulic properties of Malawian agricultural systems. Soil and Tillage Research. 2020;**201**:104639

[93] Jurado-Guerra P et al. The grasslands and scrublands of arid and semi-arid zones of Mexico: Current status, challenges and perspectives. Revista Mexicana de Ciencias Pecuarias. 2021;**12**:261-285

[94] Hou D et al. Sustainable soil use and management: An interdisciplinary and systematic approach. Science of the Total Environment. 2020;**729**:138961

[95] Bollati IM, Cavalli M. Geomorphic systems, sediment connectivity and geomorphodiversity: Relations within a small mountain catchment in the Lepontine Alps. In: Geomorphometry 2020. Padova, Italy: CNR; 2020. pp. 50-54

[96] Wei Z et al. Mitigation of nitrogen and phosphorus leaching from black soil croplands in Northeast China. 中国生态农业学报 (中英文). 2021;**29**(1):113-118

[97] Kaluarachchi Nartallo Y. Potential advantages in combining smart and green infrastructure for future cities. Frontiers of Engineering Management. 2020;**8**:98-108

Chapter 14

Recent Changes in Temperature and Maximum Snow Cover Days over the Northern Hemisphere with a Focus on Alaska

Vinay Kumar and Robert Ross

Abstract

One aspect of climate variability is the shift in seasonal change, with a given season arriving early or late. However, this shift in season is location-dependent and affects local ecology. Over subpolar regions, the change in temperature is very much associated with the regional and local variability of snow-caps, sea ice near the pole, pole-ward transportation of heat, cloud cover, and wind circulation. Based on a 36-year analysis of skin temperature, we found that the lowest temperature occurred in March rather than in February. Additionally, the maximum snow cover day has shifted from March 12 to March 17 in the last 3 to 4 decades. A plausible reason for the late accumulation of ice/snow over the Arctic/Alaskan region may be due to the multi-scale interactions between multi-decadal oscillations, for example, North Atlantic Oscillations (NAO) and climate change.

Keywords: seasonal shift, snow caps, multi-decadal oscillations, Alaska region, climate change

1. Introduction

Since 1978, NASA satellites have been monitoring the variability of polar ice and snow. Around and over Alaska, seasonal accumulation of ice and snow is an important aspect of the seasons. The Arctic region is experiencing climate change faster than anywhere else on Earth [1]. Arctic ice melt is the complex feedback process that involves poleward movement of heat (from the ocean and atmosphere), moisture transportation, cloud formation, and surface albedo. An enhanced moisture transport and latent heat from the subpolar region to the polar region drive sea-ice retreat and low-level cloud formation for surface warming [2]. The rapid change in Arctic ice impacts the environment, ecosystems, economies, and local communities. In fact, sea ice starts to melt if the temperature rises above freezing point, while undersurface water remains above freezing in Alaska and the Arctic region. After winter, the sun starts to migrate from the southern hemisphere to the Northern Hemisphere. The Arctic region carries seasons around 9 months of winter (long, icy, and dark), around

3 months of summer (short, very cool, and bright), and spring and Autumn for a few weeks. In the transition months (February and March), the heat from the growing temperature is utilized in melting the sea-land ice. It lowers the warmth in summer over the region. The melting and thawing of ice levels led to the formation of new wetlands and ponds over the Tundra region. Climate change might alter the frozen regions into warmer, less frozen regions, thus producing an uncertain future for the Arctic [3]. In addition, March and September are the months of maximum and minimum snow cover over the Arctic Circle regions. The seasonal variation of ice extent over Arctic ice is more than Antarctic region, although the Northern Hemisphere gets more snow than the counterpart southern hemisphere.

The variability of temperature is one of the intriguing crises and worries of the world community. The underlying arctic amplification feedback includes ice-albedo feedback, cloud feedback, and others [4]. This feedback of the greenhouse effect and ice-albedo contribute to around 75 and around 25%, respectively [5]. Basically, the influence of greenhouse gases and other feedback mechanisms on climate change was foresighted long back by many researchers [6, 7]. Screen and Simmonds noted warming over the Arctic faster than the rest of the globe in the last few decades and evident warming in seasons in conjunction with sea ice decline [8]. The surface temperature warming appears to be during the season change and is linked with atmospheric and ocean heat, circulation, and moisture transport [9]. Increasing air temperature and sea temperatures continue to trigger changes in the Arctic region. The Arctic is warming at twice the rate of anywhere else on Earth, which is also known as Arctic amplification. In Global oceans, the problem of coral bleaching is linked to an increase in ocean temperature. Another issue is related to Polar bears, who depend on sea ice extent and thickness for their basic needs, for example, dens, food, and mating. The loss of sea ice is affecting populations of polar bears, and walruses' populations, and their health, as mentioned by Kathy Crane (NOAA).

The rise in temperature is attributed to greenhouse gases, pollution, and anthropogenic reasons [5]. The variation in the surface temperature greatly influences the sea ice and land snow cover. In turn, variability of the snow cover and sea ice extent influences the temperature variation on seasonal and local scales. Every year the sea ice and snow cover are reduced to a minimum extent and the surface temperature is increasing. Every year there is a time when the snow cover is maximum, which has a great influence on the winter season of a region nearby.

In response to climate change, the Arctic region is experiencing a great rise in surface temperature [4, 10]. The climate and ecological aspect of the region is susceptible to tipping points due to the loss of sea ice [11]. Most of the models underestimated the projection of the fast decline of sea ice [12]. Data from NASA show that average surface temperatures across the Arctic Ocean and air increased an average of 0.16 and 0.09°C/yr. from 2003 to 2013 [13]. In another study, it is revealed that the Arctic has been experiencing warmer and wetter since 2001 [14]. The reason for such an environment is linked to sea-ice melt and evaporation of water from ice-free ocean (open-water). They further added that due to moisture and heat feeding into the Arctic-wide, the trend of early melt and later freeze will likely continue.

Owing to temperature changes, the observed and projected fluctuations in seasonal temperature might have a deep impact on animals, flora and fauna, economic activities, and ecosystems [15]. Increasing air temperatures, sea surface temperatures, declining reflectivity at the surface, shrinking spring land snow cover, summer sea ice, and declining populations of polar bear populations are among many observations being made in the Arctic region (Arctic Report card 2014). Along with the influence of Ocean-Atmospheric variability on Arctic ice melt, decadal-scale models of variability, for example, Atlantic Multidecadal Oscillation and Pacific Decadal Oscillation were also implicated in Arctic sea-ice variability [4].

The significance of this work is quite large since we are addressing an atmospheric signal, from the recent Arctic melt. The largest warm temperature in the tropical troposphere resides in the monsoon near 25oN. The circulations from this region can reach Central Asia and eventually the Arctic, with a substantial increase in the meridional heat flux over the Arctic Circle [16]. The objective of this research work is to assess and relate the decrease in skin temperature and snow/ice cover over and around the Alaska region.

2. Datasets and methodology

In total, 35 years of data have shown effective variability over polar regions and surroundings using skin temperature, geopotential height, winds, and snow cover [17]. This study had the limitation of snow cover datasets from NSDIC. Thus, all the other datasets (ERA-Interim) also showed up to 2014. A standard climatology duration of 1979–2008 was considered in all the figures. The correlation coefficient between skin temperature, and snow cover over the Arctic region, and Alaskan region is calculated. The datasets of snow cover and multi-decadal oscillations (e.g., AO/NAO, North Atlantic Oscillations) are acquired from online data portal [18].

3. Results

3.1 Northern hemisphere

3.1.1 Skin temperature anomalies

Keeping climate change in the background, as sea ice withdraws in summer, SST in the Arctic Ocean region has risen. Among all, the most significant linear trend was observed in the Chukchi Sea (northwest of Alaska), where skin temperature has increased at a rate of 0.5°C/decade. In August 2014, SST was as much as 7.2°F higher than the 1982–2010 average in the Laptev Sea (north of Russia), and in the Bering Strait region, while SST in the Barents Sea, north of Norway, was close to the 1982–2010 average (NOAA Media release, 2014). North of Alaska in the Beaufort Sea there is clockwise churning of the ice which replenishes old ice with new, known as a nursery for older ice.

The variability of temperature on a monthly to yearly scale shows complex variations over the Alaska region. Thus, we considered the decadal variability of skin temperature. **Figure 1** shows the skin temperature anomalies of March, over the North Pole region (in stereo-polar format). Anomalies of positive skin temperature can be seen over the north polar region from 1979 to 1984 to 1985–1994. However, after that temperature anomalies remain positive (warming). March is the month when the season starts to change from winter to summer. In the latest decade 2005–2014, positive temperature anomalies can be seen over the USA, Canada, Russia, and Middle East regions. For 1982–2014, we found a correlation of −0.36 between the NAO index and March surface temperature averaged over a box (**Figure 1a**).

There are many other features that are hidden in **Figure 1**, for example, abnormal warnings over the Middle East and Saudi Arabia region. However, such abnormalities are quite local and hardly have an impact on Arctic ice melt.

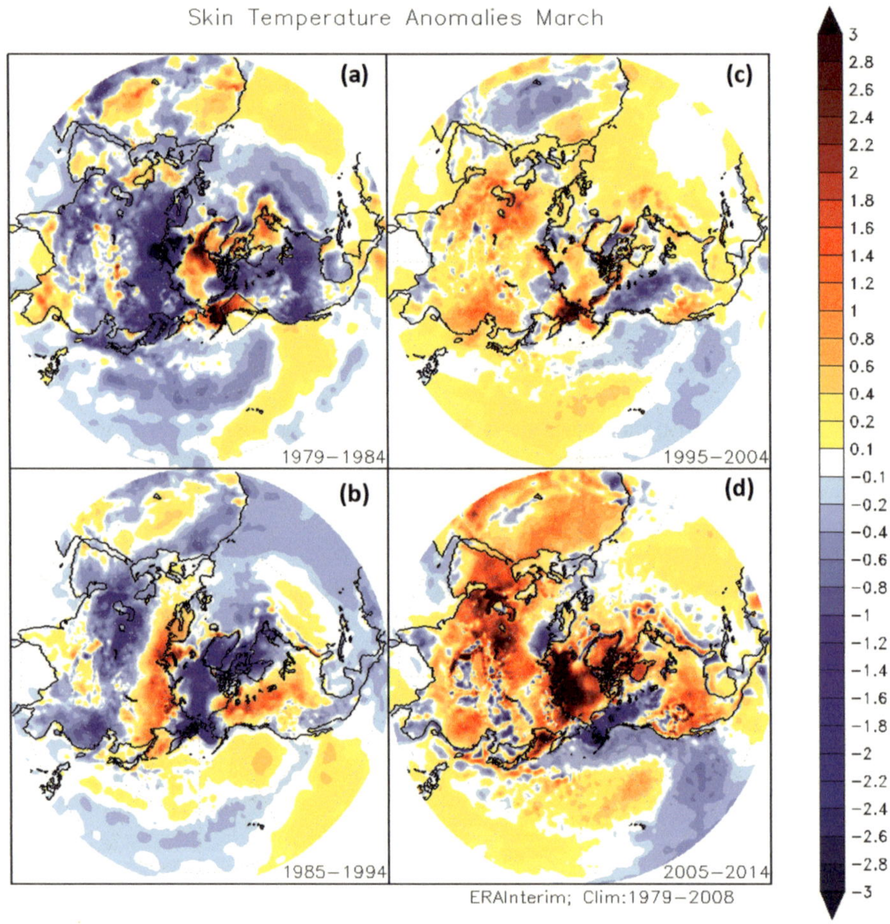

Figure 1.
Skin temperature anomalies for the month of march showing the decadal variability from some decades (a–c) and the remaining 6 years (1979–1984, d).

The Alaska region remains a valuable resource of natural resource for the USA. Any change in the atmospheric conditions might affect the economy and ecosystems of the region. **Figure 2** shows the decrease in March skin temperature over a region (135 W–168 W, 60 N–70 N; shown by a black box in **Figure 1a**) in Alaska. March is a month when the Sun starts to move toward the Northern Hemisphere from the southern hemisphere. This spring warming (heat) is used to melt the snow and surrounding sea ice over the Alaskan region, which is one of the reasons that the skin temperature goes down in the region. Other possible factors might be the oscillation of NAO and AO over the decadal scale. A decrease of 3°C is noted over selected regions in the last 36 years. Additionally, a fluctuation of 7°C between 2005 and 2007 was unique in several decades.

3.1.2 Wind anomalies

In America and Canada, regions where the jet stream plays a crucial role in producing freezing and warm weather. Such a pattern of cold and warm weather extends to Europe and Russia via these fast-moving wind streams. In 2014, the jet stream brought very cold air into eastern North America and central Russia, while

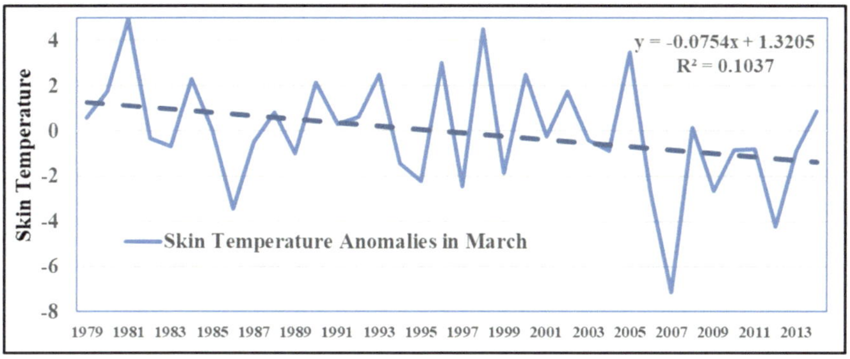

Figure 2.
March month skin temperature anomalies (°C) for 35 years averaged over (135 W–168 W, 60 N–70 N) region. Data source: Era-interim.

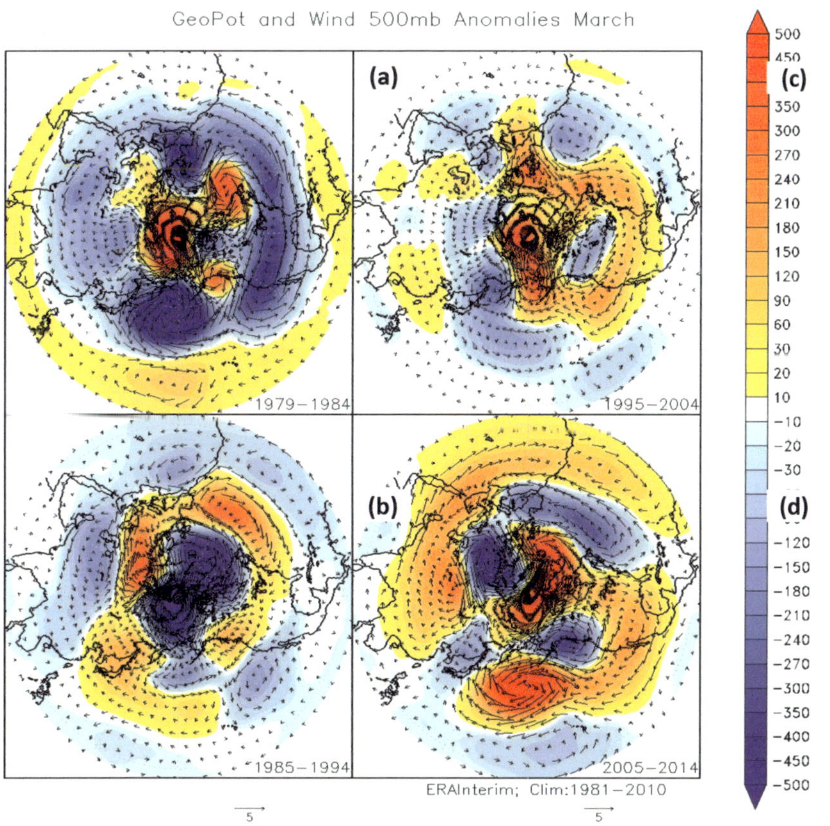

Figure 3.
Geopotential and wind anomalies at 500mb for the month of march showing the decadal variability from some decades (a–c) and the remaining 6 years (1979–1984, d).

extremely warm air into Alaska and Northern Europe [19]. Additionally, temperature anomalies of 10 DegC higher than the average for January were taped over Alaska. A difference in the temperature over different parts of the earth influences pressure and then wind pattern. **Figure 3** shows the Rossby wave pattern of low, high, and low. As a decadal average, the wind pattern has changed dramatically over different parts of the northern pole.

3.1.3 Snow cover and variability

Wind and air temperatures curate the snow cover of the surface of Earth. In the era of climate change warning, every next year seems to carry less snow cover [12]. Advancing with this reality, in spring 2014, snow cover across the Arctic was reduced below the long-term mean. Further, due to above-normal temperatures and below-normal snow aggregation, the snow vanished 3 to 4 weeks earlier than normal over the Canadian subarctic, and western Alaska [19]. **Figure 4** shows snow cover over the North Pole, which shows the maximum snow cover from March 12 to March 17 in the last 36 years. A shift of 5 days for maximum snow covering the Northern Hemisphere is notable for ecology, economy, tourist, and environmental purposes. These variations are observed from February 24 to April 1, a range of 37 days.

It is now very clear that every year maximum snow cover (in millions of square kilometers) is decreasing from many studies and datasets demonstrated in **Figure 5**.

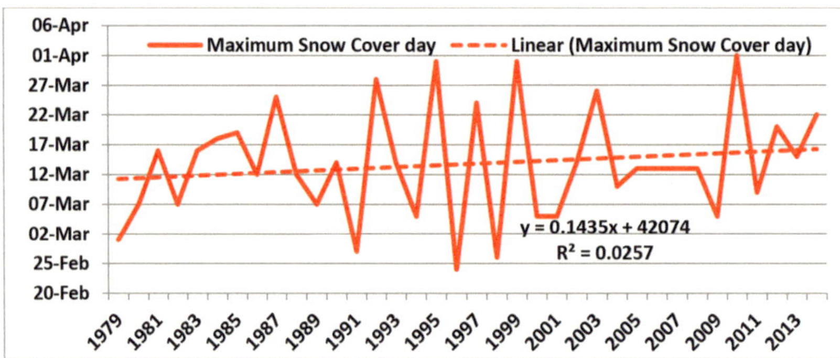

Figure 4.
The line diagram shows a day in a year when snow cover was maximum (million square kilometers) over the Arctic region. Data source: NSDIC.

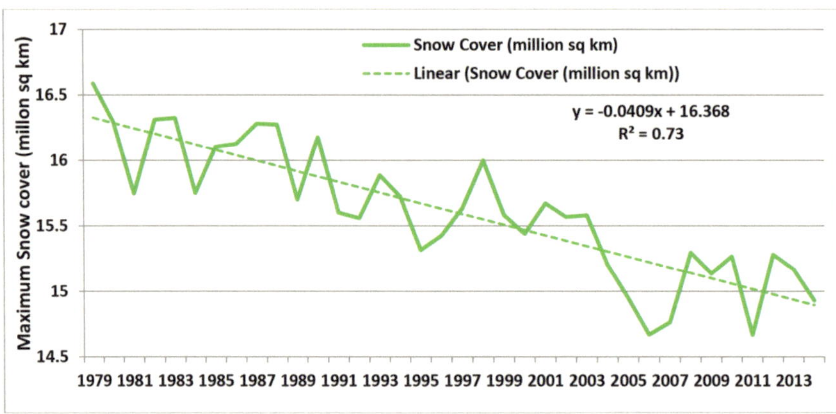

Figure 5.
The line chart shows the details of the maximum snow cover (million square kilometers) over the Arctic region. Data source: NSDIC.

Mid-March is the time when snow and ice cover start to melt over the northern pole region. In the year 2015, maximum coverage of ice started melting around February 25 as stated by Jeff Key (NOAA's Satellite and Information Service Center). It seems snow and ice cover started to melt earlier and faster than in previous years in general. In the recent year, 2009 (April 1) witnessed lower snow cover than in 2010, with a difference of 0.21 square km (**Figure 5**). However in 2009, the maximum accumulation of snow was observed on April 1, as compared to 2010, when maximum snow was observed on March 6 (**Figure 4**). Arctic-wide, the start of the melt of snow proceeded by around 6 days from 2003 to 2013, while the yearly refreeze was waited by around 11 days [13].

4. Alaska region

4.1 Temperature and snow cover in February and march

In the context of climate change, temperature has shown an increasing trend over almost all parts of Earth. Researchers, forecasters, and the public have good trust that heat and temperature have been uniformly increasing since the initial time, which is known as the Hockey Stick. The average daily maximum temperature in recent decades (2005–2014) for 2 months (February 1 to March 31) declined over many regions (cities) in Alaska (**Figure 6a–e**) while increasing over many parts of the Northern Hemisphere (**Figure 7**). It is interesting to note that in decadal graphs, March shows relatively lower temperatures in recent decades (1994 onwards) than February in previous decades (before 1994). We analyzed daily maximum temperatures in many other parts of the Northern Hemisphere. For example, Alta (Norway) has shown an increase in temperature in recent decades. The warming and cooling in the Northern Hemisphere fluctuated quite a decade in March, though. Some of the local and regional atmospheric and oceanic interests play an important role along with remote influences.

The overall trend of northern hemispheric ice is decreasing, while the Bering Sea has experienced an accumulation of ice due to a short-lived atmospheric pattern: drifting snow from the North Pole. **Figure 8** demonstrates the accumulation of ice, which peaked in the days of February and March in recent years over the Bering Sea. Here, in more recent years, the maximum ice extent/cover is showing a shift toward March.

From 2007 to 2014, sea ice extent was the lowest since satellite observations started in 1979 [19]. To extend our support for snow-ice cover over bigger parts than just the Bering Sea, here in **Figure 9**, we showed that the snow and ice growth over the Alaskan region shifted in March as compared to February (**Figure 9**). The minimum temperature and maximum snow cover are moved to the days of mid-March and later days, over the Northern Alaskan region. Such a shift in the snow cover will influence the ecology and other common traits of the region.

5. Conclusions and discussions

Humans are having a significant impact on the Arctic because of greenhouse gas emissions on the multi-decadal to century time scales, with impacts on long-wave radiation and SSTs. In recent decades, minimum temperatures and maximum snow cover show a shift in the month of March, compared to earlier decades. The accumulation of maximum snow cover has shifted by 5 days in the last 36 years (1979–214).

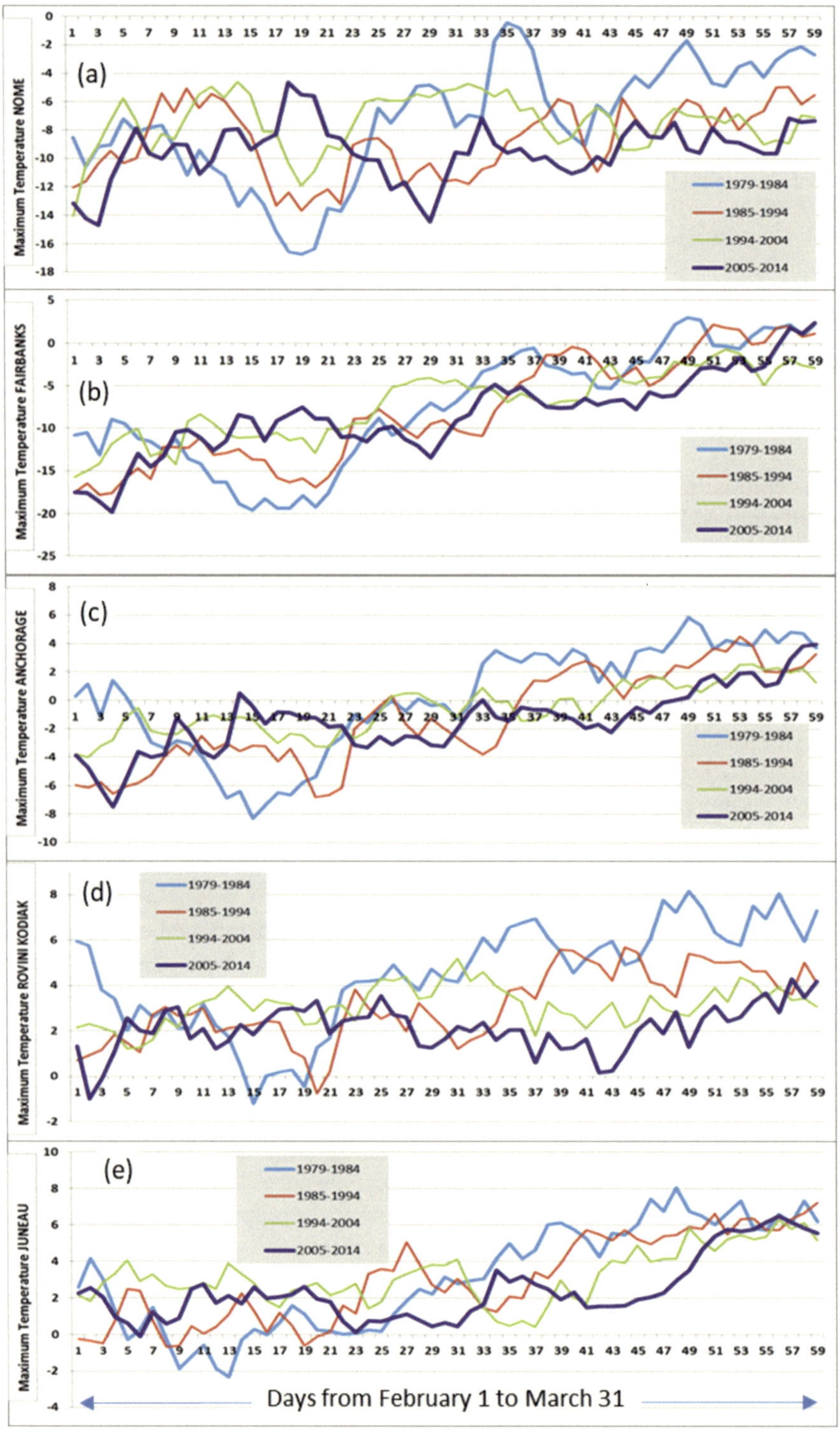

Figure 6.
Line chart shows the variation of maximum temperature over the Alaska region (a–e) for three recent decades and the remaining 6 years.

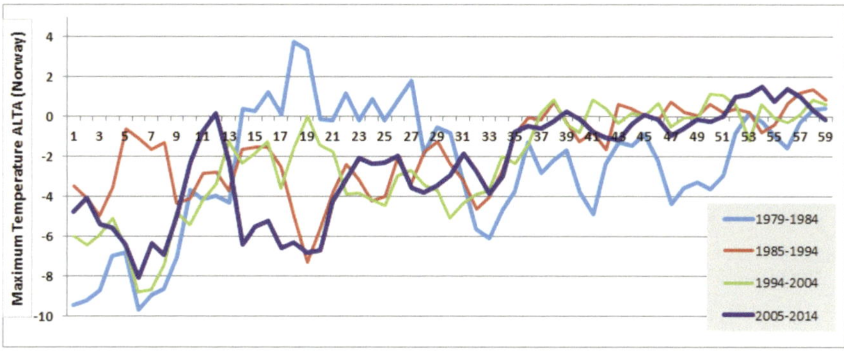

Figure 7.
Line chart shows the variation of maximum temperature over Alta (Norway) for three recent decades and the remaining 6 years.

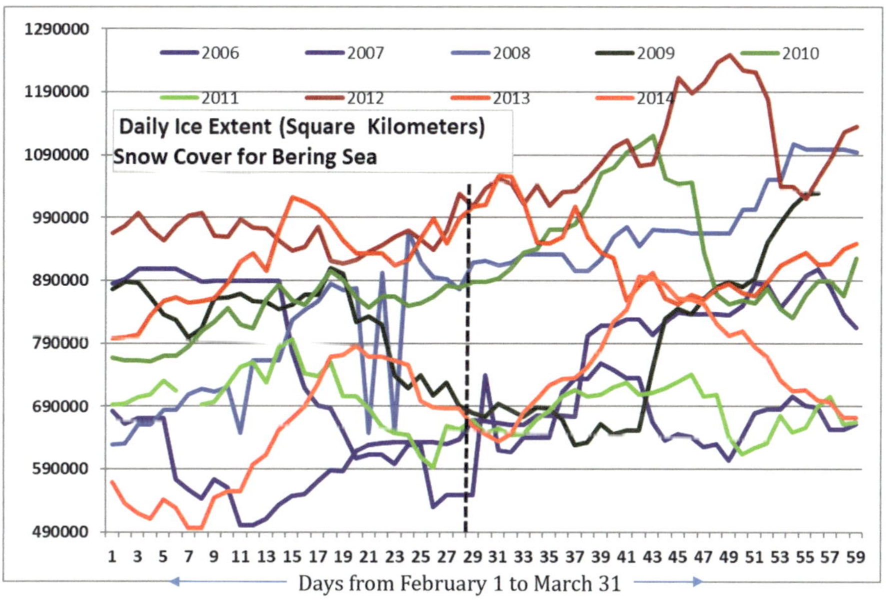

Figure 8.
The line chart shows the variation of daily ice extent over Bering Sea.

During 1980, maximum accumulation was reached by March 12, and now it is around March 17. In the north and northeast of Alaska, the Bering Sea shows the abnormal accumulation of sea ice, that is, a reduction over the North Pole region. March is regarded as colder than February due to the accumulation and melting of snow covering the Alaskan region. We did not consider the timescale of more than a decade; thus, no consideration was given to changes in astronomical factors such as changes in the earth's orbit and rotational axis, which, of course, occur on much longer time scales. Additionally, we did not calculate energy balance, rate of melting, or heat transfer in our analysis to reach conclusions. Is the shift in minimum temperature due to climate change alone or due to other prominent oscillations prevailing over the region?

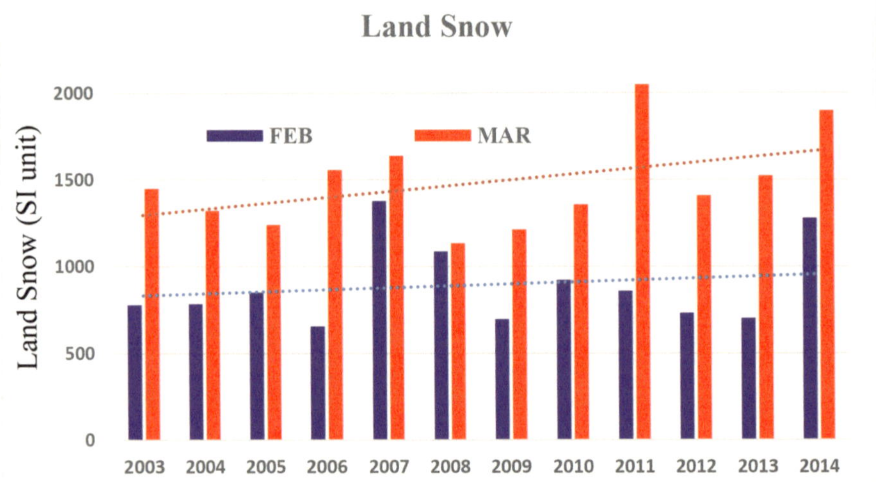

Figure 9.
Land snow and lake ice averaged over the Alaskan region and the adjoining north-western part of Canada (135 W–168 W, 60 N–70 N). Data source MODIS datasets for land snow cover at 0.05-degree horizontal resolution.

Acknowledgements

The authors are thankful to GrADS, ECMWF, and NSCID center for publicly available packages and datasets. The authors are very much thankful to the anonymous reviewer.

Author details

Vinay Kumar[1*] and Robert Ross[2]

1 Department of Atmospheric Science, Environmental Science, and Physics, University of the Incarnate Word, San Antonio, TX, USA

2 Department of Earth Sciences, Millersville University of Pennsylvania, Millersville, PA, USA

*Address all correspondence to: vkumar@uiwtx.edu

IntechOpen

© 2024 The Author(s). Licensee IntechOpen. This chapter is distributed under the terms of the Creative Commons Attribution License (http://creativecommons.org/licenses/by/4.0), which permits unrestricted use, distribution, and reproduction in any medium, provided the original work is properly cited.

References

[1] Anchorage Daily News. 2015. Available from: https://www.adn.com/arctic/article/arctic-getting-warmer-and-wetter-sea-ice-shrinks-study-says/2015/08/09/

[2] Praetorius S, Rugenstein M, Persad GG, Caldeira K. Global and Arctic climate sensitivity enhanced by changes in North Pacific heat flux. Nature Communications. 2018;**9**:3124. DOI: 10.1038/s41467-018-05337-8

[3] NOAA 2021 Annual Report. Monthly Global Climate Report for Annual 2021. NOAA 2021 Annual Report; Asheville, North Carolina, USA: National Center for Environmentla Prediction; 2022. Available from: https://www.ncei.noaa.gov/access/monitoring/monthly-report/global/202113

[4] Chylek P, Folland C, Lesins G, Dubey M, Wang M. Arctic air temperature change amplification and the Atlantic multidecadal oscillation. Geophysical Research Letters. 2009;**36**:L14801

[5] Bintanja R, van der Linden CE. The changing seasonal climate in the Arctic. Scientific Reports. 2013;**3**:1556. DOI: 10.1038/srep01556

[6] Holland MM, Bitz CM. Polar amplification of climate change in coupled models. Climate Dynamics. 2003;**21**:221-232

[7] Manabe S, Stouffer RJ. Sensitivity of a global climate model to an increase of CO2 concentration in the atmosphere. Journal of Geophysical Research. 1980;**85**(C10):5529-5555

[8] Screen JA, Simmonds I. The central role of diminishing sea ice in recent Arctic temperature amplification. Nature. 2010;**464**:1334-1337

[9] Graversen RG. Do changes in the midlatitude circulation have any impact on the Arctic surface air temperature trend? Journal of Climate. 2006;**19**:5422-5438

[10] Graversen RG, Burtu M. Arctic amplification enhanced by latent energy transport of atmospheric planetary waves. Quarterly Journal of the Royal Meteorological Society. 2016;**142**(698):2046-2054

[11] Lenton TM. Arctic climate tipping points. Ambio. 2012;**41**:10-22

[12] Stroeve JC et al. The Arctic's rapidly shrinking sea ice cover: A research synthesis. Climatic Change. 2012;**110**:1005-1027

[13] Hahn LC, Armour KC, Battisti KC, Donohoe A, Fajber R. Seasonal changes in atmospheric heat transport to the Arctic under increased CO_2. Geophysical Research Letters. 2023;**50**(20). Available from: https://agupubs.onlinelibrary.wiley.com/doi/epdf/10.1029/2023GL105156

[14] Boisvert LN, Strove JC. The Arctic is becoming warmer and wetter as revealed by the atmospheric infrared sounder. Geophysical Research Letters. 2015;**50**(20):4439-4446. DOI: 10.02/2015GL063775

[15] Hodson A et al. Glacial ecosystem. Ecological Monographs. 2008;**78**(1):41-67

[16] Krishnamurti TN, Kumar V. Prediction of a thermodynamic wave train from the monsoon to the Arctic following extreme rainfall events. Climate Dynamics. 2017;**48**:2315-2337

[17] Richter-Menge J, Overland JE, Mathis JT. Arctic Report Card 2016. NOAA; 2016. Available from: http://www.arctic.noaa.gov/Report-Card

[18] Riggs GA, Hall DK, Salomonson VV. MODIS Snow Products User Guide for Collection 4 Data Products. Greenbelt, USA: Goddard Space Flight Center; 2006

[19] Jeffries MO, Ritcher-Menge J, Overland JE. Eds. Arctic Report Card 2014. 2014. Available from: https://arctic.noaa.gov/wp-content/uploads/2023/04/ArcticReportCard_full_report2014.pdf

Section 5

Climate Change: Emission Reduction and Uncertainties

Chapter 15

Towards the 1.5°C Climate Scenario: Global Emissions Reduction Commitment Simulation and the Way Forward

Joseph Akpan and Oludolapo Olanrewaju

Abstract

This work presents an analysis of the impact of nationally determined contributions (NDC) under the Paris Agreement on global temperature rise. With the use of a climate simulation tool based on the concept of system dynamics, the study constructs a framework to project global temperature changes under other policy scenarios. The hypothesis is formulated based on the analysis of current, announced and best-case global/national policy scenarios. The research aims to address critical questions regarding the effectiveness of the ongoing NDC commitments in limiting global temperature rise to well below 2°C, in alignment with the Paris Agreement's goals. The simulation results offer a roadmap by presenting possible grey areas for optimising the current NDCs in global and national energy policies and treaties, fostering international collaboration and reinforcing the global commitment to combating climate change. In addition, this study also presents other potential strategies for decarbonisation associated with facilitating the implementation of just and fair NDCs.

Keywords: national determined contributions, environmental and energy policy, global temperature rise, climate change pathways, 1.5°C climate scenario, emissions mitigation

1. Introduction

The Paris Agreement's requirements for lowering greenhouse gas emissions and adapting to climate change require countries to be mandated to facilitate this movement by building policies to support the national determined contribution (NDC) initiative for each country. The Paris Agreement requires all nations to contribute 'nationally determined contributions' (NDCs) to the global effort to mitigate climate change. Countries publicly disclose their post-2020 climate change actions of reducing greenhouse gas emissions, establishing a balance between richer countries and developing nations and promoting equality, sustainable development and poverty eradication [1]. These goals are vital development priorities for many emerging nations. A

IntechOpen

simplified Earth system model was used in [2] to evaluate the global temperature slowdown in the NDC scenario (T = 0.6°C) and identify the causes in certain locations. The organisation for economic cooperation and development (OECD) and Asian Nations (R5ASIA) were the top two contributors with 39.3 and 36.8 per cent, respectively. The next two largest contributors were the Latin American and Caribbean (R5LAM) and Middle Eastern and African (R5MAF) areas, with 11.5 and 8.9%, respectively. The Reforming 5 Economies Forum (R5REF) is the remaining 3.5%. The extent to which a region pitches in to help cut carbon emissions is a major factor. Short-lived aerosols' influence on lowering SO_2 levels in R5ASIA was modest but significant [2].

In the study by Koven C. et al. [3], CO_2 concentrations fall below pre-industrial values when cumulative CO_2 emissions approach zero through negative CO_2 emissions, yet long-term climate change persists, guided by multi-century dynamical processes. Even if commitment to maintaining a consistent atmospheric composition and a steady stream of emissions, the global mean temperature and sea level would continue to increase [4]. A study by Grigoroudis et al. [5] explored the best emissions policies using the emission plans of only China and the USA that are compatible with temperature limitations. The findings indicate that negative emissions and severe cuts can maintain temperature stability of 2.5°C. However, relying on future technical advancements to make negative emissions feasible might lead to ongoing carbon releases and irreversible climatic implications. Jung et al. [6] examined the financial cost, risk and feasibility of lowering greenhouse gas emissions using a general equilibrium model and many burden-sharing schemes. Evaluations include one extended NDC scenario and three 2050 scenarios. The modelling results suggest that a GHG reduction in the Korean economy might cost between USD 100 and USD 350 in 2050, compared to the NDC extended scenario. Without a major economic and energy infrastructure overhaul, reducing greenhouse gas emissions in Korea would be costly.

Using a shared socio-economic pathway (SSP) of 1–2.6°C, the study by Vakilifard et al. [7] evaluated the advantages of introducing negative emission technologies in the global warming response to cumulative carbon emissions beyond 2050. The effective zero emissions commitment (eZEC) and the global warming response were evaluated over 86 unique model realisations. After net-zero emissions are achieved, the capacity to fulfil climate objectives and avoid further warming is improved by including negative emissions.

The findings from these studies highlighted above and many others, such as the ones in [8–12], have shown that reducing emissions should not be a delayed option in energy and climate policies or the effort to only selected countries. For instance, the effect of the USA's withdrawal attempt from the Paris Agreement in 2017 was evaluated to increase strain on the global average of emissions reduction as well because of high costs on other countries [8]. The urgent need to combat climate change has led to the development and implementation of various policies and commitments worldwide. Among these, the national determined contributions (NDCs) under the Paris Agreement stand as a crucial framework where countries pledge their efforts to reduce greenhouse gas emissions [13].

This study uses a system dynamic approach to examine the updated NDCs and classify all the countries under China, the USA, India, the EU, the rest of advanced countries and rest of the developing countries to study the effect of the current NDC on global temperature levels, and what best reduction would result in the 1.5°C scenario.

2. Study rationale

The ongoing discourse on climate legislation underscores the importance of enhancing and executing NDCs to achieve the objectives outlined in the Paris Agreement. It emphasises the need for meticulously tailored policies and strategies for individual sectors to successfully attain the ambitious objectives of carbon mitigation, as delineated in global accords, particularly the Paris Agreement. Examining the effectiveness of nationally determined contributions (NDCs) in reducing global temperatures to below 2°C as stipulated by the Paris Agreement goals is imperative. Using fractional integration and cointegration methods, the research by Gil-Alana [14] examines the connection between CO_2 emissions and global temperatures. The results for the short-term panel dataset contradict the hypothesis of cointegration since the orders of integration are different for the two variables. However, long-term time series data indicates a lasting positive correlation between emissions and temperatures [14]. It is also important to note that the emissions accumulation in the atmosphere within the short-term outlives into a long period, enabling the noticeable effects on global temperatures.

A dynamic simulation framework is useful in predicting variations in global temperatures under various policy scenarios of emissions reduction, which is the aim of this study with the use of a just and fair transition approach. The just transition concept is a philosophical paradigm that promotes social and economic justice in the transition to a low-carbon economy [15–17]. It prioritises equity, inclusivity, job creation, environmental justice, worker support and sustainable development by employing strategies for reducing emissions. These strategies include investing in green jobs, transitioning to renewable energy, carbon pricing, regulatory policies, infrastructure investment, community engagement, education and training. These strategies aim to create sustainable employment opportunities, reduce pollution and protect vulnerable communities. By embracing these principles, policymakers, businesses and communities can work together to transition to a low-carbon economy that promotes environmental sustainability and social justice, making the journey towards a more sustainable future fair and inclusive.

3. Progresses in global emissions, average temperature rise, renewable energies and commitment to emissions reduction

3.1 Historical emissions versus average temperature rise level

The continuous increase in the emissions level is shown in **Figure 1(a)**, with experience in increasing average temperature as well, beyond the pre-industrial level, as indicated in **Figure 1(b)**. Consequently, there has been resulting global warming and adverse effects that distort the natural ecosystem.

Greenhouse gas emissions (GHG) can be categorically divided mainly into emissions from CO_2 and other gases such as CH_4, NO_x, hydrofluorocarbons (HFCs) and sulphur hexafluoride (SF6) and perfluorocarbons (PFCs) based on the Kyoto Protocol [21]. The first constitutes most of the emissions from energy production and consumption, often found in the power, industrial processes, transport and building sectors. Understanding the effects of the emissions in terms of global warming potential (GWP) is often made with reference to its equivalence with CO_2. For instance, 1, 25, 298 are regarded as the equivalency factor for CO_2, CH_4 and N_2O in

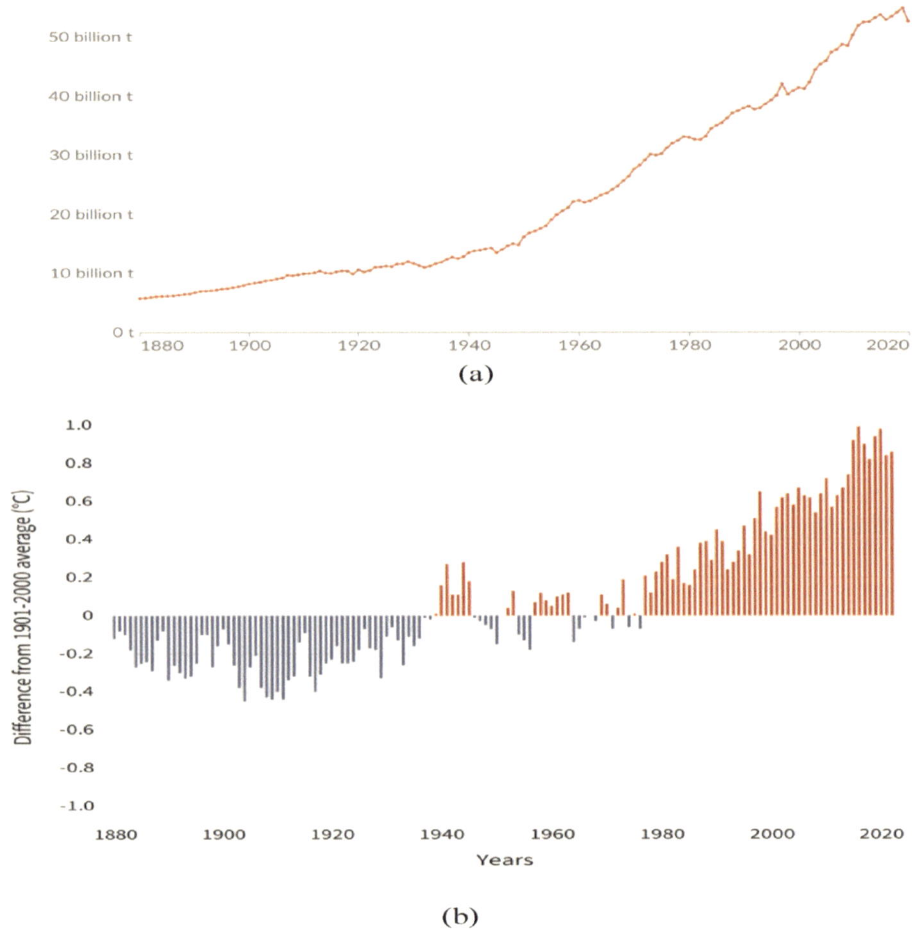

Figure 1.
Historical emissions versus average temperature rise level. (a). Emissions growth, according to Ritchie H. et al. [18] and Jones M. et al. [19]. (b). Global average changes in surface temperature 1880–2022, according to NAOO [20].

terms of 100-year GWP. Similarly, HFCs (675 and 14,800 for CH_2F_2 and CHF_3), PFCs: 7390, 12,200, 8830, 8860, 10,300, 13,300, 9300 for CF_4, C_2F_6, C_3F_8, C_4F_{10}, c-C_4F_8, C_5F_{12} and C_6F_{14}, respectively, SF_6: 22,800.

All the gases have significant impacts on global weather patterns, which last for years; hence, they are determined by adding the radiative forcing caused by a gas's pulse emission over a certain amount of time. However, an evaluation of the final implications of climate change is not directly tied to these calculations. As a result, Kirschbaum M. [22] developed s new metric called the climate-change impact potential (CCIP), which evaluates the significance of pulse emissions of different gases, such as CO2, methane and nitrous oxide. Three categories of consequences are identified: warming rate, cumulative warming and temperature rise. According to the CCIP, long-lived nitrous oxide has a greater impact than short-lived methane over 100 years.

In the next section of this study, the emissions from the two categories of GHG are presented within the last 5 years.

3.2 Short-term level emissions trajectory

Figure 2 shows the emissions trends from energy (i.e., CO_2 emissions) during the last 5 years, while **Figure 3** compares emissions of **Figure 2** with those obtained from other gases.

Over half of all emissions (i.e., 60.5, 55.6, 54.1, 55.64 and 56% between the years 2018 to 2022, respectively) come from the energy processes, where the burning of fossil fuels is the norm in the power sector with the highest value of CO_2 emissions contributions for all the years presented in **Figure 1**. Followed by industrial operations, such as cement manufacturing and chemical manufacturing, also contribute significantly to methane emissions, as do agriculture and land use. Emissions from the combustion of fossil fuels are a huge problem and transportation modes, including cars, planes, ships and trains, all contribute to this problem. Buildings contribute to emissions from heating, cooling and electrical use because of the energy used for these purposes (**Figure 4**).

3.3 Global renewable levels Progress

The use of these renewable energy technologies not only helps the economy, society and the environment but also advances the cause of sustainable development. Renewables account for less than two-thirds of total energy consumption and 85% of total power output [24]. The amount of renewable energy that must be deployed must

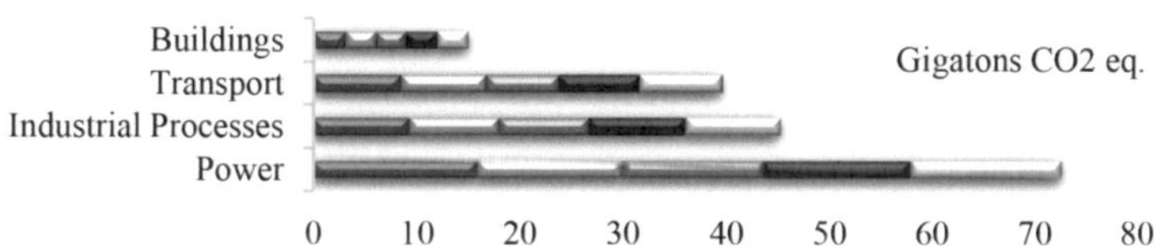

	Power	Industrial Processes	Transport	Buildings
2018	15.88	9.19	8.42	3.11
2019	14.07	8.76	8.26	2.91
2020	13.57	8.71	7.09	2.85
2021	14.39	9.31	7.73	3.01
2022	14.65	9.15	7.98	2.97

Figure 2.
Global CO_2 emissions by sector 2018–2022. Data from the IEA report and Ritchie, H. et al. in [9], and [23], respectively.

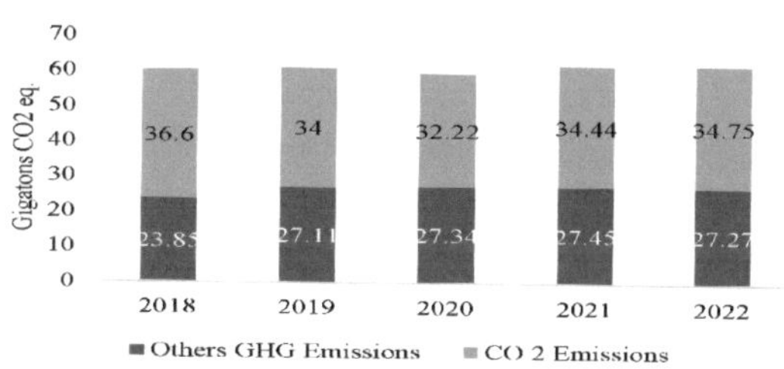

Figure 3.
Global GHG emissions from 2018 to 2022. Data from the IEA report and Ritchie, H. et al. in [9], and [23], respectively.

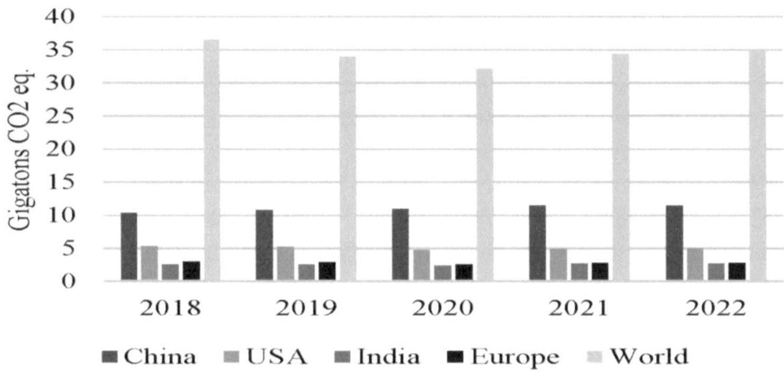

Figure 4.
Top CO_2 emitting countries with emission values from 2018 to 2022. Data from the IEA report and Ritchie, H. et al. in [9], and [23], respectively.

be increased at least six times over what is now planned to keep global warming to far below 2°C; hence, decarbonisation strategies, such as the ones highlighted by Akpan et al. in [25], for carbon emission, control is critical. If the global energy system is transformed, everyone should have access to less expensive and reliable energy at a higher level of security. Theory, ideologies, innovations, global needs and policy conceptions, as well as implementation, continue to influence the adoption of renewable energy technology in a wide range of applications and sectors. Countries striving to achieve a high level of reliance on renewable energy sources employ a diverse range of renewable energy technologies. Factors, such as geographical location, resource availability, technological capacity, governmental aims, policy frameworks, infrastructure development and economic situations, influence the variability of renewable energy sources. About 30 countries that have achieved nearly 100% renewable energy (RE) utilisation exhibit a predominant reliance on a specific RE source such as wind, solar or geothermal [26]. Notably, the most prevalent RE source among these countries is hydropower, accounting for a minimum of 70% of its RE generation. China, the United States, India and the European Union, which are the largest emitters of CO_2, are yet to reach even a 40% RE in their national electricity mix, as shown in **Figure 5**. However, these countries exhibit a diverse range of renewable energy sources within their overall energy portfolio, including hydropower, wind power and solar energy. The United States and India contain a diverse range of renewable energy sources in

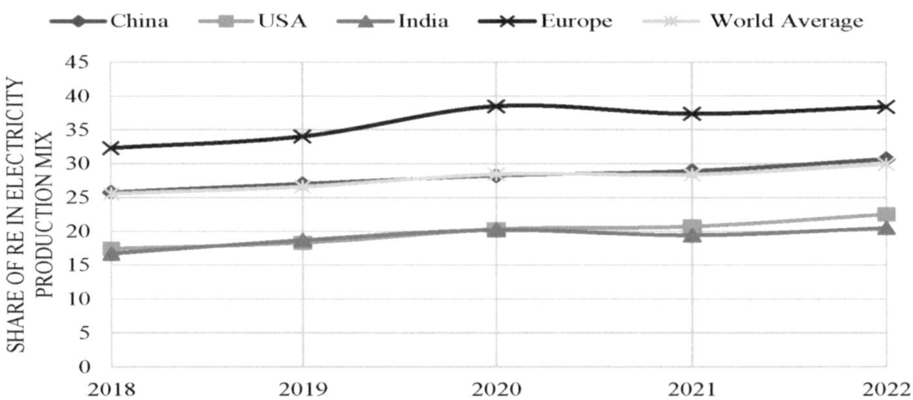

Figure 5.
Share of RE in electricity production 2018–2022. Data from Ritchie, H. et al. [27].

their respective portfolios, including wind, solar, hydropower and biomass. There is a variance in the proportion of renewable energy sources among member states of the European Union, with significant progress recorded in Germany and Denmark as they transition towards renewable energy. Unfortunately, geothermal energy has received relatively limited attention in many nations, potentially attributable to the significant upfront costs associated with infrastructure development and execution.

To keep this value well within the 1.5–2°C range and mitigate the consequences of climate change, such as drought, flood and heat wave, existing and projected policies have been made by several countries. Meanwhile, the current finance set aside to facilitate clean energy projects to curb the emissions from the different contributing sectors and countries is not sufficient to achieve the Paris Agreement of COP21 [28, 29].

3.4 Updated National determined contribution of top CO_2-emitting countries

The use of these renewable energy technologies not only helps the economy, society and the environment but also advances the cause of sustainable development [5]. And because of the urgent need to keep the global temperature within the 1.5°C scenario, many countries, herein top emitting countries, have proposed and are working towards this goal by introducing national energy policies that could drive this change. Mostly the transition to the use of renewable energies. These policies are summarised in **Table 1**.

Category	Country/ Region	Summary of RE Targets	NDC (Emissions Reduction)	To Address 1.5 °C scenario issues	Ref.
A	China	Increased RE Target in the National Grid • The fourteenth five-year plan raises the target for renewable energy to 30 per cent of total electricity consumption by 2025 (18 per cent for non-hydro renewables). • Energy Storage/Hydrogen Roadmap Development • 50GW new added battery energy storage capacity by 2025 • 1.3GW of annual hydrogen production per year by 2025	65%	Partial	[28, 30, 31]
A	USA	Approval of the Inflation Reduction Act • Per unit-energy and investment tax credits for solar PV and wind energy systems are extended. • Battery storage and zero-emission nuclear power can qualify for an investment tax credit. • Investment in sustainable energy infrastructure and technology production • Energy Storage/ Hydrogen Roadmap Development • 20.8GW of battery storage by 2025, in addition to the 7.8GW capacity at present	50–52%	Partial	[28, 32]

Category	Country/ Region	Summary of RE Targets	NDC (Emissions Reduction)	To Address 1.5 °C scenario issues	Ref.
D	India	Expansion of the Production-Linked Incentive (PLI) Scheme • 40 GWh of capacity to produce batteries • 50 GWh of capacity to produce solar photovoltaic cells is expected to be added in the next 3 years. • Reduction of 50 MM Tons Annual Emissions of CO_2 by 2030 • Hydrogen Roadmap Development • 125GW Capacity of RE for green hydrogen by 2030	45%	Partial	[28, 33]
A	Europe	Commitment to Increasing Offshore Wind Capacity • Nine EU member states have pledged that more than 120 GW of offshore wind capacity is expected to be installed by 2030 and more than 300 GW by 2050. • Announcements by the European Commission-REPowerEU Plan, Net-Zero Industry Act Proposal and other Potential Reforms • The European Union has proposed a few changes, including a faster permitting process. • An increase in the EU's 2030 renewables target to 45% by 2030 (total energy matrix, not just power). • An increase of around EUR 225 billion in loans for grids • Hydrogen Roadmap Development	40%	Partial	[28, 34]

A-Advanced, D-Developing.

Table 1.
Updated policies of selected top GHG emitters by energy in response to the 1.5°C scenario issues (between 2020 to 2023).

4. Methodology

4.1 Overview

The process of simulating pathways towards limiting global warming to 1.5 degrees celsius requires the use of complex climate models and scenario studies. These models are designed to incorporate various factors influencing the climate system such as greenhouse gas emissions, land use changes and aerosols. They help predict potential climatic impacts over time resulting from various emission reduction strategies and policies. The intergovernmental panel on climate change (IPCC) frequently assesses diverse emission scenarios and their corresponding implications for global temperature increases. Climate models are sophisticated tools used to simulate the potential effects of various emission scenarios on the Earth's climate system, considering complex interactions among the atmosphere, oceans, land and ice. Mitigation methods include renewable energy sources, energy efficiency enhancements, reforestation efforts, carbon capture and storage initiatives and modifications in agricultural practices. Simulations also consider potential adaptation actions required to address climate change consequences such as uncertainties related to human behaviour, technological improvements, natural variability and policy changes within the climate system. Policy assessments evaluate the viability and efficacy of various policies and measures designed to mitigate global warming impacts.

4.2 Existing and related climate models and model selection for study

Long-term data (temperature and GHG concentrations) spanning thousands of years is required for studying climate change. The study of current climate change and the development of projections for the future requires a solid foundation in the study of previous climate change. Based on historical findings and trends, many tools have been used to reconstruct the temperature record and predict GHG values and vice versa. The key tools, models and their corresponding findings are presented in **Table 2**.

The three models and tools (C-ROAD, MAGIC and GEC) highlighted in **Table 2** are all integrated assessment tools, which extract insights and data from several existing and conventional climate models, meteorological, macroeconomic and sector-by-sector data. The GEC model is used mainly to study possible future states of

Year of Development	Climate Model (s)	Organisation/ Country	Baseline Emissions CO_2 eq. (GTons)	Baseline Temperature Rise Level (°C)	Other Scenarios	Period	Ref.
2023 (Most updated version)	C-ROAD	MIT/Ventana System/ Climate Interactive	66.8	3.32	U/D	2023–2100	[35]
2021	Model for the Assessment of Greenhouse Gas-Induced Climate Change (MAGICC)	Climate Resource	126.29	< 4.0	1.5 °C 2.0 °C	2020–2100	[36]
2023	Global Energy-Climate (GEC)	IEA	—	A	NZE Scenario AP Scenario STEP Scenario	2023–2100	[37, 38]

U/D – User dependent, NZE – Net Zero, Announced Pledges (AP), STEP (Stated and Planned), A – Assumed based on temperature projections from IPCC reports.

Table 2.
Key emerging integrated climate assessment models.

Capability	Description
Accessibility	The model provides real-time features through a user-friendly graphical interface.
Consistency	The findings of the Intergovernmental Panel on Climate Change's Fifth Assessment Report (AR5), along with other organisations and knowledge derived from comprehensive models, align consistently with the outcomes of the simulator.
Flexibility	The model can accommodate a diverse range of user-defined scenarios, which can vary in terms of their complexity.
Robustness	The model effectively encompasses the inherent uncertainty around the potential climatic effects that are linked to decisions regarding emissions.
Transparency	There is access to the equations; they may be checked, and they are displayed graphically.
Understanding	The causes of the model pattern's expression can be identified and traced to real behaviours.
Limitation	No economic assessment considerations, less spatial resolution and less detailed assessment of climate impacts.

Table 3.
C-ROAD model capabilities and limitations.

transition into net-zero carbon emissions from energy systems only while keeping in view several sets of assumptions consistent with the sixth assessment report by the IPCC to limit global warming to less than 1.5°C (with a minimum of 50% probability) with a negligible possibility of exceeding that objective [37]. In the temperature projection part, IPCC scenarios are used. The IPCC scenarios are majorly derived from the MAGICC model [36]. The MAGICC is a less complicated climate system model, which anticipates future climate change and Earth's component interactions with uncertainty. MAGICC simulates the carbon cycle, methane cycle and anthropogenic aerosol emissions in four boxes representing the Earth's land and ocean. Hence, climate change, greenhouse gas concentrations, effective radiative forcing, temperature change, Earth system heat absorption and sea-level rise are projected [36]. The drawback with the MAGICC is its set temperature projection limitation, making it difficult to assess the exact estimate values of other constituting factors that influence temperature changes. Also, the MAGICC does not allow for user definitions to ascertain possible feedback based on inputs. Therefore, C-ROAD, with several additional capabilities highlighted in **Table 3**, has become more pertinent and open-source for user interaction and feedback for policy simulation and decisions.

As already mentioned in sections 1 and 2 of this study, the objective of the study is to simulate different policy measures available and may be required to transit towards the 1.5°C scenario. The goal of this simulation procedure is to ascertain the different NDC composition pathways under the context of climate justice needed by all countries and or regions to reduce emissions. The simulation is done to ascertain the possible reduction value at critical years up to the year 2100. **Figure 5** shows a schematic of how the model's process flow is represented, as embedded in the tool used, which is described in the next section.

4.3 Description of tool used for the study and governing equations

The tool used in this study integrates climate science, economics and policy analysis to develop a comprehensive and reliable simulation framework. Using fundamental principles, C-ROADS is a real-time system dynamics model that allows users to

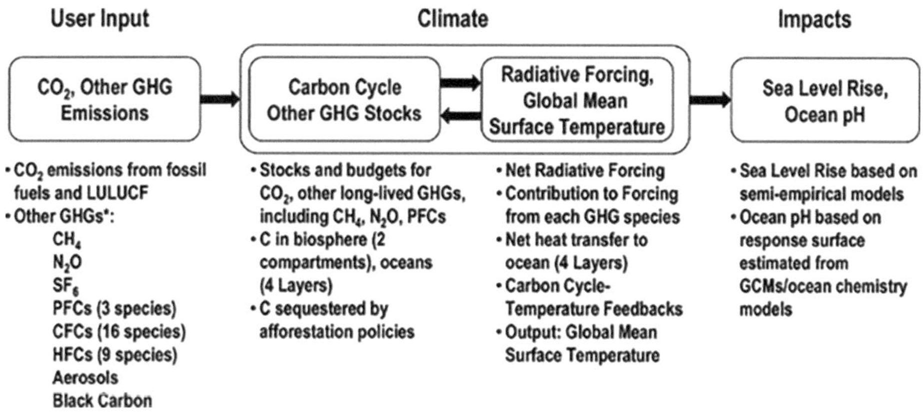

Figure 6.
C-ROADS model architecture [45]. Note: Key stocks (carbon cycle and other GHG stocks), key flows (CO_2 and other GHG emissions), and feedback (temperature feedback based on carbon cycle from ocean-atmosphere interaction owing to radiative forcing, cooling feedback and heat contents).

input actions to reduce GHG emissions from land use and forestry. For studying complex systems across time, Jay Forrester developed system dynamics in the 1950s [39]. The system dynamics employs system thinking to analyse the dynamic interplay and feedback between various system components and factors [40, 41]. A system dynamics model's structure and behaviour may be shown using stock and flow, as in **Figure 6** for this study. Users can see the system's responses through simulations utilising system dynamics models. They also investigate the impact of dynamic complexity on system behaviour and unintended outcomes and can use it to develop a case for business [42], technology, market, carbon dynamics emission [43], strategic scenario forecasting [44] and as a guide for other policy decisions.

The C-ROADS integrates national and regional emission and land use sectors, aiming to understand how national and regional commitments could contribute to climate objectives. It generates findings efficiently and is used to support other integrated assessment models, calibrating results using larger disaggregated models [45]. The capabilities and limitations of the C-ROADS based on the model technical reference in [45] are presented in **Table 3**.

Figure 6 depicts the C-ROAD architecture for the emissions versus global temperature transition.

The C-ROADS has mainly three phases, as shown in **Figure 6**. The user input, climate model and the impact phase.

The user input phase allows users to set the year of peak emissions, start reduction and level of effort to prevent deforestation and promote afforestation. The reduction levels are a function of the summation of the aggregate percentages of greenhouse gas emissions. In the second phase, the climate model is a fifth-order linear system with three negative feedback loops that control deep ocean warming ($\boldsymbol{H_d}$, and $\boldsymbol{R_d}$), as well as heat content in the atmosphere and surface ocean ($\boldsymbol{H_s}$, and $\boldsymbol{R_s}$). Radiative forcing $\boldsymbol{F_r}$, cooling feedback ($\boldsymbol{F_o}$, and $\boldsymbol{F_d}$) and heat content (H and R) all have an impact on the first-order ocean warming response. Based on the climate interactive technical reference in [45], the equation governing the variables can be represented in the Eqs. (1)–(6) as follows:

$$\boldsymbol{T_s} = \frac{\boldsymbol{H_s}}{\boldsymbol{R_s}} \tag{1}$$

$$T_d = \frac{H_d}{R_d} \tag{2}$$

$$H_s = \int (F_r(t) - F_o(t) - F_d(t))\, dt + H_s(0) \tag{3}$$

$$H_d = \int (F_d(t))\, dt + H_d(0) \tag{4}$$

Where (T_s, and T_d) are the surface and deep ocean temperatures, respectively. (F_o, and F_d) are the cooling feedback, outgoing radioactive flux and heat flux to the ocean, respectively.

$$F_o(t) = \lambda * T_s \tag{5}$$

$$F_d(t) = R_d * \frac{T_s - T_d}{\tau} \tag{6}$$

Where λ and τ are climate feedback parameters and heat transfer constant.

The radiative forcing F_r from CO_2 is a logarithmic function of atmospheric CO_2 concentration [46, 47], while the total F_r, required to quantify CH_4 and N_2O forcings, is smaller than the sum of the F_r for either gas alone. F-gas forcings are the product of CO_2 concentration and radiative forcing coefficient, while other forcings from aerosols and tropospheric ozone are exogenous time-varying parameters. While CO_2 concentrations in Eq. (7) are a function of the ratio of atmospheric CO_2 concentrations to pre-industrial atmospheric CO_2 concentrations.

$$CO_2\ concentration = \frac{CO_2\ atm}{CO_2\ pre - industrial\ atm} \tag{7}$$

The equilibrium temperature response T_E is determined by the radiative forcing coefficient and climate feedback parameters. Therefore, T_E is represented in Eq. (8) as follows:

$$T_E = \frac{\kappa}{\lambda} * \frac{In\left(\frac{CO_2\ atm}{CO_2\ pre-industrial\ atm}\right)}{In\ (2)} \tag{8}$$

On running the model, the output shows the temperature changes with the emissions, which can be extracted for further analysis. The last phase, the impact phase, produces the consequence effect of temperature changes on sea level rise with flood risk map, pH level, ocean acidification, crop yield decrease, possible death from extreme heat and animal and plant species loss.

5. Dataset and definition of scenarios

5.1 Dataset

Under full yearly resolution up to 2010, six regions modes were assumed, categorised into China, US, India, EU, rest of advanced countries and the rest of

Input	Description	References
Historical Country-level data	1. CO_2 emissions from fossil fuels 2. CO_2 emissions from land use 3. GDP 4. Population growth prospects 5. Other GHGs	[35]
Carbon cycle and climate sector data	The carbon cycle modelling approach uses historical country data for its estimation. Following Eqs. (1)–(8) and other relevant approaches	[35, 45]
NDC	The nationally determined contributions (NDC) are extracted from the UNFCCC registry, and with Eq. (9), the emissions reduction rate is determined per annum for the baseline and other scenarios, as in Section 5.4.	[13]

Table 4.
Model input data and sources.

developing countries. Both the techno-economic and socio-economic factors play a role in forecasting CO_2 emissions and temperature level changes. The main input variables for estimating emissions in Gigatons per CO_2 equivalent and global temperature changes are based on the features of Eqs. (1)–(8).

The model input baseline values of countries and regional generation profiles used are extracted from recent data from the following sources, as stated in **Table 4**.

$$e_r = \left[\frac{NDC_i}{100}\right]^{1/n} \tag{9}$$

Where e_r is the annual emissions reduction (expressed in percentage), NDC is the national determined contribution for each country *i*, and n is the number of periods (in years) to fulfil that commitment, starting from the year the NDC was submitted to the UNFCCC under the registry in [13]. The Announced Policies Type 1 emission reduction rate per annum, as calculated, is presented in **Table 5**.

Category	Country/Region	Annual reduction (%)	NDC (Emissions Reduction Targets)	NDC Target Year
A	China	0.94	65%	2030
A	USA	0.928	50–52%	2030
D	India	0.915	45%	2030
A	Europe	0.903	40%	2030
D	Rest of Advanced Countries (RAC)	NU	—	P/D
A	Rest of Developing Countries (RDC)	NU	—	P/D

**NU – Non-Uniform, P/D – Partly defined and not same across all the developing and advanced countries.*

Table 5.
Annual emissions reduction based on updated policies of selected top GHG emitters (Announced Policies Type 1, calculated as per eq. (9)).

Scenarios	Hypothesis	Ref.
Baseline Case	Without NDC	Based on historical country-level data available through [35]
Announced Policies Type 1	With NDC only for China, Europe, USA and India	**Table A1**
Announced Policies Type 2	With the same NDC of China, Europe, USA, India, the rest of the advanced countries (RAC) and the rest of developing countries (RDC)	**Table A1**
2.0 °C Scenario	Increased NDC but equal shares across all regions	**Table A1**
1.5 °C Scenario	Increased NDC but equal shares across all regions	**Table A1**

Table 6.
Scenario definitions and hypothesis.

5.2 Scenario definitions

The definition of the scenarios is presented in **Table 6**. The different simulation input data in the form of NDC based on each scenario is made as a hypothesis as in **Table 6** and described in the Appendix.

6. Results and discussion

6.1 Findings

The results of the simulation are presented in this section under two categories: Policy versus global temperature rise in **Figure 7** and policy versus emissions in **Figure 8**. The detailed emissions values per country or region (i.e., USA, EU, Rest of Developed Countries, China, India, and the Rest of Developing Countries) are included in the Appendix, **Figures A1–A5** for all the global temperature rises.

The simulation results of the emissions reduction scenarios of **Figures 7** and **8** provide insights into how the five scenarios (baseline, announced type 1 and 2, 2 and 1.5°C scenarios) might influence future temperature levels. Simulations show that reducing emissions results in lower temperatures compared to scenarios where emissions continue to rise at the baseline scenarios. The extent of reduction significantly influences the simulated temperature outcomes, with more substantial reductions of 0.9, 2.5 and 6% having a more pronounced impact on limiting global temperature increase to 2.62, 2.0 and 1.5°C, respectively. In all the scenarios, the temperature rises from 2023 till 2100. However, the average annual temperature growth rate is different, with 1.15, 1.04, 0.84, 0.50 and 0.12%, for the baseline, announced type 1 and 2, 2 and 1.5°C scenarios, respectively.

Similarly, the average annual emissions reduction growth rate is −0.09, 0.50, 0.90, 2.28 and 4.12%, respectively. The baseline scenario has a negative emissions reduction growth rate, which implies that emissions may continually be on the increase. Emissions are expected to peak between the years 2070–2080 owing to the envisaged world's population peak at that time, too; hence, there may be a reduced need for high energy usage in the building and other sectors, resulting in fewer emissions. Ideally,

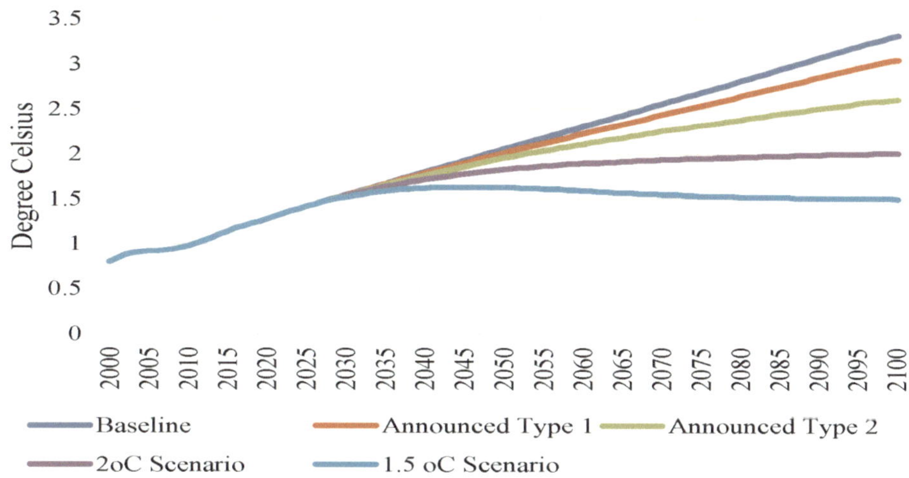

Figure 7.
Different simulated policies versus global temperature rise.

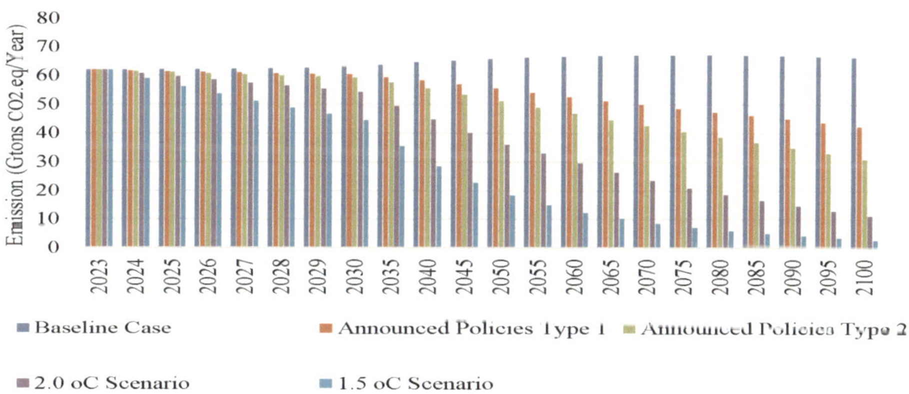

Figure 8.
Different simulated policies versus emission.

the output of average annual emissions reduction growth rate in, particularly, the announced 2, 2 and 1.5°C scenarios are supposed to be equal to the input emission reduction commitment per annum of 0.9, 2.5 and 6%, respectively. However, this is not the case with the simulation results as the model considers other factors, such as the population-economic growth dynamics of countries. Also, the world economic growth measured in annual GDP growth rate is expected to decrease, based on the OECD data for real GDP forecast [48]. The economies of industrialised nations, such as China, are expected to peak, while those of the developing nations may continue to rise as they transition into being industrialised and with more accessibility to electricity for all population. These dynamics resulted in discrepancies between emission reduction commitment and the output of the average annual emissions reduction growth rate in the announced 2, 2 and 1.5°C scenarios. The cost of delaying the decision to reduce emissions is not to be overlooked as the consequences outweigh the immediate cost of action.

Given a short time interval between 2023 and 2030, the average annual temperature growth rates are 1.76, 1.67, 1.67, 1.67 and 1.48%, while the average annual emissions reduction growth rates are 0.03, 0.06, 0.17, 0.43% for the baseline, announced type 1 and 2, 2 and 1.5°C scenarios, respectively. The percentage changes within a short time series appear very little compared with the case of the long-term difference, even with the large emission reduction commitments. This is in line with the work by L. A. Gil-Alana [14], which has shown that global temperature changes are hardly noticeable until a long-term span. The degree of uncertainty in determining how much temperature rise can be ascertained in the short term is owing to many variables that result in these changes. Some of these variables could be attributed to various factors such as carbon cycle dynamics, complex interactions and other uncertainties such as future human behaviour in terms of the release of anthropogenic gases, technological advancements and natural system responses.

Overall, simulations consistently show that reducing emissions is crucial for mitigating global temperature rise, with even moderate reductions of 0.9% in the announced type 1 scenario slowing the pace of temperature rise.

Figure 9 shows the results of all five scenarios in the years 2030, 2050 and 2100 for the different countries and regions where the commitments were made. In the baseline scenario of **Figure 9(a)**, only the EU shows emission reduction to 4.06, 3,63 and 3,03 Gtons CO_2 eq in the years 2030, 2050 and 2100, respectively. At the same time, China showed a continuous increase to 15.98 Gtons CO_2 eq in 2050 from the 15.29 Gtons CO_2 eq value of 2030, with a corresponding reduction to 12.25 Gtons CO_2 eq by 2100. The rest of the advanced countries showed a value of 8.4 and 7.67 Gtons CO_2 eq in 2030 and 2050, respectively, with an increase to 7.76 Gtons CO_2 eq by 2100. For the USA, India and the rest of developing countries, emissions value continually increased across the baseline scenario. In the rest of the other four scenarios, as in **Figure 9(b)–(d)**, all six regions show progress in emissions reduction due to the introduction of policies to support emissions reduction. However, the results show an envisaged high growth in emissions value of the rest of developing countries. This change is an indication of the anticipated high growth in industrialisation and increase in energy accessibility across developing countries, being a part of the UN Sustainable Development Goal (SDG) 7 of providing clean and modern energy for all. Though substantial progress has been made in this regard, there is still a huge gap in the population's access to electricity in developing countries. For instance, only about 60% of Africa's population has access to electricity [49] and yet, there is also an increasing population growth compared with the other regions or nations [50]. Therefore, in closing this gap while reducing the probability of high emissions from this form of industrialisation, energy from renewable sources should be highly utilised as this could drastically reduce further emissions from the power sector. Mitra S. et al. [51] emphasised the importance of such energies to be affordable towards increasing universal accessibility. The issues and challenges to achieving these are numerous and diverse, with the key ones being discussed by Akpan J. et al. in [26] and Batinge B. et al. in [52] providing a roadmap for reducing energy poverty in African electricity markets.

Consequently, considering climate justice, the NDC initiative proposed during COP26 is left in the hands of each country to determine its contribution to the global value in the drive to meet the UN SDG 2030 target. The results from sections 6.1 and 6.2 indicate that in the baseline case, the announced policies, types 1 and 2, are insufficient to mitigate global warming and climate change effects. Whereas, in the

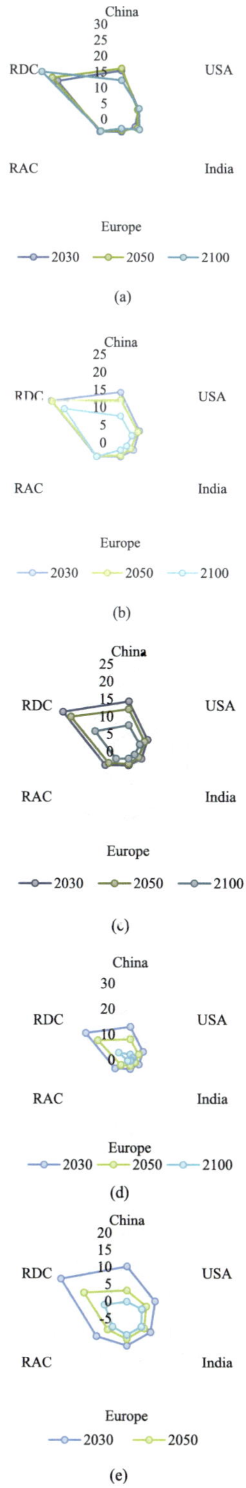

Figure 9.
Summary of simulation results at strategic period across countries/regions' (all scenarios). (a) Baseline scenario; (b) announced policy type 1; (c) announced policy type 2; (d) (d) 2.0°C scenario; (e) 1.5°C scenario.

other two scenarios (i.e., 1.5 and 2.0°C scenarios), the world needs to increase its current NDC of emissions reduction by about 6.5% and 2.5%, respectively, and concurrently across the six regions used in the hypothesis of this study has been defined in **Table A1**. In this way, reduction commitments can be. However, this measure is not without a challenge. The next section presents a discussion on this constraint and then proposes potential pathways that could help overcome the challenge with the strategy of equal percentage reduction commitment.

6.2 The way-forward

The recent turnout from each country for the submission of her NDC has been poor; for instance, to reduce emissions in accordance with the Paris Agreement, ten nations updated their nationally determined contributions (NDCs) in 2022. Of the 195 nations that have ratified the Paris Agreement, just about 30 have nevertheless set clear goals for achieving net-zero emissions [13]. While regulatory policy announcements remained the same, there were around 80 new renewable energy programmes made on the demand side. Eight announcements came from Latin America and the Caribbean, seven from Asia, four from Africa, two from Oceania and just one each from North America, the Middle East and North Africa. Most announcements came from Europe. At the end of the year 2022, 94 countries either had goals or policies in at least one end-use sector, but only three (Spain, Portugal and Turkey) had plans or objectives in all four end-use categories [13].

It is well understood that the complexity of ascertaining NDC is high, as it requires many sectors' inputs and data. For many countries, arriving at this data that presents the peculiar situation of the country is more cumbersome and would also involve regulation of existing energy policies and regulatory frameworks. Commitment to research findings in preferring holistic pathways in addressing these current issues is very timely. It cannot be overemphasised as many governments and international organisations are actively looking out for the best measures to facilitate the 100% RE vision, UN SDG and climate change agenda of 2030.

Accurate data and long-term estimates of energy demand, the need to create adaptable and responsive market and regulatory frameworks, international collaboration and coordinated activities are mostly highlighted as the necessary strategies to decarbonise the current global systems. In this study, we emphasise just and fair sharing as an important consideration for developing strategic emissions reduction, energy and climate policy for progressive achievements through planning and implementation. Hence, leaving no room for prejudice in the sharing. For instance, this reduction measure employed in this study to achieve the 2.0 to 1.5°C scenarios was fairly shared among the regions and countries used in the simulation.

Suggested in **Table 7** below are other possible areas that can help have a more responsive NDC needed to facilitate the rapid deceleration of global emissions increase and to stay within the right temperature rise threshold.

To help reduce carbon emissions, policy alternatives are continuously being widened to preserve the environment and improve society are complementary goals. The claims in [53, 54] show that Africa generally has historically had a small impact on global warming in terms of countries' contribution to global CO_2 and GHG emissions. But these assertions cannot be used to justify continuing to pollute at current rates for these countries. Why? Because the consequence effect does not exclude fewer CO_2 emitters, as the impact can be direct or indirect. Africa, today, can make huge

Concepts	Description of proposed pathway
Regional NDC Stock-taking and Equally Shared Globally Determined Commitment (GDC) Concept	This initiative should be aimed at replacing the current pattern of NDC global stock-taking in a 5-year interval. Rather, stock-taking should be a 2-year interval carried out at the regional or continental level, at least twice before the main global forum. This way, countries' efforts are measured earlier, and appraisal for immediate support in the continuous reduction of emission synonymous with the requirements for the 1.5°C - 2.0°C Scenario is done well in time. This initiative acknowledges the efforts of a few countries that have made the commitments. But because this commitment is made at the will of individual countries, there is bound to be a possibility of no prioritisation of commitments. Therefore, finding the right emissions reduction sharing ratio that considers other factors such as developmental issues, renewable resource constraints and the economic situation of developing countries is pertinent.
Reducing the CO_2 emissions per wealth class per population	The wealthiest individuals and highest-income nations produce a disproportionate amount of the world's CO_2 emissions. In order to reduce emissions, both high-income individuals and rapidly expanding places should be the focus, with each group's unique challenges and opportunities being carefully taken into account. This initiative should be driven by a global effort that takes these complex dynamic issues, such as behavioural changes in the emissions disparities, into account to ensure a just and fair decrease in CO_2 emissions.
Equality with the Drivers of Climate Change Acts and Decisions	This initiative is poised towards the inclusiveness and even appropriation of global climate decision makers as most key representation of climate change decisions is made from a selected few and from countries not representing full global coverage representation.
Emissions Budgeting Framework for Attractiveness of Investment Towards 1.5°C	To meet the 1.5°C objective and reduce global emissions, a huge financial investment commitment is required. In addition to government and public support, private investment is highly required. To ensure that financial commitment in this region is attractive, an emissions budgeting framework is important. This budgeting framework should allow the predictability of carbon emissions cost per investment options, mitigate high-carbon asset risks, align with global climate goals and induce green capital through carbon pricing and tax incentives. Organisations can commit to decarbonisation since clean technology investment and climate action are promoted *via* low-carbon technology innovation. Hence, promoting social and environmental responsibility, as well as increasing stakeholders and public support. With this approach, the investing community can help create a more sustainable and climate-resilient future.

Table 7.
The way forward in reaching the 1.5°C - 2.0°C scenario.

technological leaps that would contribute to flattening the curves to net-zero emissions by pushing into policies, research and implementation of projects from renewable energy sources to meet each country's electricity consumption.

The EU NDC commitment, done at a regional level [55], with noticeable progress even with the advent of the recent Russia – Ukraine war that kept the EU's energy dependence on imports at a disadvantaged position, proves the resiliency of the EU towards the NDC emissions goal. The EU has had a joint effort to reduce GHG emissions, being evidenced in their NDC proposal. In the coming years, the bulk of the emission is envisaged to come from developing countries due to the anticipated increase in energy access by the population and the growing industrialisation agenda.

Therefore, it is pertinent to consider ways of both managing and mitigating this emission through a determined commitment at a centralised level. The current NDCs are done in a decentralised pattern, leaving the individual countries with a wavering will and dwindling decision over the implementation of announced strategies.

From the perspective of this study, it is believed that the most effective way of reducing emissions without impeding other developmental agendas could be best done through a just fair approach yet at a regional commitment level. Furthermore, most developing countries are energy poverty-driven yet full of one of the world's largest renewable energy resources; therefore, in the commitment to reaching the 1.5° C scenario, this inequality in energy access should be accounted for in any further NDC commitment, proposed pathways in **Table** 7, and climate science models, as these are often overlooked in many existing frameworks. Hence, this study emphasises the unity in diversity towards reducing emissions burden and mitigating climate change.

7. Conclusion

There is a lot of technical and political depth to the current conversation on how to evaluate a country's NDC progress towards the 1.5°C scenario. This complexity makes it challenging to have a central consensus on the 'right' or 'objective' technique to follow. Many models and climate systems have been developed to provide platforms to determine the effects of the current practices on global warming and climate change. The challenges with the complexity of these existing tools have been supported by the introduction of feedback loops to allow for the simulation of the different actions that can either enhance or mitigate the impacts of climate change. In this study's simulations, feedback processes are incorporated in the simulation tool used, which accounts for uncertainties already established by existing complex integrated assessment models. The process involves studying variables such as greenhouse gas emissions, energy consumption patterns, technological deployment, land use changes and policy adoption. Key components include implementing significant reductions in greenhouse gas emissions, transitioning to sustainable energy sources, implementing renewable energy infrastructure and implementing carbon reduction technologies. All these processes were assumed to have been accounted for in the NDC reduction percentages used in this study. The model considers data from historical emissions, current NDC commitments, technological advancements and socio-economic factors.

The findings from this study promote the initiatives of the intergovernmental panel on climate change (IPCC) reports to keep the global temperature level below 1.5°C and suggest that the comprehensive adoption and serious commitment towards NDCs of reducing emissions at about 2.5 and 6.0% can significantly mitigate global temperature rise to 2.0 and 1.5°C, respectively as against the current practices. The overall results produced projections of Earth's climate responses to temperature changes based on the emissions values at the strategic periods, as shown in **Figure 9** from each policy scenario could help governments, corporations and communities make informed decisions regarding mitigation techniques, adaptation measures and policies to limit global temperature rise to 1.5°C. The findings and perspectives presented in Section 6, with the proposed pathways in **Table 7**, provide valuable insights for policymakers, stakeholders and the broader scientific community, offering a strategic framework for improving NDCs, fostering international collaboration and bolstering global commitment to tackling climate change complexities.

Acknowledgements

The authors are grateful to the Durban University of Technology for funding the manuscript's APC and the grant provided under the PG RFA-Energy research theme.

Author's contribution

The first author conceptualised, investigated, analysed and developed the manuscript, while the second author supervised, reviewed and validated the manuscript.

Data availability

All data used are publicly available, and the ones from the result can be made available in full detail upon request.

A. Appendix

See **Table A1**.
Policies versus Emissions Outcome

Peak of Emissions (Year)	Start of Emissions Reduction (Year)	Emissions Reduction Rate (% per annum)	Countries
		Scenario-Baseline	
2100	2100	0	China
2100	2100	0	US
2100	2100	0	India
2100	2100	0	EU
2100	2100	0	The rest of the advanced countries

Peak of Emissions (Year)	Start of Emissions Reduction (Year)	Emissions Reduction Rate (% per annum)	Countries
2100	2100	0	Rest of the developing countries
		Announced Policies Type 1	
2023	2023	0.94	China
2023	2023	0.928	US
2023	2023	0.915	India
2023	2023	0.903	EU
2100	2100	0	The rest of the advanced countries
2100	2100	0	Rest of the developing countries
		Announced Policies Type 2	
2023	2023	0.94	China
2023	2023	0.928	US
2023	2023	0.915	India
2023	2023	0.903	EU
2023	2023	0.9	The rest of the advanced countries
2023	2023	0.9	Rest of the developing countries
2023	2023	0.94	China
		2.0 °C Scenario	
2023	2023	2.5	China
2023	2023	2.5	US
2023	2023	2.5	India
2023	2023	2.5	EU
2023	2023	2.5	The rest of the advanced countries
2023	2023	2.5	Rest of the developing countries
		1.5 °C Scenario	
2023	2023	6	China
2023	2023	6	US
2023	2023	6	India
2023	2023	6	EU
2023	2023	6	The rest of the advanced countries
2023	2023	6	Rest of the developing countries

Table A1.
Hypothesis definition for each of the scenarios.

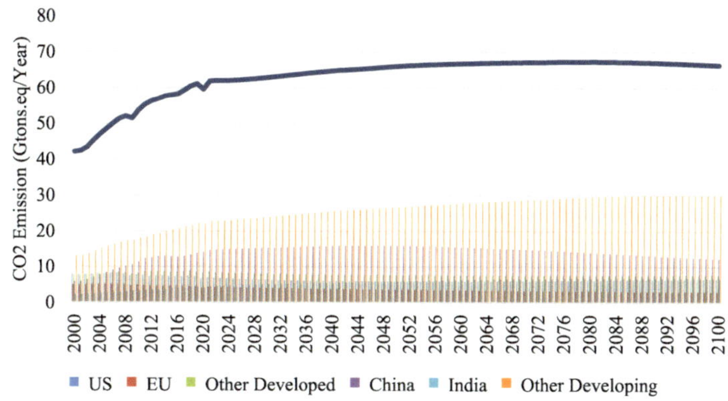

Figure A1.
Global CO_2 emissions with baseline path.

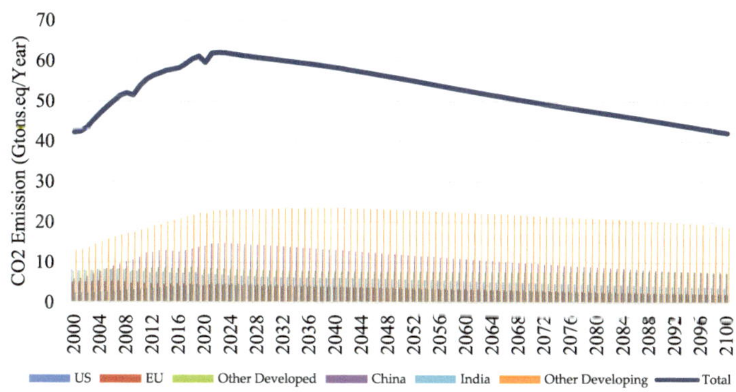

Figure A2.
Global CO_2 emissions with announced policies type I.

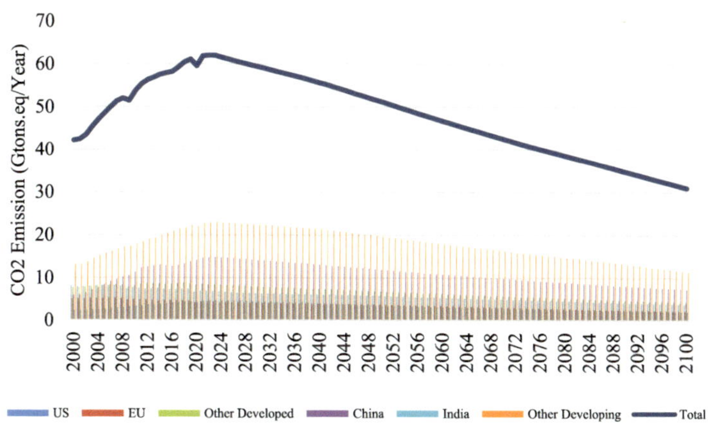

Figure A3.
Global CO_2 emissions with announced NDC policies type 2.

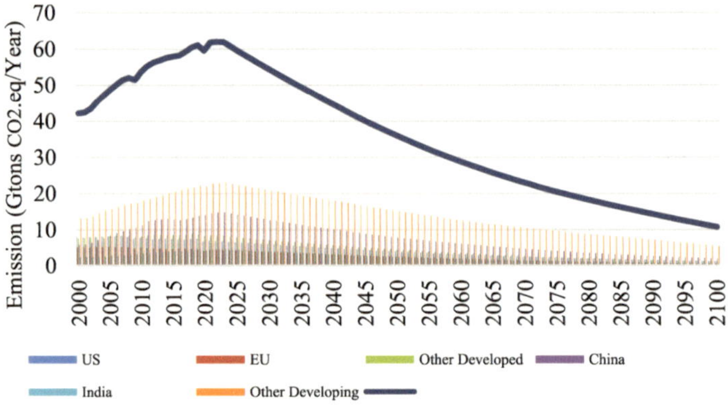

Figure A4.
Global CO_2 emissions with 2.0°C scenario.

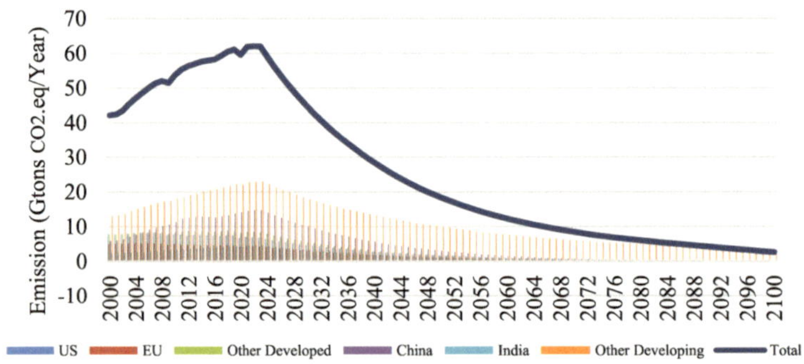

Figure A5.
Global CO_2 emissions with 1.5°C scenario.

Author details

Joseph Akpan* and Oludolapo Olanrewaju
Industrial Engineering, Durban University of Technology, Durban, South Africa

*Address all correspondence to: 22176142@dut4life.ac.za

IntechOpen

© 2023 The Author(s). Licensee IntechOpen. This chapter is distributed under the terms of the Creative Commons Attribution License (http://creativecommons.org/licenses/by/4.0), which permits unrestricted use, distribution, and reproduction in any medium, provided the original work is properly cited.

References

[1] UNFCCC. Nationally Determined Contributions (NDCs). Available from: https://unfccc.int/NDCREG. [Accessed: May 24, 2023]

[2] den Elzen MGJ et al. Updated nationally determined contributions collectively raise ambition levels but need strengthening further to keep Paris goals within reach. Mitigation and Adaptation Strategies for Global Change. 2022;**27**(6):33. DOI: 10.1007/s11027-022-10008-7

[3] Koven CD, Sanderson BM, Swann ALS. Much of zero emissions commitment occurs before reaching net zero emissions. Environmental Research Letters. 2023;**18**(1):014017. DOI: 10.1088/1748-9326/acab1a

[4] Wigley TML. The climate change commitment. Science (1979). 2005; **307**(5716):1766-1769. DOI: 10.1126/science.1103934

[5] Grigoroudis E, Kanellos F, Kouikoglou V, Phillis YA. The challenge of the Paris agreement to contain climate change. Intelligent Automation and Soft Computing. 2018; **24**(2):319-330

[6] Jung TY, Park C. Estimation of the cost of greenhouse gas reduction in Korea under the global scenario of 1.5°C temperature increase. Carbon Management. 2018;**9**(5):503-513. DOI: 10.1080/17583004.2018.1476587

[7] Vakilifard N, Williams RG, Holden PB, Turner K, Edwards NR, Beerling DJ. Impact of negative and positive CO_2 emissions on global warming metrics using an ensemble of earth system model simulations. Biogeosciences. 2022;**19**(17):4249-4265. DOI: 10.5194/bg-19-4249-2022

[8] Dai H, Xie Y, Zhang H, Yu Z, Wang W. Effects of the US withdrawal from Paris agreement on the carbon emission space and cost of China and India. Frontiers in Energy. 2018;**12**(3): 362-375. DOI: 10.1007/s11708-018-0574-y

[9] IEA. Greenhouse Gas Emissions from Energy Highlights. Paris, France: IEA; 2022. Available from: https://www.iea.org/data-and-statistics/data-product/greenhouse-gas-emissions-from-energy-highlights. [Accessed: August 24, 2023]

[10] Morfeldt J et al. Emission pathways and mitigation options for achieving consumption-based climate targets in Sweden. Communications Earth & Environment. 2023;**4**(1):342. DOI: 10.1038/s43247-023-01012-z

[11] van den Berg NJ et al. Implications of various effort-sharing approaches for national carbon budgets and emission pathways. Climatic Change. 2020; **162**(4):1805-1822. DOI: 10.1007/s10584-019-02368-y

[12] Patrick P et al. Greenhouse gas emission budgets and policies for zero-carbon road transport in Europe. Climate Policy. 2023;**23**:343-354. DOI: 10.1080/14693062.2023.2185585

[13] UNFCCC. Nationally Determined Contributions (NDCs). New York, United States: UNFCCC; 2020. Available from: https://unfccc.int/NDCREG. [Accessed: July 7, 2023]

[14] Gil-Alana LA, Monge M. Global emissions and global temperatures: Are they related. International Journal of Climatology. 2020;**40**(15):6603-6611. DOI: 10.1002/joc.6601

[15] Wang X, Lo K. Just transition: A conceptual review. Energy Research and Social Science. 2021;**82**:102291. DOI: 10.1016/j.erss.2021.102291

[16] Heffron RJ, McCauley D. What is the 'just transition? Geoforum. 2018;**88**: 74-77. DOI: 10.1016/j.geoforum. 2017.11.016

[17] Stevis D, Felli R. Planetary just transition? How inclusive and how just? Earth System Governance. 2020;**6**: 100065. DOI: 10.1016/j.esg.2020.100065

[18] Ritchie H, Rosado P, Roser M. Greenhouse Gas Emissions. Oxford, England: Our World Data; 2023. Available from: https://ourworldindata.org/greenhouse-gas-emissions

[19] Jones MW, Peters GP, Gasser T, et al. National contributions to climate change due to historical emissions of carbon dioxide, methane, and nitrous oxide since 1850. Scientific Data. 2023; **10**:155. DOI: 10.1038/s41597-023-02041-1

[20] Lindsey R, Dahlman L. Climate Change: Global Temperature [Online]. New York, United States: National Oceanic and Atmospheric Administration (NOAA); 2023. Available from: https://www.climate.gov/news-features/understanding-climate/climate-change-global-temperature. [Accessed: June 22, 2023]

[21] UN. KYOTO Protocol to the United Nations Framework Convention on Climate Change [Online]. New York, United States: UN; 1998. Available from: https://unfccc.int/process-and-meetings/the-kyoto-protocol/history-of-the-kyoto-protocol/text-of-the-kyoto-protocol. [Accessed: October 24, 2023]

[22] Kirschbaum MUF. Climate-change impact potentials as an alternative to global warming potentials. Environmental Research Letters. 2014; **9**(3):034014. DOI: 10.1088/1748-9326/9/3/034014

[23] Ritchie H. Who has contributed most to global CO_2 emissions? 2019. Available from: https://ourworldindata.org/contributed-most-global-co2. [Accessed: July 18, 2023]

[24] OECD. OECD Green Growth Studies. Paris, France: OECD iLibrary; 2012. DOI: 10.1787/22229523

[25] Akpan J, Olanrewaju O. Sustainable energy development: History and recent advances. Energies (Basel). 2023;**16**(20): 7049. DOI: 10.3390/en16207049

[26] Akpan J, Olanrewaju O. Towards a common methodology and modelling tool for 100% renewable energy analysis: A review. Energies (Basel). 2023;**16**(18): 6598. DOI: 10.3390/en16186598

[27] Ritchie H, Roser M, Rosado P. Our World in Data. Renewable Energy. 2023. Available from: https://ourworldindata.org/renewable-energy. [Accessed: May 22, 2023]

[28] IEA. World Energy Investment 2023 [Online]. Paris, France: IEA. Available from: https://iea.blob.core.windows.net/assets/8834d3af-af60-4df0-9643-72e2684f7221/WorldEnergyInvestment2023.pdf; 2023. [Accessed: September 25, 2023]

[29] Daniel R, Yuqi Z, Richard GN, Brian CP, Aaron B. Global Energy Outlook 2023: Sowing the Seeds of an Energy Transition. Washington: Resources for the Future; 2023

[30] Government of China. China Nationally Determined Contribution (Reducing Greenhouse Gases in China: A 2030 Emissions Target) [Online].

Beijing, China: Government of China; 2022. Available from: https://unfccc.int/NDCREG. [Accessed: August 28, 2023]

[31] World Economic Forum. Green Hydrogen in China: A Roadmap for Progress. Cologny, Switzerland: World Economic Forum; 2023. Available from: https://www.weforum.org/whitepapers/green-hydrogen-in-china-a-roadmap-for-progress. [Accessed: July 31, 2023]

[32] Government of USA. The USA Nationally Determined Contribution (Reducing Greenhouse Gases in the United States: A 2030 Emissions Target) [Online]. New York, United States: Government of USA; 2021. Available from: https://unfccc.int/NDCREG. [Accessed: August 28, 2023]

[33] Government of India. India's Updated First Nationally Determined Contribution under Paris Agreement (2021–2030) [Online]. New Delhi, India: Government of India; 2022. Available from: https://unfccc.int/NDCREG. [Accessed: August 28, 2023]

[34] Germany (on behalf of Europe). Europe Nationally Determined Contribution (Reducing Greenhouse Gases in the Europe: A 2030 Emissions Target) [Online]. Berlin: Germany (on behalf of Europe); 2021. Available from: https://unfccc.int/NDCREG. [Accessed: August 28, 2023]

[35] Climate Interactive. Climate Interactive [Online]. Available from: https://www.climateinteractive.org/. [Accessed: June 14, 2023]

[36] Climate Resource, MAGICCS. 2021. Available from: https://live.magicc.org/scenarios/9742acaa-4b02-4f1a-b9bb-507066c06ff5/overview. Climate Resource PTY LTD

[37] IEA. Global Energy and Climate Model Documentation 2023 [Online]. Paris, France: IEA; 2023. Available from: https://www.iea.org/reports/global-energy-and-climate-model. [Accessed: October 28, 2023]

[38] IEA. Global Energy and Climate Model-Techno-Economic Inputs [Online]. Paris, France: IEA; 2023. Available from: https://www.iea.org/reports/global-energy-and-climate-model. [Accessed: October 28, 2023]

[39] Forrester JW. Industrial Dynamics. Cambridge, USA: Massachusetts Institute of Technology (MIT) Press; 1961

[40] Sapiri H, Zulkepli Hew J, Ahmad N, Zainal Abidin N, Hawari NN. Introduction to System Dynamic Modelling and Vensim Software. Kedah, Malaysia: UUM Press; 2017. DOI: 10.32890/9789672064084

[41] Mihailovs N, Cakula S. Dynamic system sustainability simulation modelling. Baltic Journal of Modern Computing. 2020;**8**(1):192-201. DOI: 10.22364/BJMC.2020.8.1.12

[42] Cosenz F, Rodrigues VP, Rosati F. Dynamic business modeling for sustainability: Exploring a system dynamics perspective to develop sustainable business models. Business Strategy and the Environment. 2020; **29**(2):651-664. DOI: 10.1002/bse.2395

[43] Pillay NS, Brent AC, Musango JK, van Geems F. Using a system dynamics modelling process to determine the impact of ECAR, EBUS and etruck market penetration on carbon emissions in South Africa. Energies (Basel). 2020; **13**(3):575. DOI: 10.3390/en13030575

[44] Ahmad SA, Tahar RM. Strategic forecasting of electricity demand using

system dynamics approach. International Journal of Environmental Science and Development. 2012;**3**(4): 328-333. DOI: 10.7763/IJESD.2012.V3.241

[45] Lori SS, Chris C, Tom F, Andrew PJ, Charles J, John S. C-ROADS Technical Reference [Online]. 2023. Available from: support.climateinteractive.org. [Accessed: August 19, 2023]

[46] IPCC. AR5 Synthesis Report: Climate Change 2014The Synthesis Report (SYR) of the IPCC Fifth Assessment Report (AR5) [Online]. Geneva, Switzerland: IPCC; 2014. Available from: https://www.ipcc.ch/report/ar5/syr/. [Accessed: October 26, 2023]

[47] Lan XKWT, Dlugokencky EJ. Trends in Globally-Averaged CH4, N2O, and SF6 Determined from NOAA Global Monitoring Laboratory Measurements, Version 2023–09. Boulder, Colorado, United States: National Oceanic and Atmospheric Administration; 2022

[48] OECD. Real GDP Long-Term Forecast [Online]. Paris, France: OECD; 2023. Available from: https://data.oecd.org/gdp/real-gdp-long-term-forecast.htm. [Accessed: October 26, 2023]

[49] Ritchie R, Mispy OO. SDG Tracker-Measuring Progress towards the Sustainable Development Goals [Online]. Oxford, England: Our World in Data; 2023. Available from: https://sdg-tracker.org/energy. [Accessed: June 15, 2023]

[50] World Bank. Population [Online]. Washington, D.C., United States: World Bank; 2023. Available from: https://data.worldbank.org/indicator/SP.POP.TOTL?end=2022&locations=DZ&start=2002. [Accessed: October 26, 2023]

[51] Mitra S, Buluswar S. Universal access to electricity: Closing the affordability gap. Annual Review of Environment and Resources. 2015;**40**:261-283. DOI: 10.1146/annurev-environ-102014-021057

[52] Batinge B, Kaviti Musango J, Brent AC. Perpetuating energy poverty: Assessing roadmaps for universal energy access in unmet African electricity markets. Energy Research and Social Science. 2019;**55**:1-13. DOI: 10.1016/j.erss.2019.05.004

[53] Tadasse D. The Impact of Climate Change in Africa. Pretoria, South Africa: Institute for Security Studies; 2010. Available from: https://api.semanticscholar.org/CorpusID:17135311 [Accessed: July 30, 2023]

[54] Odimegwu F. "An Analysis of Climate Change in Africa". The Effects of Climate Change in Developing Areas: Assessment of the African Situation [Online]. Mauritius: LAP LAMBERT Academic Publishing; 2020. Available from: https://www.researchgate.net/publication/344070153

[55] Wolkinger B et al. Implementing Europe's climate targets at the regional level. Climate Policy. 2012;**12**(6): 667-689. DOI: 10.1080/14693062.2012.669096

Chapter 16

Measuring the Environmental Impact and Uncertainty Analysis of Portland Cement Production in South Africa: A Recipe 2016 v 1.04 Endpoint Method Approach

Oluwafemi E. Ige, Kevin J. Duffy, Oludolapo A. Olanrewaju and Obiora C. Collins

Abstract

The cement industry is among the growing industries globally that negatively impact human health and global warming due to various substances released into the water, air and soil. This impact and potential damage have been studied in several ways to understand their effects, but more details are still needed. This study examines the damage done by producing 1 kg of cement in South Africa using the Recipe 2016 endpoint method. It also conducted an uncertainty analysis using the Monte Carlo method to confirm and establish its credibility. The results showed that the clinkering stage causes the most damage to human health (49%) and ecosystems 60% due to large amounts of carbon dioxide emissions. The result showed high uncertainty in Water consumption, Human health, Water consumption, Terrestrial ecosystem, Aquatic ecosystems, Human carcinogenic toxicity and Ionizing radiation. These results align with existing literature but highlight the specific contributions of clinkering.

Keywords: global warming, endpoint method, portland cement, damage impact, uncertainty

1. Introduction

The global state of energy is categorized by reducing fuel sources and environmental worries related to fossil fuel usage [1]. Although the global COVID-19 pandemic reduced carbon dioxide (CO_2) emissions by 5.8% in 2020, carbon emissions are estimated to attain roughly similar atmospheric levels by 2023 [2]. In line with a three-year delay created by the economic recovery situation, energy demand will lag behind compared to pre-pandemic projections [3, 4]. Environmental protection has grown in popularity recently and has become an essential measure for social and political settings [5, 6]. The most sought-after goal is the reduction of greenhouse gas (GHG) emissions, for example, carbon dioxide (CO_2), nitrous oxide (N_2O) and methane (CH_4) which contribute to the greenhouse effect in an energy-intensive industry such as cement [7]. Cement production is a multiplex process that uses large raw materials such as limestone, marl, clay and iron ore and various fuels such as coal, natural gas, fuel oil, petroleum coke and waste, heat and electricity. Cement production contributes to global warming through CO_2 emission during the clinker production stage [8]. Cement is an essential material in the construction industry with high environmental impacts [9]. The global cement demand has risen due to rising population and urbanization. Cement production in 2016 was about 809 × 10^6 t, making it the most-used produced substance globally [10]. Also, global cement demand was expected to climb 4.5% per year to reach 5.2 Bmt in 2019 [11]. The cement production growth rate is projected to rise in the following years.

The main impact of cement production is GHGs emissions, excess resource use and energy (fossil fuel and electricity) [12]. The main components in Portland cement are calcium (limestone), silicate, iron and alumina. Gypsum is used to improve the setting time of cement [13]. CO_2, CH_4 and NOx are cement production's most common GHG emissions [14]. Several Studies have indicated that the clinkering stage has the most significant impact, with CO_2 being one of the most released gases [15–17]. Coal is the primary fuel (thermal) used in the combustion process, contributing to these GHGs. The impact of cement production is affected by the raw material specifications, fuel type, and technology used. Cement production contributes to about 5% of CO_2 emissions, producing 0.81 kg CO_2-eq for one ton of cement [18]. Due to this, cement production faces tremendous challenges in meeting global demand while reducing CO_2 emissions. In modern cement industries, both wet and dry rotary kilns are used [19]. Growing concern over the capacity of the Earth to sustain economic development arose due to the increasing threat of harmful global environmental change and mismanagement problems. Many international groups and developed nations have recommended different strategies and measures to mitigate these effects, including green economy, green growth, green transformation, green industrial blueprint and sustainable transformation [20]. Since cement is one of the most vital products in the world, the damage caused by this industry must be critically analyzed. To properly recommend and provide mitigation strategies, it is necessary to measure cement production and its impact on human health, ecosystems and resources in the production stages. Various mitigation strategies were recommended, including partial clinker replacement, alternative fuel use, etc. [21]. Also included in the suggestions was adopting the BAT in cement production [15, 22, 23].

Furthermore, while it is crucial to accurately quantify and communicate uncertainties in Life cycle assessment (LCA) regarding input results to ensure accurate interpretation and use, LCA specialists still do not consider uncertainty when conducting their studies. Uncertainty analysis is a commonly used analytical method in LCA tools. It mainly addresses the uncertainty associated with measured parameters, such as inaccuracies in emissions measurements or normalization data and the variability observed among different sources and objects, such as diverse emissions from similar processes [24]. These methods are appropriate for assessing the impact of the uncertain parameter in the model's output, thereby determining the overall accuracy of the model. Therefore, an effective LCA tool integrates uncertainty and variability analyses. Furthermore, a few studies have conducted an uncertainty analysis using a Monte Carlo simulation to determine the importance of the results in South African cement plants.

2. Literature review

LCA is a critical assessment tool for analyzing the effects of all stages of product production, from raw extraction to the waste disposal finish [25]. This tool is of great significance when it is necessary to know the environmental impact of various production stages. The LCA tool can provide a comprehensive overview of the whole product life cycle [26]. It also serves as a decision-supporting tool for cement producers, allowing them to improve the production process. The cement industry's environmental impact can be assessed using an LCA. Several studies have been investigated using LCA to evaluate the environmental impact of cement production in various countries, including Spain [22, 27, 28], Egypt and Switzerland [29], China [30, 31], the European Union, the USA [15, 16, 32, 33] and others [17, 34–36]. Tun et al. [35] evaluated the environmental impacts of the cement industry in Myanmar using LCA. They identify major environmental impacts such as climate change, photochemical oxidant formation, fine particulate matter formation, terrestrial acidification, and fossil resource scarcity, with CO2, NOx, SO2, and PM2.5 emissions from clinker production being the key contributors, suggesting the need for mitigation options such as energy-saving measures, alternative fuels and materials and process upgrades to promote sustainable development in the industry. Çankaya and Pekey [34] used SimaPro 8.0.4 software to conduct a comparative LCA of cement production using traditional raw materials and fuels (Traditional Scenario) versus alternative ones (Alternative Scenario). The results showed that the alternative scenario reduced the overall environmental impact of clinker production by approximately 12%, with positive effects on climate change, human health, ecosystem quality, and resources, particularly for cement types CEM IV and CEM II that used trass. Thwe et al. [36] examined the environmental impacts of cement production in Myanmar, focusing on the Max Myanmar Cement Plant in Naypyitaw. The results showed that conventional cement production has significant negative environmental impacts, with the calcination stage being the primary contributor. However, switching from coal to natural gas as a fuel source can substantially reduce these impacts, with potential reductions of 68% for climate change, 83% for acidification, and 96% for eutrophication. Olagunju and Olanrewaju [17] assessed the environmental impact of 1 kg of Ordinary Portland Cement (OPC) using both midpoint and endpoint approaches of life cycle impact assessment (LCIA). The results showed that the clinker production phase had the highest impact, with CO_2 emissions being the highest pollutant emitter, contributing to global warming and posing risks to human health and ecosystems. Additionally, the study highlighted the emission of high copper, which has toxic effects on both the ecosystem and humans and the high consumption of fossil resources, particularly crude oil, which raises concerns about possible scarcity.

Stafford et al. [16] assessed the environmental impacts of using waste materials as partial replacements for fossil fuels in a cement plant in Southern Europe. The study uses a life cycle assessment approach to evaluate the environmental profile of cement manufacturing, focusing on indicators such as abiotic depletion, acidification, eutrophication, photochemical oxidant formation, and global warming. The main contributors to these impact categories were atmospheric emissions in the kiln, consumption of fossil fuels, and electricity usage in the mills. Moretti and Caro [15] investigated the environmental impact of the Italian gray cement and clinker industry, specifically focusing on the upstream and core phases of the production processes. The analysis reveals that the core production phase contributes the most to greenhouse emissions, acidification and eutrophication potential, emphasizing the need to implement BAT in the cement industry to achieve environmental benefits. Therefore, the aim of this paper is as follows: to investigate the environmental impact of Portland cement production (CEM1) in South Africa at an endpoint method using Recipe 2016 v 1.04 of the SimaPro LCA; to conduct a comprehensive uncertainty analysis throughout the entire life cycle to enhance the credibility of the results and to compare the environmental performance of cement production in South Africa with that of other regions globally.

3. Materials and methods

The LCA method has been used to evaluate the environmental impacts and resource use across a product's lifetime, i.e., from raw materials extraction to the production line and use stages to waste [25]. The LCA method evaluates the environmental impact of products, services, and processes from "cradle to grave." Based on this data, the LCA analysis assesses the potential impact on the environment, natural resources and human health [37]. The methodology is one of the most crucial components of LCA tools. The analysis and the data validation follow the ISO/TS 14072 guidelines for organizational LCA [26] using the endpoint method of the SimaPro 9.1.1 software [38, 39]. The LCA consists of four stages: definition of goal and scope, inventory analysis, impact assessment and interpretation, following the guidelines of the ISO 14040 [25], 14,044 [40] and ISO/TS 14072 [26].

3.1 The definition of goal and scope

The goal and scope definition gives a detailed assessment objective along with system boundaries, assumptions, and functional units of a product or process [41]. This study will use 1 kg of Portland cement as a functional unit. This will include all datasets, analyses and interpretations.

3.2 Inventory data

The LCI stage covers all the input and output inventory data required to evaluate a product's production process [41]. The environmental impact database includes all emissions produced during the production process. This study uses inventory data from the Ecoinvent v3.7.1 [42] database of SimaPro, one of the most highly regarded databases for construction materials [43], for 1 kg Portland cement production. The basic materials model describes the production of various materials used in the life cycle of the South African cement plant. The input data for 1 kg of Portland cement is included in **Table 1**.

The data includes all energy used and raw materials in the production process. The figures are computed based on original data collected in South African cement plants. The data comes from five typical cement plants in South Africa, representing 90% of the cement market share.

3.3 Impact analysis

LCIA is a multiple-faced evaluation method used to reveal all the potential environmental impact categories based on the environmental resources data provided

Inputs from technosphere	Amount
Cement factory	5.36e−11 unit
Clinker	0.902 kg
Gypsum, mineral	0.0475 kg
Limestone	0.05 kg
Ethylene glycol	0.00019 kg
Steel, low-alloyed	5.25e−05 kg
Electricity	0.0376 kWh
Output	
Emissions to air (Heat, waste)	0.135 MJ
Products (Portland Cement)	1 kg

Table 1.
The inputs and outputs to produce 1 kg of Portland cement.

in the LCI. Several environmental issues are addressed in this assessment, including energy use, global warming, water pollution, etc., offering a comprehensive evaluation of the product's environmental impacts [25]. Therefore, this study describes the LCA methods for the various stages of cement production, i.e., clinkering, raw material usage, fuel usage, transportation and electricity usage explained by [14]. Since SimaPro uses different analysis methods, the method selected for this study is ReCiPe 2016 Endpoint (H) V1.04, which has three main impact categories; human health, ecosystem and resources. In the LCIA stage, all inventories are grouped into various impact categories, then the results of the LCIA and LCI are interpreted [44].

3.4 Interpretation

This stage explains the results of the LCIA based on the LCI [45]. This stage comprehensively presents processes and substances with significant impacts with a clear presentation, after which we will formulate recommendations. This study will analyze the environmental impact of 1 kg of Portland cement (CM1) using the endpoint LCIA method to measure the impact and correctly provide appropriate recommendations. The investigation was carried out from raw material extraction to cement production, i.e., from cradle to gate system boundary.

3.5 Damage assessment

As previously stated, the endpoint method employs three major impact categories, each of which includes specific impacts, as described below:

3.5.1 Human health

- Global warming is measured in Disability-adjusted Life Years (DALY), caused by increased death and diseases due to climate change.

- Stratospheric ozone depletion is measured in DALY, resulting from increased UV radiation caused by ozone-depleting substances in the atmosphere.

- Ionizing radiation; expressed in DALY, causing by radioactive radiation.
- Ozone formation, Human health; the damage is measured in DALY. This effect is emissions of organic substances into the air caused by summer smog.
- Fine particulate matter formation: the damage is measured in DALY. This effect is produced by sulfur, dust and nitrogen oxide emissions into the atmosphere caused by winter smog.
- Human carcinogenic and non-carcinogenic toxicity; the damage is measured in DALY. This results from carcinogenic effects due to carcinogenic emissions of substances to air, soil and water.

3.5.2 Ecosystem

- Global warming; Terrestrial ecosystems and Freshwater ecosystems, Ozone formation; Terrestrial ecosystems, the damage to the ecosystem is measured in species/year resulting from the emission of ecotoxic substances to air, soil and water.
- Terrestrial acidification, Freshwater and Marine eutrophication, Terrestrial ecotoxicity. Freshwater and Marine ecotoxicity; the damage to the ecotoxicity is measured in species/year caused from the emission of acidifying substances into the air.
- Land use: the damage is expressed in species/year. Either land conversion or land occupation causes this damage.

3.5.3 Resources

- Mineral resource: the damage is measured in USD2013 due to decreasing mineral grades.
- Fossil resource: the damage is expressed in USD2013, resulting from lower quality fossil fuel extraction.

3.6 Uncertainty analysis

Uncertainty analysis is a systematic method employed to measure the degree of uncertainty that emerges from a life cycle inventory analysis due to the interaction between model imprecision impact, input uncertainty, and data variability [25]. In 1998, Huijbregts [24, 46] identified six types of uncertainties and variabilities, these include (i) spatial variability; (ii) parameter uncertainty; (iii) uncertainty due to choices; (iv) model uncertainty; (v) temporal variability; and (vi) variability between objects and sources. Multiple methods can be employed to conduct this analysis, but Monte Carlo simulation, a statistical modeling method, is widely used during the LCIA stage. Commercial LCA software such as SimaPro, developed by PRéConsultants, now includes Monte Carlo simulations but is only used in a few LCA studies [34, 47]. In this study, the uncertainty analysis will center around evaluating the degree of uncertainty present in the LCI. This uncertainty arises from several factors, including uncertain inputs, inconsistent data and model imprecision [48]. This study used a Monte Carlo simulation to conduct an uncertainty analysis, aiming to evaluate the level of uncertainty of the result since we used secondary data for the analysis. The uncertainty analysis used numerical data and 1000 iterations and a 95% confidence interval (CI) to meet its requirements.

4. Results and discussion

4.1 The characterization result (endpoint)

The endpoint method measures the damage-oriented impacts of the processes and shows several impact categories that are further grouped into damage categories. According to the endpoint method, the flows are divided into 22 impact categories, as shown in **Table 2**. Endpoint methodologies assess human health, ecosystems,

Human health (DALY)
Global warming, Human health
Stratospheric ozone depletion
Ionizing radiation
Ozone formation, Human health
Fine particulate matter formation
Human carcinogenic toxicity
Human non-carcinogenic toxicity
Water consumption, Human health
Ecosystems (species.yr)
Ozone formation, Terrestrial ecosystems
Terrestrial acidification
Freshwater eutrophication
Marine eutrophication
Terrestrial ecotoxicity
Freshwater ecotoxicity
Marine ecotoxicity
Global warming, Terrestrial ecosystems
Land use
Global warming, Freshwater ecosystems
Water consumption, Terrestrial ecosystem
Water consumption, Aquatic ecosystems
Resources (USD2013)
Mineral resource scarcity
Fossil resource scarcity

Table 2.
The damage category investigated (endpoint method).

and resource damage categories. In addition, the endpoint method shows impacts within various categories but removes other aspects, not considering the emissions factors [49].

Based on the endpoint method, **Table 2** shows the endpoint impact category, which explains each impact in the damage category. This method presents 22 different impact indicators with three damage units, i.e., DALY, species/yr. and USD2013, according to their impacts. The various impacts listed in **Table 2** are grouped according to the area of significance to life (AoSL) damage categories, as shown in **Table 3**.

Figure 1 shows a graphical illustration of the impact categories of the five production processes. The characterization result of the endpoint method shows a similar pattern as the midpoint method with four additional impact categories [14].

The contribution of the five production processes, i.e., clinkering, raw material usage, fuel usage, transportation and electricity usage, to the damage categories, as shown in **Figure 2**. These production processes were assessed and interpreted according to damage categories: Human health, ecosystems and resources. The evaluation and interpretation of the production processes are outlined as follows.

4.1.1 Raw material usage stage

Raw material consumption contributed to all three impact categories, as shown in **Figure 2**. Human health (5%), ecosystem (4%), and resources (4%). Overall, raw

Damage category	Unit	Portland cement, Production
Human health	DALY	1.6176×10^{-6}
Ecosystems	Species. yr	3.9×10^{-9}
Resources	USD2013	1.686×10^{-2}

Table 3.
Damage assessment results of 1 kg Portland cement using endpoint method.

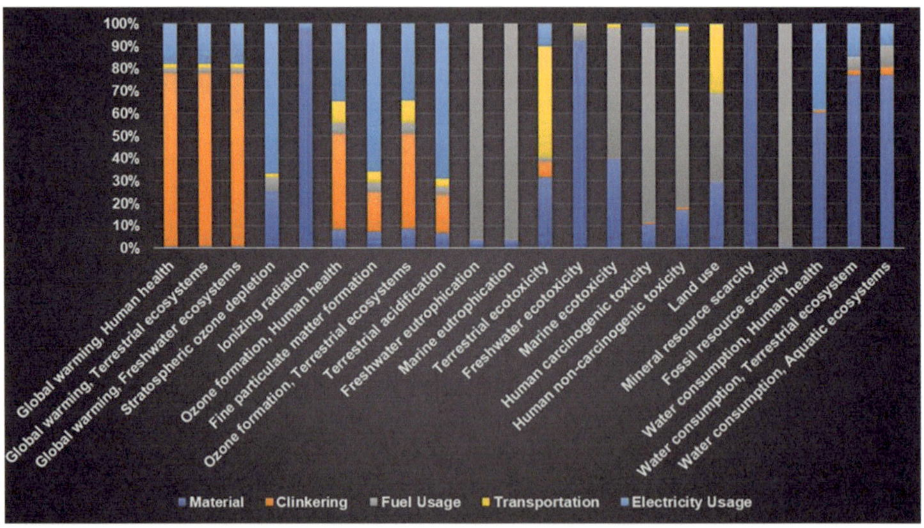

Figure 1.
Endpoint environmental impact of cement production in South Africa.

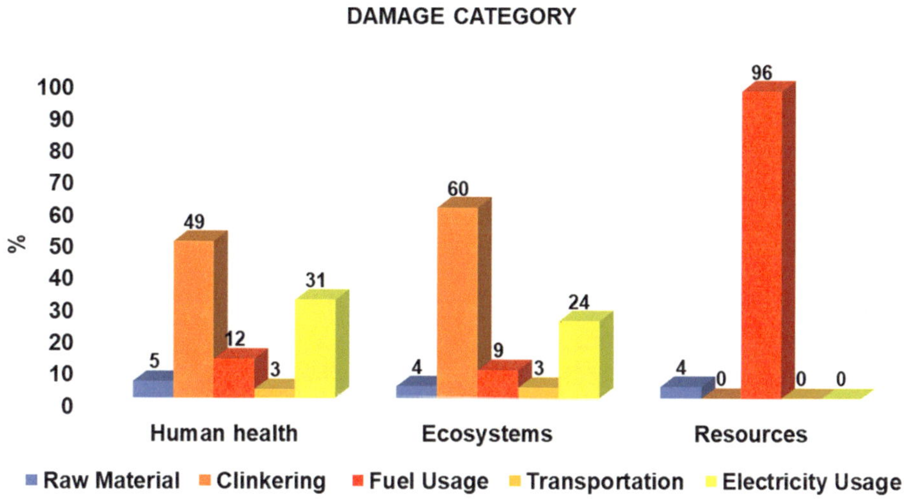

Figure 2.
Contribution of substances to human health.

material consumption had a minor impact on the damage categories. Therefore, raw material usage did not contribute significantly to damage categories.

4.1.2 Clinkering stage

In **Figure 2**, the clinkering stage is responsible for 49% of the damage to human health. It contributes to the ecosystem (60%) due to primary gas emissions such as CO_2, SO_2, NO_x and particulate matter, but clinkering stage had no impact on Resources.

4.1.3 Fuel usage stage

Figure 2 showed that Fuel usage contributed to all damage categories with human health damage (12%), ecosystem (9%) and significantly impacted resource damage (96%). Most of these impacts were related to fossil resource consumption, direct emissions from cement production and transport materials during clinker production.

4.1.4 Transportation stage

As shown in **Figure 2**, transportation usage had a negligible impact on the damage categories. There was a 3% contribution to human health and the ecosystem had no impact on resource damage.

4.1.5 Electricity usage stage

Electricity usage did not impact damage to Resources, as shown in **Figure 2**, but caused damage to human health (31%) and the ecosystem (24%). The following section further examined the damage category and the findings are detailed below.

4.2 Damage assessment result

The endpoint analysis classifies the various impact categories into the relevant damage categories based on their effects. **Table 3** summarizes the damage assessment under which each impact category falls: Human Health with a value of 1.6176×10^{-6} DALY, Ecosystems has 3.9×10^{-9} species/yr. and Resources with 1.686×10^{-2} USD2013. More in-depth analysis was done to determine the exact substance that contributed to these damage categories and their contribution level in the production process stage. **Figure 3** compares the three damages in five production process stages, i.e., raw material, clinkering, fuel usage, electricity usage and transportation.

4.2.1 Human health

Damage to human health is measured in DALY and determined by many categories, including carcinogens, radiation, respiratory organics, climate change, respiratory inorganics and the ozone layer. The World Health Organization (WHO) defines the unit of human health damage disability-adjusted life years as the total annual number of potential life lost or that a person is disabled due to a pandemic, disease, or accident. **Table 3** shows the damage to human health was 1.6176×10^{-6} DALY. **Figure 2** shows the substances emitted during the clinkering stage that cause this human health damage.

Figure 3 shows that 49% of the damage caused to human health is due to the clinkering stage. At this stage, clinker production causes massive damage to human health. Others are raw material usage (5%), fuel usage (12%), electricity usage (31%) and energy generation and transportation (3%). The substances resulting during the production stages are CO_2, CH_4, SO_2, NOx, Particulates, <2.5 μm, etc. Therefore, a detailed damage assessment was conducted on human health. CO_2 emissions have a high contribution, as shown in the midpoint method result from [14]. In **Figure 2**, the result showed that 64% of the damage is caused by CO_2 emissions, with other substances like SO_2 (20%), NO_X (10%), PM2.5 (4%), NH_3 (1%), CH_4 (0.4%), As (0.6%). All of which have different effects on human health. The reaction between the raw material and coal in the clinker causes the sulfur content to produce SO_2, one of the major gases released into the air and water.

4.2.2 Ecosystem

Damage to the ecosystem is calculated in species/yr.; it refers to the number of species lost in a year due to emissions to the environment, water bodies, etc. **Table 3**

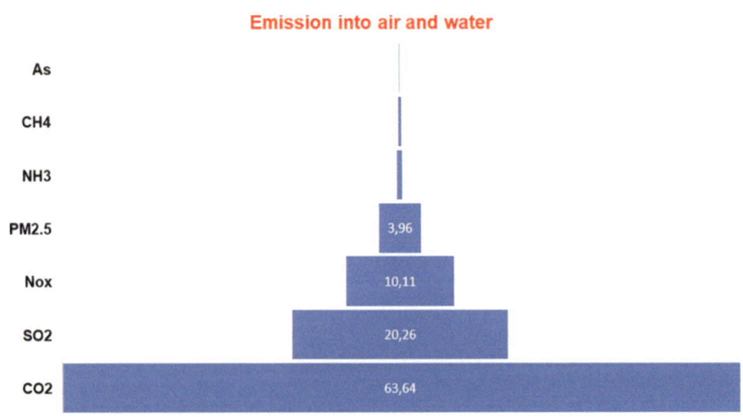

Figure 3.
Damage assessment results for five production processes.

shows the ecosystem's damage is 3.9×10^{-9} species. yr. per 1 kg of Portland cement produced. This damage was the average of species threatened calculated each year. In other words, for every 1 kg of cement produced, 3.9×10^{-9} species likely die yearly. The South African cement industry requires approximately 34.2 MT of cement annually, which endangers 133 species.

Figure 3 shows that the clinkering stage is responsible for 60% of the damage to the ecosystem, as seen in the case of human health. However, the remaining percentage comes from electricity usage (24%), fuel usage (9%), raw material usage (4%) and transportation (3%). An in-depth analysis was done on the ecosystem damage according to the substances that cause this damage. **Figure 4** shows the analysis results of the substances that cause ecosystem damage. The result showed that CO_2 has the highest emission (76%) due to energy generation, followed by NO_X (12%), SO_2 (9%), CH_4 (2%) and NH_3 (1%). Again, this proved that whatever affects the ecosystem affects human health and vice versa.

4.2.3 Resources

Damage to resources is expressed in USD2013. **Table 3** shows the resource damage is 1.686×10^{-2} USD2013. The potential marginal increase in the price of resources per kg of Portland cement produced was due to scarcity. Therefore, for each resource used to produce 1 kg of Portland cement, the cost of that resource will increase by 0.01685 USD in 2013.

Figure 3 shows the details of the total damage to resources. The main resource damage is fuel usage (96%) and raw materials (4%). An in-depth damage analysis was performed on the resources according to the substances that caused this damage. According to **Figure 5**, the result showed that coal (60%), crude oil (32%), gas (6%) and aluminum (2%) contributed to resources.

The endpoint analysis's results are consistent with those found in the literature, with CO_2 emissions and the clinkering stage being the most significant contributors [30, 35]. The resources used by Chen et al. [30] and Tun et al. [35] differ because coal was the primary fossil fuel used to produce cement. According to **Figure 2**, the SO_2 emitted from clinkering stage, as shown in **Figures 3** and **4**, was due to the coal combustion in the

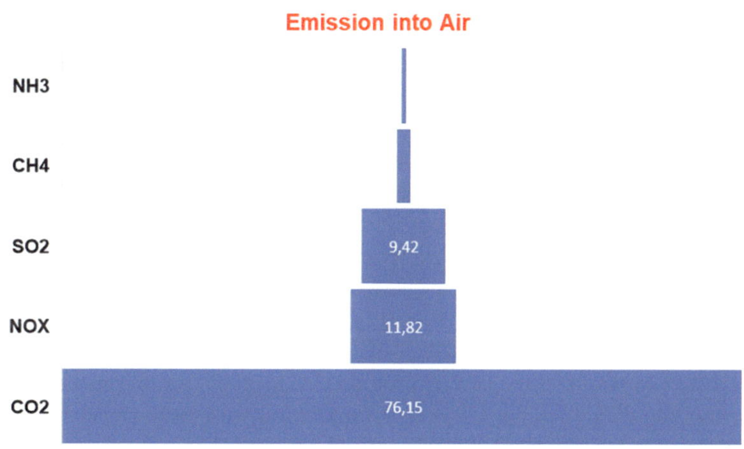

Figure 4.
Contribution of substances to the ecosystem.

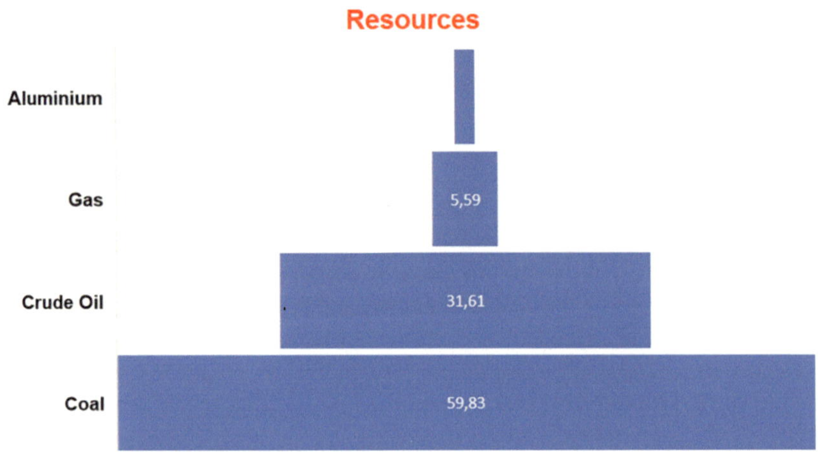

Figure 5.
Contribution of substances to resources.

South African cement plant, which established the production of SO_2 [29, 50]. The sulfur content of coal heavily influences the SO_2 emissions. Meanwhile, SO_2 emissions threaten human health and the ecosystem while harming people's chances of receiving clean air, as previously stated [18, 30, 51]. In the clinkering stage (calcination process), NO_X is emitted during fuel burning. These emissions were emitted into the air and water.

4.3 Uncertainty analysis result

The uncertainty analysis results given in **Table 4** represent the uncertainty in the endpoint, while **Table 5** represents damage assessment, respectively. The Monte Carlo method was used to conduct an uncertainty analysis with a confidence interval of 95%.

According to the uncertainty analysis, Water consumption, Human health, Water consumption, Terrestrial ecosystem, Water consumption, Aquatic ecosystems, Human carcinogenic toxicity and Ionizing radiation have a high degree of uncertainty. Human non-carcinogenic toxicity, Freshwater eutrophication, Marine ecotoxicity and Freshwater ecotoxicity have a relatively high degree of uncertainty. On the contrary, all other uncertainties at endpoint impact categories are relatively low. Also, the uncertainty was averagely low in the damage assessment uncertainty analysis result. In addition to the range of data sources, the uncertainty of the results is also influenced by variations in the system scope. CO_2 emissions from cement production activities primarily come from two main components: fuel consumption and cement clinker production. When limestone mainly contains calcium carbonate ($CaCO_3$) and is heated or burned to produce lime (CaO) during cement clinker production, it releases CO_2 emissions.

5. Conclusions

In this study, we assessed the environmental impact of 1 kg of Portland cement at the production stages, raw materials usage, clinkering, fuel usage, transportation and electricity usage caused by South African cement plants. This study used the endpoint (damage approach) method of LCIA. The study conducted an uncertainty analysis to confirm its results and establish its credibility. The assessment used data from five

DOI: http://dx.doi.org/10.5772/intechopen.1004055

Impact category	Unit	Mean	Median	SD	CV	2,50%	97,50%	SEM
Fine particulate matter formation	DALY	5,04E–07	4,68E–07	1,99E–07	39,5	2,36E–07	1,03E–06	6,30E–09
Global warming, Human health	DALY	9,33E–07	8,51E–07	4,15E–07	44,5	3,73E–07	2,09E–06	1,31E–08
Human carcinogenic toxicity	DALY	8,05E–08	4,22E–08	1,35E–07	168,0	1,29E–08	4,26E–07	4,28E–09
Human non-carcinogenic toxicity	DALY	1,16E–07	8,28E–08	1,14E–07	98,2	2,76E–08	4,02E–07	3,62E–09
Ionizing radiation	DALY	8,01E–11	4,49E–11	1,19E–10	148,1	8,46E–12	3,70E–10	3,75E–12
Ozone formation, Human health	DALY	1,93E–09	1,78E–09	8,32E–10	43,1	8,11E–10	3,96E–09	2,63E–11
Stratospheric ozone depletion	DALY	1,05E–10	9,31E–11	5,65E–11	53,5	3,73E–11	2,41E–10	1,79E–12
Water consumption, Human health	DALY	2,02E–09	4,79E–09	3,42E–08	1695,2	–8,52E–08	6,64E–08	1,08E–09
Freshwater ecotoxicity	species.yr	1,09E–11	8,98E–12	7,72E–12	71,1	3,56E–12	3,06E–11	2,44E–13
Freshwater eutrophication	species.yr	2,20E–10	1,57E–10	2,15E–10	97,7	3,85E–11	7,45E–10	6,79E–12
Global warming, Freshwater ecosystems	species.yr	7,69E–14	7,01E–14	3,42E–14	44,5	3,08E–14	1,73E–13	1,08E–15
Global warming, Terrestrial ecosystems	species.yr	2,81E–09	2,57E–09	1,25E–09	44,5	1,13E–09	6,32E–09	3,96E–11
Land use	species.yr	6,97E–11	6,26E–11	3,27E–11	46,9	2,88E–11	1,55E–10	1,03E–12
Marine ecotoxicity	species.yr	2,24E–12	1,84E–12	1,63E–12	73,1	7,23E–13	6,45E–12	5,17E–14
Marine eutrophication	species.yr	3,34E–14	2,99E–14	1,69E–14	50,6	1,20E–14	7,74E–14	5,35E–16
Ozone formation, Terrestrial ecosystems	species.yr	2,76E–10	2,55E–10	1,19E–10	43,1	1,16E–10	5,66E–10	3,76E–12
Terrestrial acidification	species.yr	5,24E–10	4,88E–10	2,06E–10	39,3	2,40E–10	1,07E–09	6,51E–12
Terrestrial ecotoxicity	species.yr	1,19E–11	1,08E–11	5,83E–12	49,0	4,48E–12	2,72E–11	1,84E–13
Water consumption, Aquatic ecosystems	species.yr	1,94E–15	2,33E–15	1,58E–14	814,3	–3,09E–14	3,17E–14	4,99E–16
Water consumption, Terrestrial ecosystem	species.yr	2,37E–11	3,77E–11	2,07E–10	875,1	–4,88E–10	4,05E–10	6,55E–12
Fossil resource scarcity	USD2013	0,016518	0,015107	0,007362	44,6	0,006978	0,036469	0,000233
Mineral resource scarcity	USD2013	0,000504	0,000469	0,000211	41,9	0,000217	0,001041	6,67E–06

Table 4.
Uncertainty results for 1 kg of cement analysis endpoint impacts.

Human health	DALY	Mean	Median	SD	CV	2,50%	97,50%	SEM
Ecosystems	species.yr	1,64E−06	1,50E−06	7,09E−07	43,2896	6,80E−07	3,44E−06	2,24E−08
		3,95E−09	3,64E−09	1,72E−09	43,59,447	1,60E−09	8,40E−09	5,45E−11
Resources	USD2013	0,017022	0,015595	0,007564	44,43,417	0,007175	0,037425	0,000239

Table 5.
Uncertainty results for endpoint impacts of 1 kg of cement analysis: Damage assessment.

South African cement plants obtained from the Ecoinvent database v3.7.1. This data was analyzed using the Endpoint Methods in SimaPro 9.1.1. software. The characterization results presented 22 impact categories from the endpoint method. The results of this study will be beneficial in increasing South Africa's LCI database and provide policymakers in South Africa with valuable scientific-based insights for making decisions related to cement production.

Based on AoSL, these impact categories were divided further into damage to human health, ecosystem and resources. Damage to human health was measured as 1.6176×10^{-6} DALY, which means that for every 1 kg of Portland cement produced, potential 1.6×10^{-6} human lives are damaged. Damage to ecosystems measured by the method shows that 3.9×10^{-9} species are potentially lost annually for every 1 kg of Portland cement produced. Damage to resources measured by the method indicates that 1.686×10^{-2} USD of resources are potentially wasted for every 1 kg of Portland cement produced. Among the three main damage categories, human health is the most affected by releasing substances into the air during Portland cement production. Also, these emissions have significant adverse impacts on global warming. The most released substances to air and water from all emissions are CO_2, As, CH_4, NH_3, $PM_{2.5}$, SO_2 and NO_X.

The clinkering stage contributes 49% of the damage to human health and 60% to ecosystems, affecting the health of humans and other species due to the amount of CO_2 released at this production stage. These results align with the literature [12, 15, 22, 28, 31, 34, 52]. CO_2 emissions contribute to global warming, with possible effects on climate change and other impacts on humans and ecosystems. Human health and the ecosystem damage from cement production results in other gas emissions such as A_s, CH_4, NH_3, $PM_{2.5}$, SO_2 and NO_X into the air and water; however, they are minor compared to CO_2. In conclusion, the clinkering stage is the most harmful production stage for human health and ecosystems since it produces the highest amounts of CO_2 gas. For every 1 kg of Portland cement produced, approximately 0.74 kg of CO_2 gas is emitted. It is advisable to select natural gas as the source of electricity generation and optimize the raw materials consumption during clinker production are effective methods to reduce the overall environmental impact in the cement industry.

Author details

Oluwafemi E. Ige[1,2]*, Kevin J. Duffy[2], Oludolapo A. Olanrewaju[1]
and Obiora C. Collins[2]

1 Department of Industrial Engineering, Durban University of Technology, Durban, South Africa

2 Institute of System Science, Durban University of Technology, Durban, South Africa

*Address all correspondence to: phemmyigoh@yahoo.com

IntechOpen

© 2024 The Author(s). Licensee IntechOpen. This chapter is distributed under the terms of the Creative Commons Attribution License (http://creativecommons.org/licenses/by/4.0), which permits unrestricted use, distribution, and reproduction in any medium, provided the original work is properly cited.

References

[1] Gençer E, Torkamani S, Miller I, Wu TW, O'Sullivan F. Sustainable energy system analysis modeling environment: Analyzing life cycle emissions of the energy transition. Applied Energy. 2020;**277**:115550

[2] Reaño RL, de Padua VAN, Halog AB. Energy efficiency and Life Cycle Assessment with system dynamics of electricity Production from Rice straw using a combined gasification and internal combustion engine. Energies. 2021;**14**(16):4942. Available from: https://www.mdpi.com/1996-1073/14/16/4942

[3] IEA, "Net Zero by 2050: A Roadmap for the Global Energy Sector," 2021. Available from: https://www.iea.org/reports/net-zero-by-2050. [Accessed: February 2022]

[4] IEA. World energy outlook 2020-summary. In: International Energy Agency. Paris, France: IEA; 2020. Available from: https://www.iea.org/reports/world-energy-outlook-2020. [Accessed: February 2, 2022]

[5] Miccoli S, Finucci F, Murro R. Assessing project quality: A multidimensional approach. In: Advanced Materials Research. Vol. 1030. Switzerland: Trans Tech Publ; 2014. pp. 2519-2522

[6] Miccoli S, Finucci F, Murro R. Criteria and procedures for regional environmental regeneration: A European strategic project. In: Applied Mechanics and Materials. Vol. 675. Switzerland: Switzerland Trans Tech Publ; 2014. pp. 401-405

[7] Anderson TR, Hawkins E, Jones PD. CO2, the greenhouse effect and global warming: From the pioneering work of Arrhenius and Callendar to today's earth system models. Endeavour. 2016;**40**(3):178-187

[8] Nta S, Olorunnisola A. Experimental production and evaluation of cement-bonded composite pipes for water conveyance. International Journal of Composite Materials. 2016;**6**(1): 9-14

[9] Feiz R, Ammenberg J, Baas L, Eklund M, Helgstrand A, Marshall R. Improving the CO2 performance of cement, part I: Utilizing life-cycle assessment and key performance indicators to assess development within the cement industry. Journal of Cleaner Production. 2015;**98**:272-281

[10] WBCSD. GNR project reporting CO2. In: Cement Sustainability Initiative. 2015. Available from: https://www.wbcsdcement.org/GNR-2016/index.html. [Accessed: November 2, 2022]

[11] Green J. Global demand for cement to reach 5.2 billion t. Worldcement. 2015. Available from: https://www.Worldcom/europe-cis/27082015/global-demand-cement-billion-tons-449/. [Accessed: August 18, 2022]

[12] Stafford FN, Raupp-Pereira F, Labrincha JA, Hotza D. Life cycle assessment of the production of cement: A Brazilian case study. Journal of Cleaner Production. 2016;**137**:1293-1299. DOI: 10.1016/j.jclepro.2016.07.050

[13] Caillahua MC, Moura FJ. Technical feasibility for use of FGD gypsum as an additive setting time retarder for Portland cement. Journal of Materials Research and Technology. 2018;7(2):190-197

[14] Ige OE, Olanrewaju OA, Duffy KJ, Collins OC. Environmental impact analysis of Portland cement (CEM1) using the midpoint method. Energies. 2022;**15**(7):2708. Available from: https://www.mdpi.com/1996-1073/15/7/2708

[15] Moretti L, Caro S. Critical analysis of the life cycle assessment of the Italian cement industry. Journal of Cleaner Production. 2017;**152**:198-210

[16] Stafford FN, Dias AC, Arroja L, Labrincha JA, Hotza D. Life cycle assessment of the production of Portland cement: A southern Europe case study. Journal of Cleaner Production. 2016;**126**:159-165. DOI: 10.1016/j.jclepro.2016.02.110

[17] Olagunju BD, Olanrewaju OA. Life Cycle Assessment of ordinary Portland cement (OPC) using both problem oriented (midpoint) approach and damage oriented approach (endpoint). In: Product Life Cycle-Opportunities for Digital and Sustainable Transformation. London, UK, London: Intech Open; 2021

[18] Huntzinger DN, Eatmon TD. A life-cycle assessment of Portland cement manufacturing: Comparing the traditional process with alternative technologies. Journal of Cleaner Production. 2009;**17**(7):668-675

[19] Madlool N. Assessment of waste preheater gas and dust bypass systems: Al-Muthanna cement plant case study. Case Studies in Thermal Engineering. 2016;**8**:330-336

[20] Lü Y-L, Geng J, He G-Z. Industrial transformation and green production to reduce environmental emissions: Taking cement industry as a case. Advances in Climate Change Research. 2015;**6**(3-4):202-209

[21] Ige OE, Duffy KJ, Olanrewaju OA, Collins OC. An integrated system dynamics model and Life Cycle Assessment for cement production in South Africa. Atmosphere. 2022;**13**(11):1788. Available from: https://www.mdpi.com/2073-4433/13/11/1788

[22] García-Gusano D, Garraín D, Herrera I, Cabal H, Lechón Y. Life Cycle Assessment of applying CO2 post-combustion capture to the Spanish cement production. Journal of Cleaner Production. 2015;**104**:328-338

[23] Holt SP, Berge ND. Life-cycle assessment of using liquid hazardous waste as an alternative energy source during Portland cement manufacturing: A United States case study. Journal of Cleaner Production. 2018;**195**:1057-1068

[24] Huijbregts M. Uncertainty and variability in environmental life-cycle assessment. The International Journal of Life Cycle Assessment. 2002;7(3):173-173

[25] ISO. 14040: International organization for standardization. Environmental Management: Life Cycle Assessment; Principles and Framework. In: International Organization for Standardization. ISO; 2006. Available from: https://www.iso.org/obp/ui/#iso:std:iso:14040:ed-2:v1:en. [Accessed: July 31, 2023]

[26] ISO/TS. Environmental management — Life cycle assessment — Requirements and guidelines for organizational life cycle assessment. ISO/TS,14072: 2014. In: International Organization for Standardization. Switzerland: ISO Geneva; 2014. Available from: https://www.iso.org/obp/ui/#iso:std:iso:ts:14072:ed-1:v1:en. [Accessed: July 23, 2023]

[27] García-Gusano D, Herrera I, Garraín D, Lechón Y, Cabal H. Life cycle assessment of the Spanish cement industry: Implementation of

environmental-friendly solutions. Clean Technologies and Environmental Policy. 2015;**17**(1):59-73

[28] Valderrama C, Granados R, Cortina JL, Gasol CM, Guillem M, Josa A. Implementation of best available techniques in cement manufacturing: A life-cycle assessment study. Journal of Cleaner Production. 2012;**25**:60-67

[29] Ali AAM, Negm AM, Bady MF, Ibrahim MG, Suzuki M. Environmental impact assessment of the Egyptian cement industry based on a life-cycle assessment approach: A comparative study between Egyptian and Swiss plants. Clean Technologies and Environmental Policy. 2016;**18**(4):1053-1068

[30] Chen W, Hong J, Xu C. Pollutants generated by cement production in China, their impacts, and the potential for environmental improvement. Journal of Cleaner Production. 2015;**103**:61-69

[31] Li C, Cui S, Nie Z, Gong X, Wang Z, Itsubo N. The LCA of Portland cement production in China. The International Journal of Life Cycle Assessment. 2015;**20**(1):117-127

[32] Josa A, Aguado A, Cardim A, Byars E. Comparative analysis of the life cycle impact assessment of available cement inventories in the EU. Cement and Concrete Research. 2007;**37**(5):781-788

[33] Somoza-Tornos A, Guerra OJ, Crow AM, Smith WA, Hodge B-M. Process modeling, techno-economic assessment, and life cycle assessment of the electrochemical reduction of CO2: A review. iScience. 2021;**24**(7):102813. DOI: 10.1016/j.isci.2021.102813

[34] Çankaya S, Pekey B. A comparative life cycle assessment for sustainable cement production in Turkey. Journal of Environmental Management. 2019;**249**:109362. Available from: https://www.sciencedirect.com/science/article/pii/S0301479719310710?via%3Dihub

[35] Tun TZ, Bonnet S, Gheewala SH. Life cycle assessment of Portland cement production in Myanmar. The International Journal of Life Cycle Assessment. 2020;**25**(11):2106-2121

[36] Thwe E, Khatiwada D, Gasparatos A. Life cycle assessment of a cement plant in Naypyitaw, Myanmar. Cleaner Environmental Systems. 2021;**2**:100007

[37] Nigri EM, Rocha SDF, Romeiro Filho E. Portland cement: An application of life cycle assessment. Product: Management and Development. 2010;**8**(2):167-172

[38] B. Pré Consultants. SimaPro. In: https://network.simapro.com/esuservices. 2016. Available from: https://network.simapro.com/esuservices. [Accessed: August 13, 2022]

[39] PRé Sustainability. SimaPro. In: Software to Measure and Improve The Impact of Your Product Life Cycle. Available: Pre-sustainability; 2017

[40] ISO. 14044: International organization for standardization. Environmental Management: Environmental management: Life cycle assessment; requirements and guidelines. In: International organization for standardization. Switzerland: ISO Geneva; 2006. Available from: https://www.iso.org/obp/ui/#iso:std:iso:14044:ed-1:v1:en. [Accessed: July 31, 2022]

[41] Marinković S. Life cycle assessment (LCA) aspects of concrete. In: Eco-efficient Concrete. Sawston, Cambridge: Elsevier; 2013. pp. 45-80

[42] Moreno Ruiz E et al. Cement Production: Documentation of Changes Implemented in Ecoinvent Database v3. 7 & v3. 7.1. Documentation Cement Production, Portland - ZA, Ecoinvent Association, Allocation, cut-off ed. Identifying Improvement Potentials in Cement Production with Life Cycle Assessment 2010. Zurich, Switzerland: Ecoinvent Association; 2019

[43] Martínez-Rocamora A, Solís-Guzmán J, Marrero M. LCA databases focused on construction materials: A review. Renewable and Sustainable Energy Reviews. 2016;**58**:565-573

[44] Ige OE, Olanrewaju OA, Duffy KJ, Collins OC. A review of the effectiveness of Life Cycle Assessment for gauging environmental impacts from cement production. Journal of Cleaner Production. 2021;**324**:129213. DOI: 10.1016/j.jclepro.2021.129213

[45] Andersson K, Ohlsson T, Olsson P. Life cycle assessment (LCA) of food products and production systems. Trends in Food Science & Technology. 1994;**5**(5):134-138

[46] Huijbregts MA. Application of uncertainty and variability in LCA. The International Journal of Life Cycle Assessment. 1998;**3**(5):273-280

[47] Güereca LP, Torres N, Juárez-López CR. The co-processing of municipal waste in a cement kiln in Mexico. A life-cycle assessment approach. Journal of Cleaner Production. 2015;**107**:741-748

[48] I. O. F. Standardization. Environmental Management: Life Cycle Assessment; Principles and Framework (no. 2006). Geneva, Switzerland: ISO; 2006

[49] Goedkoop M, Heijungs R, Huijbregts M, De Schryver A, Struijs J, Van Zelm R. ReCiPe 2008. In: A Life Cycle Impact Assessment Method Which Comprises Harmonised Category Indicators at the Midpoint and the Endpoint Level. Vol. 1. 2009. pp. 1-126. Available from: https://web.universiteitleiden.nl/cml/ssp/publications/recipe_characterisation.pdf

[50] Mittal ML, Sharma C, Singh R. Decadal emission estimates of carbon dioxide, sulfur dioxide, and nitric oxide emissions from coal burning in electric power generation plants in India. Environmental Monitoring and Assessment. 2014;**186**(10):6857-6866

[51] Song D, Yang J, Chen B, Hayat T, Alsaedi A. Life-cycle environmental impact analysis of a typical cement production chain. Applied Energy. 2016;**164**:916-923

[52] Georgiopoulou M, Lyberatos G. Life cycle assessment of the use of alternative fuels in cement kilns: A case study. Journal of Environmental Management. 2018;**216**:224-234